Methods in Molecular Biology

Series Editor
John M. Walker
School of Life Sciences
University of Hertfordshire
Hatfield, Hertfordshire, AL10 9AB, UK

For further volumes:
http://www.springer.com/series/7651

Hox Genes

Methods and Protocols

Edited by

Yacine Graba

IBDML, Marseille, France

René Rezsohazy

Institut des Sciences de la Vie, Université Catholique de Louvain, Louvain-la-Neuve, Belgium

Humana Press

Editors
Yacine Graba
IBDML
Marseille, France

René Rezsohazy
Institut des Sciences de la Vie
Université Catholique de Louvain
Louvain-la-Neuve, Belgium

ISSN 1064-3745 ISSN 1940-6029 (electronic)
ISBN 978-1-4939-1241-4 ISBN 978-1-4939-1242-1 (eBook)
DOI 10.1007/978-1-4939-1242-1
Springer New York Heidelberg Dordrecht London

Library of Congress Control Number: 2014944736

Printed on acid-free paper

Humana Press is a brand of Springer
Springer is part of Springer Science+Business Media (www.springer.com)

Preface

Hox Genes: A Fertile Interplay of Concepts and Methods

Why a *Methods in Molecular Biology* issue on Hox genes? "Methods" books usually focus on a set of techniques suited to a given experimental approach. Here, it was chosen to present techniques and methods centered on a research topic: Hox genes. The study of Hox genes has been at the heart of the emerging molecular embryology, molecular genetics of development, and evolutionary-developmental biology (evo-devo). Key steps have been the identification of "A gene complex controlling segmentation in *Drosophila*" reported by Ed Lewis in 1978, the cloning of bithorax in the early 1980s, the discovery of homeobox genes in vertebrate genomes, and the discovery that Hox gene functions in patterning embryos are conserved [1–6].

Hox genes have been initially identified as crucial determinants to pattern the main anteroposterior body axis of animal embryos [1]. Their functions however rapidly emerged as extending beyond axial patterning, with involvement in multiple organogenetic processes, including but not restricted to limb, heart, gut, respiratory, urogenital, or endocrine organogenesis, as well as in later neuronal circuit connectivity or in adult hematopoiesis [7, 8]. The repertoire of Hox functions has become immensely vast, and their recognized role in a number of pathologies is likely only an under-representation of the consequences of their dysfunction [9]. Hox genes encode transcription factors [10], and their extensive structural and functional conservation allowed apprehending how variations in gene sequence and activity underlie morphological evolution [11–15].

The study of Hox genes has been paradigmatic for the unification of biology by the end of the twentieth century, when the emergence of molecular biology promoted the interconnection and cross-influence of most fields of biological sciences. Formerly, experimental embryology, genetics, biochemistry, physiology, evolutionary biology, and other life sciences were loosely cross-talking due to a lack of unifying concepts, as well as difficulties in appraising the biological reality in a holistic way, from the bottom molecular scale to the top, organismic (or ecosystemic) scale. Equally responsible for the limited cross talk between disciplines were the technological difficulties and deficit to appraise questions from a multiscale point of view. In the 1990s, these challenges had been taken up with the convergence of genome sequencing projects, reverse molecular genetics, descriptive genetics, embryology, and histology of model organisms like *Drosophila*, the mouse, the worm *Caenorhabditis elegans*, *Xenopus*, or the zebrafish. The study of Hox genes has accompanied this unifying movement, feeding and being fed by this forwards-backwards movement between methodological development and concept evolution, questions raised, and strategies to address them.

In this issue of *Methods in Molecular Biology*, we aim at reviewing techniques and methodologies which arose from or were successfully applied to the study of Hox genes and Hox proteins. This overview does not pretend, by far, to be exhaustive and only provides a sample of possible experimental approaches that have accompanied 35 years of research in

Hox biology. By essence, the techniques and strategies described in the 21 chapters constitutive of this *Methods in Molecular Biology* issue on Hox genes are also relevant beyond the Hox field.

Extracting Information from Genome and Gene Sequences

Comparative studies permit insights into the degree of freedom in molecular evolution that results from a balance between permissiveness and restrictiveness to changes. This depends on the biochemical, structural, and functional constraints acting on molecules as well as on the robustness and plasticity of the molecules and of the networks of interacting actors involved. Such comparative studies, notably focused on Hox and homeobox genes more broadly, are facilitated by the availability of fully sequenced and assembled genomes, which is the case for classical animal models but lacking for non-model organisms. With the now powerful capacity of high-throughput sequencing platforms, the perspective of sequencing new animal genomes is conceivable for most research projects. Marlétaz et al. describe the rationale and methods to proceed with genome sequencing and assembly. They provide relevant reference to bioinformatic tools, which allow homeobox gene extraction and classification, thereby enabling the reconstruction of the phylogenetic history of homeobox genes.

Comparative studies not only focus on protein-coding sequence but also on *cis*-regulatory elements governing Hox gene expression. This is essential as, in many instances, the functional specificity of Hox genes is driven by their expression levels and patterns rather than by the specificity of the corresponding Hox proteins. This has been shown to especially stand for Hox genes belonging to a same paralogy group in vertebrates [16, 17]. Furthermore, evolutionary changes in Hox gene regulation have been identified to be at the heart of body plan diversification, as it has been illustrated in chelicerates, myriapods, and insects or in snakes, lacertilians, and mammals for example [18, 19]. In their contribution, Matharu and Mishra discuss the challenging issue of identifying Hox cis-regulatory sequences and suggest strategies to mine the most plausible Hox regulatory information.

Manipulating Genomes

Once genes are identified and putative regulatory elements inferred, their functional analysis will heavily rely on genome manipulation. In the mouse, since the initial gene recombination strategies that led to the first gene knockouts in the late 1980s to early 1990s [20, 21], a tremendous number of tools have been developed relying on the use of site-specific recombination systems and on the generation of vectors allowing to carry hundreds of kilobases. Genome manipulation strategies and techniques nowadays are versatile and permit inverting, deleting, and exchanging from single genes to entire Hox clusters, thereby allowing the study of long-distance mechanisms of gene regulation. In their chapter, Tschopp and Duboule review the vast repertoire of alleles displaying lox recombination sites suited to engineer modified Hox gene clusters and provide guidance on how to design and to proceed with site-specific recombination-based genome engineering techniques.

Working with bacterial artificial chromosomes (BACs) as recombination vectors allows the manipulation of large sequences encompassing multiple genes and their associated regulatory sequences. Parrish et al. report how combining BAC transgenesis with site-specific recombination provides enough modularity to tackle the complexity of Hox gene biology by inactivating, deleting, or reassorting genes or sequence elements.

In *Drosophila*, a convenient way to manipulate the genome relies on P-element mobilization. In their chapter, de Navas et al. present a gene replacement strategy to insert P-GAL4 elements in Hox genes so as to induce UAS-responsive genes in cells where the

Hox::P-GAL4 is active, in a context mutant for the resident Hox gene. The UAS-driven gene may be chosen to promote reporter activity or to express mutant Hox variants or any other genes of interest, making the approach useful to study both protein function and *cis* -regulatory elements. While lacking the targeted precision of site-specific recombination, the methodology takes advantage of the facility to screen for phenotypic markers.

Regulatory Landscapes and Topological Organization of Regulatory Elements

Gene regulation relies on short-range as well as long-range regulatory mechanisms, highlighting the necessity to apprehend the role of regulatory landscapes and nuclear topological organization. Recently developed approaches now allow grasping the physical connection between remote regulatory elements and gene promoters. These approaches capture DNA synapses involved in enhanceosomes or silenceosomes. Generic methods include 3C, 4C, 5C, Hi-C, ChIP-loop, or ChIA-PET [22]. Tagging DNA elements involved in such distant contacts is another way to map distant chromosome interactions. Cléard et al. describe an elegant DNA tagging method in *Drosophila* based on fusing the bacterial DAM DNA methyltransferase to a DNA-binding or a chromatin-binding protein that targets the methyltransferase activity towards specific DNA elements (insulators, enhancers/silencers, promoters, ...). In the method detailed here, suited for genomes poorly methylated only, the DAM methyltransferase is fused to the DNA-binding domain of GAL4, and the long-range DNA-DNA interactions are mapped from a UAS sequence recognized by GAL4 and inserted at a locus of interest. Since DAM methylate GATC sequences which cleavage by DpnI and DpnII are methylation dependent, DNA methylation can easily be located both on a loci-specific or on a genome-wide scale.

A somewhat conceptually similar approach is presented by Agelopoulos et al. and consists in cell- and gene-specific chromatin immunoprecipitation (cgChIP). It involves the tagging of a *cis*-regulatory element of interest (providing the gene specificity) with a lacO sequence that allows the recruitment of an epitope-tagged LacI expressed in a cell-specific way using the UAS-GAL4 system (providing the cell specificity). ChIPing the tagged LacI allows fishing distant DNA elements in contact with the *cis*-regulatory element specifically in cells expressing LacI. Combining cgChIP with a second immunoprecipitation step or with western blot analyses further allows the identification of proteins involved in connecting the *cis*-regulatory element with the remote DNA element.

Long-distance interactions involving regulatory sequences serve to not only connect remote regulatory sequences to gene promoters but also locate genes within the nucleus according to their transcriptional status. This highlights the importance of viewing gene regulation in correlation with the topological organization of the nucleus. Coupling DNA fluorescent in situ hybridization (FISH) with immunostaining (FISH-I) allows visualizing DNA topology and DNA relocation in nuclear sub-compartments. Bantignies and Cavalli describe how these techniques capture Hox gene nuclear relocation upon transcriptional activation and repression in *Drosophila* embryos and cultured cells.

Recognizing Expression Patterns

The function of a gene primarily depends on its temporal and spatial deployment, making the characterization of gene expression patterns an obligatory step in understanding gene functions. Hox gene studies largely relied on but also contributed to the development and refinement of in situ mRNA hybridization techniques and immunohistochemical detection of proteins [23–25]. These expression studies uncovered the collinearity relationship whereby gene order in the Hox clusters relates to the embryonic territories where they are active.

While prominent aspects of a gene expression pattern are usually rapidly identified, expression at early developmental stages often characterized by the paucity of material, size of biological samples, and weak expression levels renders grasping some aspects of the pattern quite challenging. Illustrative of this difficulty, early Hox gene expression, a way before the onset of gastrulation in vertebrates, was only recently identified. Characterizing gene expression in single oocytes, zygotes, morula stage, or blastocyst embryos raises new technical challenges. Paul et al. describe a set of techniques useful to analyze gene expression in oocytes and early bovine embryos.

Complementary to ISH or IHC techniques which identify where and when Hox genes are active, it is also extremely informative in a functional perspective to assess the destiny of cells that have expressed Hox genes at an earlier time of their ontogeny. Lineage tracing meets this objective. As a straightforward way to trace cell lineages in mouse, Laforest et al. present a binary transgenic strategy combining cell- or tissue-specific Cre-expressing transgene with the induction of reporter genes inserted in the permissive ROSA26 locus. Once the reporter has been switched on by Cre-mediated recombination, it remains activated in all their descent. Such approaches have been instructive to better understand Hox-knockout phenotypes that impact on structures which at late stages of development do not express Hox genes. Lineage tracing provided evidence that cells having expressed the gene at earlier stages of development actually contribute to these structures and that activity of the gene is critical for the proper determination, migration, and functional allocation of these sometimes distant cells, as exemplified by migratory neural crest cells [26, 27].

Addressing Gene Function in Vertebrate Model Organisms

Addressing the functions of Hox genes is at the heart of developmental biology. There are numerous approaches to invalidate gene functions. Gene inactivation by homologous recombination has been a major approach to inactivate genes in the mouse. Recombining and engineering the mouse genome have increasingly become more amenable by the use of site-specific recombinases and BAC recombination vectors, as described by Tschopp and Duboule and Parrish et al. Such approaches require massive access to the genome, as allowed by electroporation of ES cells in mouse, which is not always the case. This for example applies to the zebrafish. Ladam and Sagerström describe how Hox gene inactivation is achieved taking the advantage of accessing the living, translucid, one- or two-cell fish embryos to inject antisense morpholinos, mRNA, or TOL2-transposon-based transgenes. Studying gene function will certainly gain efficacy with new targeted recombination systems derived from CRISPR-cas9 or TALEN [28] that appear to be usable in most living organism, from *C. elegans* to human cell lines and plants.

Addressing Gene Function in Non-model Organisms

Studying non-model organisms is central to evo-devo and constitutes a major challenge for the future of Hox gene studies. It should delineate the variations as well as the constraints underlying Hox gene and protein evolution and provide a phylogenetic perspective on Hox gene function. The study of Parhyale is illustrative of a non-model organism for which transgenesis tools have recently been developed and applied to Hox biology. Parhyale is a crustacean, a sister group of hexapods showing a direct embryonic development. Kontarakis and Pavlopoulos provide a description of this organism supporting the soundness of investing tools to study its Hox genes. They also present how to carry out transposon-based transient and stable transgenesis in Parhyale.

The bovine embryo has slow one cell-to-blastocyst stage transition, favoring expression and functional studies at this transition when compared to the mouse. Hox transcripts and proteins accumulate during oocyte maturation and varying Hox transcript abundance in early bovine embryos indicates that Hox genes could play roles at the onset of embryonic development. Paul et al. have set up RNA interference to knock down mRNA levels by microinjecting siRNAs in oocytes or zygotes in the bovine embryo, which should allow addressing this possible very early Hox gene function.

Identifying the Genomic Distribution and Target Genes of Hox Proteins

Hox proteins are transcription factors. Grasping how Hox proteins fulfill their function requires apprehending genome occupancy and genome landscapes associated with Hox protein activity as well as the identification of Hox target genes [29–31].

Recognizing Hox targets has first relied on candidate gene approaches, resulting in a slowly growing list of target genes. With the entry into the post-genomic era and the development of DNA microarrays, we now have a more comprehensive view of Hox downstream target genes. Polychronidou and Lohmann describe how to proceed to run mRNA extraction and microarray hybridization from *Drosophila* embryos. An additional layer of resolution in generated datasets can be achieved by sorting cells to increase sample homogeneity. Defaye and Perrin propose a method to FACS cells based on cell-specific GFP expression in a way that is quantitatively compatible to carry out a transcriptomic characterization of tissue-specific cell populations.

Transcriptomic approaches do not discriminate between direct and indirect target genes. Crossing transcriptomic data with the genomic distribution of Hox proteins should in principle identify direct target genes. This is now routinely achieved through chromatin immunoprecipitation (ChIP) that requires a ChIP-grade antibody directed against the protein of interest and that allows the recovery of the chromatin associated to that protein. The approach carried out at a genome-wide scale is informative to map Hox-responsive enhancers but also to predict Hox protein partners that may influence Hox DNA-binding specificities. Amin and Bobola describe how to proceed with ChIP and ChIP-Seq following microdissection of mouse embryonic territories, i.e., branchial arches, and Agrawal and Shashidara describe a method to perform ChIP analysis from dissected *Drosophila* imaginal discs.

Probing Protein-DNA Interactions

Understanding the mode of action of transcription factors requires deciphering the molecular rules governing the selectivity, affinity, and specificity in DNA binding. For homeoproteins, and Hox proteins in particular, there appears to be a paradox in that their homeodomain is extensively conserved while the genetics establishes that their functions and activities are specific. This paradox can at least partly be resolved considering that Hox proteins do not act alone but along with cofactors proposed to increase Hox DNA-binding specificities.

Techniques to appraise the DNA-binding specificity of Hox proteins are diverse. Electrophoresis mobility shift assays (EMSA) have been widely used to study Hox DNA-binding properties [32–34]. Churion et al. describe how EMSA techniques allow us to quantitatively determine DNA-binding specificities and interaction affinities. Quantitative EMSA also enables tackling cooperativity and stoichiometry in DNA binding. As pointed out by Churion et al., working with Hox proteins is especially challenging since these proteins are prone to form aggregates [35]. The authors also address this issue in the context of EMSA experiments.

A major limitation of EMSA experiments is that they are not suited to handle extensive repertoires of sequences. The SELEX-Seq method described by Riley et al. circumvents this limitation. It consists in a variant of SELEX, or "Systematic Evolution of Ligands by Exponential Enrichment," combined to high-throughput next-generation sequencing. The principle of the method is to prepare an extensive library of DNA probes and to proceed with cycles of selection-amplification of protein-bound sequences to determine, qualitatively and quantitatively, the repertoire of sequences bound by a protein or a protein complex. Deep sequencing permits informative data to emerge in the very first round of the selection procedure and allows the identification of multiple preferred binding sequences.

Identifying Protein Partners

Partnerships with other DNA-binding proteins have been central to apprehend Hox protein function, with the most illustrative example being provided by the evolutionarily conserved three amino acid loop extension (TALE) homeodomain proteins [36–39]. Protein partnership also has the potential to uncover possible novel functions, including non-transcriptional processes. Two proteome-wide interaction screens provided data suggesting such functions for Hox proteins [40, 41]. Bergiers et al. present methods to screen for interactions by high-stringency yeast two-hybrid assays, to validate candidate interactions by affinity coprecipitation, and to build interaction networks. These methods have an important heuristic value, but data nevertheless need to be anchored in biological contexts. An elegant way to validate protein interactions, to probe their structural determinants, and to follow them up in vivo is reported by Duffraisse et al. who describe how to carry out bimolecular fluorescence complementation (BiFC) in live *Drosophila* embryos. The BiFC rationale is to fuse semi-GFPs (or Venus) to two validated or candidate interactors which upon interaction will reconstitute a functional fluorescent protein. The method is sensitive and allows mapping binding interfaces, following intracellular shuttling and monitoring live interactions.

From the Bench to the Clinic

There is a wealth of evidence supporting roles for Hox proteins in pathological processes. The most documented situations relate to cancers and in particular to hematological malignancies [9]. Correlating Hox gene activities to disease development is informative to refine the typology and etiology of pathologies. Further, such correlations are also valuable to formulate prognostics and to design therapeutic strategies. In that context, Kettyle et al. describe procedures to establish correlations between transcriptomic profiling of a given disease, which in the paradigm the authors describe consists in HOX-TALE-related leukemia, and transcriptomic changes elicited by exposure to a small molecule. This approach, referred to as "connectivity mapping," is based on functional correlations rather than on structural modeling and allows screening for molecules which can be envisioned as a putative drug to treat the disease.

Marseille, France *Yacine Graba*
Louvain-la-Neuve, Belgium *René Rezsohazy*

References

1. Lewis EB (1978) A gene complex controlling segmentation in *Drosophila*. Nature 276:565–570
2. Sanchez-Herrero E, Vernos I, Marco R, Morata G (1985) Genetic organization of Drosophila bithorax complex. Nature 313:108–113
3. Scott MP, Weiner AJ (1984) Structural relationships among genes that control development: sequence homology between the *Antennapedia*, *Ultrabithorax*, and *fushi tarazu* loci of Drosophila. Proc Natl Acad Sci U S A 81:4115–4119
4. McGinnis W, Levine M, Hafen E, Kuroiwa A, Gehring WJ (1984) A conserved DNA sequence in homeotic genes of the *Drosophila Antennapedia* and *bithorax* complexes. Nature 308:428–433
5. McGinnis W, Garber RL, Wirz J, Kuroiwa A, Gehring WJ (1984) A homologous protein-coding sequence in *Drosophila* homeotic genes and its conservation in other metazoans. Cell 37:403–408
6. Malicki J, Schugart K, McGinnis W (1990) Mouse *Hox 2.2* specifies thoracic segmental identity in Drosophila embryos and larvae. Cell 63:961–967
7. McGinnis W, Krumlauf R (1992) Homeobox genes and axial patterning. Cell 68:283–302
8. Hombria JC, Lovegrove B (2003) Beyond homeosis—HOX function in morphogenesis and organogenesis. Differentiation 71:461–476
9. Shah N, Sukumar S (2010) The Hox genes and their roles in oncogenesis. Nat Rev Cancer 10:361–371
10. Winslow GM, Hayashi S, Krasnow M, Hogness DS, Scott MP (1989) Transcriptional activation by the *Antennapedia* and *fushi tarazu* proteins in cultured Drosophila cells. Cell 57:1017–1030
11. Rohs R, Jin X, West SM, Joshi R, Honig B, Mann RS (2010) Origins of specificity in protein-DNA recognition. Annu Rev Biochem 79:233–269
12. Gellon G, McGinnis W (1998) Shaping animal body plans in development and evolution by modulation of Hox expression patterns. Bioessays 20:116–125
13. Pearson JC, Lemons D, McGinnis W (2005) Modulating Hox gene functions during animal body patterning. Nat Rev Genet 6:893–904
14. Merabet S, Hudry B, Saadaoui M, Graba Y (2009) Classification of sequence signatures: a guide to Hox protein function. Bioessays 31:500–511
15. Akam ME, Averof M, Castelli-Gair J, Dawes R, Falciani F, Ferrier D (1994) The evolving role of Hox genes in arthropods. Development 120(Suppl):209–215
16. Tvrdik P, Capecchi MR (2006) Reversal of Hox1 gene subfunctionalization in the mouse. Dev Cell 11:239–250
17. Greer JM, Puetz J, Thomas KR, Capecchi MR (2000) Maintenance of functional equivalence during paralogous Hox gene evolution. Nature 403:661–665
18. Cohn MJ, Tickle C (1999) Developmental basis of limblessness and axial patterning in snakes. Nature 399:474–479
19. Hughes CL, Kaufman TC (2002) Exploring the myriapod body plan: expression patterns of the ten Hox genes in a centipede. Development 129:1225–1238
20. Thomas KR, Capecchi MR (1990) Targeted disruption of the murine int-1 proto-oncogene resulting in severe abnormalities in midbrain and cerebellar development. Nature 346:847–850
21. Koller BH, Hagemann LJ, Doetschman T, Hagaman JR, Huang S, Williams PJ, First NL, Maeda N, Smithies O (1989) Germ-line transmission of a planned alteration made in a hypoxanthine phosphoribosyltransferase gene by homologous recombination in embryonic stem cells. Proc Natl Acad Sci U S A 86:8927–8931
22. Montavon T, Duboule D (2012) Landscapes and archipelagos: spatial organization of gene regulation in vertebrates. Trends Cell Biol 22:347–354
23. Tautz D, Pfeifle C (1989) A non-radioactive in situ hybridization method for the localization of specific RNAs in *Drosophila* embryos reveals translational control of the segmentation gene *hunchback*. Chromosoma 98:81–85
24. Gaunt SJ, Miller JR, Powell DJ, Duboule D (1986) Homoeobox gene expression in mouse embryos varies with position by the primitive streak stage. Nature 324:662–664
25. Awgulewitsch A, Utset MF, Hart CP, McGinnis W, Ruddle FH (1986) Spatial restriction in expression of a mouse homoeo box locus within the central nervous system. Nature 320:328–335
26. Makki N, Capecchi MR (2010) Hoxa1 lineage tracing indicates a direct role for Hoxa1 in the development of the inner ear, the heart, and the third rhombomere. Dev Biol 341:499–509
27. Bertrand N, Roux M, Ryckebusch L, Niederreither K, Dolle P, Moon A, Capecchi M, Zaffran S (2011) Hox genes define distinct progenitor sub-domains within the second heart field. Dev Biol 353:266–274
28. Wei C, Liu J, Yu Z, Zhang B, Gao G, Jiao R (2013) TALEN or Cas9—rapid, efficient and specific choices for genome modifications. J Genet Genomics 40:281–289

29. Graba Y, Aragnol D, Pradel J (1997) *Drosophila* Hox complex downstream targets and the function of homeotic genes. Bioessays 19:379–388
30. Sanchez-Herrero E (2013) Hox targets and cellular functions. Scientifica 2013:738257. doi:10.1155/2013/738257
31. Hueber SD, Lohmann I (2008) Shaping segments: Hox gene function in the genomic age. Bioessays 30:965–979
32. Corsetti MT, Briata P, Sanseverino L, Daga A, Airoldi I, Simeone A, Palmisano G, Angelini C, Boncinelli E, Corte G (1992) Differential DNA binding properties of three human homeodomain proteins. Nucleic Acids Res 20:4465–4472
33. Damante G, Di Lauro R (1991) Several regions of Antennapedia and thyroid transcription factor 1 homeodomains contribute to DNA binding specificity. Proc Natl Acad Sci U S A 88:5388–5392
34. Catron KM, Iler N, Abate C (1993) Nucleotides flanking a conserved TAAT core dictate the DNA binding specificity of three murine homeodomain proteins. Mol Cell Biol 13:2354–2365
35. Bondos SE, Bicknell A (2003) Detection and prevention of protein aggregation before, during, and after purification. Anal Biochem 316:223–231
36. Longobardi E, Penkov D, Mateos D, De Florian G, Torres M, Blasi F (2014) Biochemistry of the tale transcription factors PREP, MEIS, and PBX in vertebrates. Dev Dyn 243:59–75
37. Mann RS, Affolter M (1998) Hox proteins meet more partners. Curr Opin Genet Dev 8:423–429
38. Chan SK, Jaffe L, Capovilla M, Botas J, Mann RS (1994) The DNA binding specificity of Ultrabithorax is modulated by cooperative interactions with extradenticle, another homeoprotein. Cell 78:603–615
39. Peifer M, Wieschaus E (1990) Mutations in the Drosophila gene extradenticle affect the way specific homeo domain proteins regulate segmental identity. Genes Dev 4:1209–1223
40. Lambert B, Vandeputte J, Remacle S, Bergiers I, Simonis N, Twizere JC, Vidal M, Rezsohazy R (2012) Protein interactions of the transcription factor Hoxa1. BMC Dev Biol 12:29. doi:10.1186/1471-213X-12-29
41. Bondos SE, Tan XX, Matthews KS (2006) Physical and genetic interactions link hox function with diverse transcription factors and cell signaling proteins. Mol Cell Proteomics 5:824–834

Contents

Part II Hox Proteins: Mode of Action and Biomedical Applications

Contributors

NAMIKO ABE • *Department of Biochemistry and Molecular Biophysics, Columbia University Medical Center, New York, NY, USA*

MARIOS AGELOPOULOS • *Department of Biochemistry and Molecular Biophysics, Columbia University, New York, NY, USA*

PAVAN AGRAWAL • *Janelia Farm Research Campus, HHMI, Ashburn, VA, USA*

YOUNGWOOK AHN • *Stowers Institute for Medical Research, Kansas City, MO, USA*

SHILU AMIN • *School of Dentistry, Faculty of Medical and Human Sciences, University of Manchester, Manchester, UK*

FRÉDÉRIC BANTIGNIES • *Institut de Génétique Humaine, CNRS UPR-1142, Montpellier, France*

ISABELLE BERGIERS • *Institut des Sciences de la Vie, Université Catholique de Louvain, Louvain-la-Neuve, Belgium*

NICOLAS BERTRAND • *Inserm, GMGF UMR-S910, Aix-Marseille Université, Marseille, France*

NICOLETTA BOBOLA • *School of Dentistry, Faculty of Medical and Human Sciences, University of Manchester, Manchester, UK; Centre for Endocrinology and Diabetes, Institute of Human Development, Faculty of Medical and Human Sciences, Manchester Academic Health Science Centre, University of Manchester, Manchester, UK; School of Dentistry, The University of Manchester, Manchester, M13 9PT, UK*

SARAH E. BONDOS • *Department of Molecular and Cellular Medicine, Texas A&M Health Science Center, College Station, TX, USA; Department of Biochemistry and Cell Biology, Rice University, Houston, TX, USA*

HARMEN J. BUSSEMAKER • *Department of Biological Sciences, Columbia University, New York, NY, USA; Department of Systems Biology, Columbia University, New York, NY, USA*

GIACOMO CAVALLI • *Institut de Génétique Humaine, CNRS UPR-1142, Montpellier, France*

KELLY CHURION • *Department of Molecular and Cellular Medicine, Texas A&M Health Science Center, College Station, TX, USA*

FABIENNE CLÉARD • *Department of Genetics and Evolution, University of Geneva, Geneva, Switzerland*

SARAH DAAKOUR • *Laboratory of Signaling and Protein Interactions, GIGA-R, University of Liege, Liège, Belgium*

ARNAUD DEFAYE • *Technologies Avancées pour le Génome et la Clinique (TAGC), UMR 1090 INSERM, CNRS, Aix Marseille Université, Marseille, France*

ISABELLE DONNAY • *Institut des Sciences de la Vie, Université Catholique de Louvain, Louvain-la-Neuve, Belgium*

DENIS DUBOULE • *National Research Centre 'Frontiers in Genetics', Department of Genetics and Evolution, University of Geneva, Sciences III, Geneva, Switzerland; School of Life Sciences, Federal Institute of Technology (EPFL), Lausanne, Switzerland*

MARILYNE DUFFRAISSE • *Institut de Génomique Fonctionnelle de Lyon, ENS de Lyon - CNRS UMR5242, Lyon, France*

DAVID FORONDA • *Centro de Biología Molecular Severo Ochoa (CSIC-UAM), Universidad Autónoma de Madrid, Madrid, Spain; Institute of Molecular and Cell Biology Singapore, Singapore*

PETER W.H. HOLLAND • *Department of Zoology, University of Oxford, Oxford, UK*

HAO-CHING HSIAO • *Department of Molecular and Cellular Medicine, Texas A&M Health Science Center, College Station, TX, USA*
BRUNO HUDRY • *MRC Clinical Sciences Centre, Imperial College London, London, UK*
FRANÇOIS KARCH • *Department of Genetics and Evolution, University of Geneva, Geneva, Switzerland*
LAURA M.J. KETTYLE • *Centre for Cancer Research and Cell Biology, Queen's University Belfast, Northern Ireland, UK*
ZACHARIAS KONTARAKIS • *Max Planck Institute for Heart and Lung Research, Bad Nauheim, Germany*
ROBB KRUMLAUF • *Stowers Institute for Medical Research, Kansas City, MO, USA; Department of Anatomy and Cell Biology, Kansas University Medical Center, Kansas City, KS, USA*
BONY DE KUMAR • *Stowers Institute for Medical Research, Kansas City, MO, USA*
FRANCK LADAM • *Department of Biochemistry and Molecular Pharmacology, University of Massachusetts Medical School, Worcester, MA, USA*
BRIGITTE LAFOREST • *Inserm, GMGF UMR-S910, Aix-Marseille Université, Marseille, France*
BARBARA LAMBERT • *Institut des Sciences de la Vie, Université Catholique de Louvain, Louvain-la-Neuve, Belgium*
FABIO G. LIBERANTE • *Centre for Cancer Research and Cell Biology, Queen's University Belfast, Northern Ireland, UK*
DAHONG LIU • *Department of Biological Sciences, New York, NY, USA*
YING LIU • *Medical Genomics Laboratory, UAB School of Medicine, University of Alabama at Birmingham, Birmingham, AL, USA*
INGRID LOHMANN • *Centre for Organismal Studies (COS) Heidelberg, University of Heidelberg, Heidelberg, Germany*
ROBERT K. MAEDA • *Department of Genetics and Evolution, University of Geneva, Geneva, Switzerland*
IGNACIO MAESO • *Department of Zoology, University of Oxford, Oxford, UK*
RICHARD S. MANN • *Department of Biochemistry and Molecular Biophysics, Columbia University Medical Center, New York, NY, USA*
FERDINAND MARLÉTAZ • *Department of Zoology, University of Oxford, Oxford, UK*
NAVNEET KAUR MATHARU • *Centre for Cellular and Molecular Biology, Council of Scientific and Industrial Research, Hyderabad, India*
KATHLEEN S. MATTHEWS • *Department of Biochemistry and Cell Biology, Rice University, Houston, TX, USA*
DANIEL J. MCKAY • *Department of Biochemistry and Molecular Biophysics, Columbia University, New York, NY, USA*
SAMIR MERABET • *Institut de Génomique Fonctionnelle de Lyon, ENS de Lyon - CNRS UMR5242, Lyon, France*
RAKESH K. MISHRA • *Centre for Cellular and Molecular Biology, Council of Scientific and Industrial Research, Hyderabad, India*
LUIS DE NAVAS • *Centro de Biología Molecular Severo Ochoa (CSIC-UAM), Universidad Autónoma de Madrid, Madrid, Spain*
CHRISTOF NOLTE • *Stowers Institute for Medical Research, Kansas City, MO, USA*
JORDI PAPS • *Department of Zoology, University of Oxford, Oxford, UK*
MARK PARRISH • *Stowers Institute for Medical Research, Kansas City, MO, USA*

Delphine Paul • *Institut des Sciences de la Vie, Université Catholique de Louvain, Louvain-la-Neuve, Belgium*
Anastasios Pavlopoulos • *Max Planck Institute of Molecular Cell Biology and Genetics, Dresden, Germany; Howard Hughes Medical Institute, Janelia Farm Research Campus, Ashburn, VA, USA*
Laurent Perrin • *Technologies Avancées pour le Génome et la Clinique (TAGC), UMR 1090 INSERM, CNRS, Aix Marseille Université, Marseille, France*
Maria Polychronidou • *Centre for Organismal Studies (COS) Heidelberg, University of Heidelberg, Heidelberg, Germany*
Chaitanya Rastogi • *Department of Biological Sciences, Columbia University, New York, NY, USA; Department of Applied Physics and Applied Mathematics, Columbia University, New York, NY, USA*
René Rezsohazy • *Institut des Sciences de la Vie, Université Catholique de Louvain, Louvain-la-Neuve, Belgium*
Todd R. Riley • *Department of Biological Sciences, Columbia University, New York, NY, USA; Department of Systems Biology, Columbia University, New York, NY, USA; Department of Biology, University of Massachusetts - Boston, Boston, MA, USA*
Charles G. Sagerström • *Department of Biochemistry and Molecular Pharmacology, University of Massachusetts Medical School, Worcester, MA, USA*
Ernesto Sánchez-Herrero • *Centro de Biología Molecular Severo Ochoa (CSIC-UAM), Universidad Autónoma de Madrid, Madrid, Spain*
Caroline Sauvegarde • *Institut des Sciences de la Vie, Université Catholique de Louvain, Louvain-la-Neuve, Belgium*
Delia del Saz • *Centro de Biología Molecular Severo Ochoa (CSIC-UAM), Universidad Autónoma de Madrid, Madrid, Spain*
L.S. Shashidhara • *Indian Institute of Science Education & Research (IISER), Pashan, Pune, Maharashtra, India*
Matthew Slattery • *Department of Biochemistry and Molecular Biophysics, Columbia University Medical Center, New York, NY, USA; Department of Biomedical Sciences, University of Minnesota Medical School, Duluth, MN, USA*
Alexander Thompson, Ph.D. • *Centre for Cancer Research and Cell Biology, Queen's University Belfast, Northern Ireland, UK*
Patrick Tschopp • *Department of Genetics, Harvard Medical School, Boston, MA, USA*
Jean-Claude Twizere • *Laboratory of Signaling and Protein Interactions, GIGA-R, University of Liege, Liège, Belgium*
Stéphane Zaffran, Ph.D. • *Inserm, GMGF UMR-S910, Aix-Marseille Université, Marseille, France*

Part I

Hox Genes: From Discovery to Functional Analysis

Chapter 1

Discovery and Classification of Homeobox Genes in Animal Genomes

Ferdinand Marlétaz, Jordi Paps, Ignacio Maeso, and Peter W.H. Holland

Abstract

The diversification of homeobox genes is of great interest to evolutionary and developmental biology. To generate a catalogue of all homeobox genes within species of interest, it is necessary to sequence complete genomes. It is now possible for small research projects and individual laboratories to determine near-complete genome sequences of animal species. We provide bioinformatic methods for assembling draft genome sequences from any animal species, including read filtering and error correction, plus methods for extracting and classifying all homeobox sequences.

Key words Metazoa, Genomics, Genome assembly, Homeodomain, Phylogeny, Gene family, Molecular evolution

1 Introduction

The comparison of Hox genes between species has made major contributions to biology. For example, the discovery that Hox gene clusters are homologous between flies and mice started a revolution in comparative developmental biology [1, 2]. In evolutionary biology, Hox genes were important in supporting a "new animal phylogeny" originally proposed from ribosomal DNA analysis [3]. Hox gene comparisons also proved pivotal in reconstructing major events in molecular evolution, such as the timing of genome duplications [4, 5]. Initially, the main methods for isolating Hox genes used degenerate PCR [6], never guaranteed to amplify all relevant genes, coupled with RACE-PCR or inverse-PCR to extend clones. More laborious, but ultimately more informative was library screening, genomic walking, and clone-by-clone sequencing [4]. In recent years, significant advances in DNA sequencing technologies have enabled new genome-scale methods that bypass degenerate PCR and library screening. Furthermore, these analyses enable survey of all homeobox genes, not just Hox genes.

Yacine Graba and René Rezsohazy (eds.), *Hox Genes: Methods and Protocols*, Methods in Molecular Biology, vol. 1196, DOI 10.1007/978-1-4939-1242-1_1, © Springer Science+Business Media New York 2014

Hox genes comprise a minority of the total number of homeobox genes in most genomes. For example, many invertebrate genomes contain around 100 homeobox genes, of which usually 10–15 are Hox genes; vertebrate genomes have over 200 homeobox genes, of which around 35–50 are Hox genes [7]. Homeobox genes are defined by possession of a DNA sequence, the homeobox, usually 180 nucleotides in length coding for a 60-amino acid domain within the encoded protein, although some homeobox sequences are longer. Homeobox sequences are extremely variable, such that sequence similarity searching using a single homeobox sequence will not detect divergent homeobox sequences. Despite this variation, each homeobox sequence shows sequence similarity to a subset of other homeobox sequences, and the encoded homeodomain has the potential to fold into three alpha helices [8].

In animals, homeobox genes can be divided into around 11 gene classes, of which the largest are the ANTP class (including Hox genes) and the PRD class [9]. There are many situations when we may wish to determine the full complement of homeobox genes in a genome. These include studies investigating links between genomic and morphological evolution, and questions in developmental biology when one needs to know if a gene under study has closely related duplicate genes. The diversity of homeobox genes is also a good indicator of overall evolutionary trends in genome evolution, such as secondary simplification or whole genome duplication. However, sequence similarity tools such as BLAST are not precise enough to delineate gene families and define groups of "related genes." We therefore recommend compiling a full list of all homeobox genes, before drawing evolutionary conclusions or designing further studies in developmental biology. Here we provide protocols for compiling complete lists of homeobox genes from animal genomes, and classifying these according to evolutionary relationships. We describe methods to use when a genome sequence is already available and we outline an economical strategy involving draft genome sequencing, assembly, and analysis (Fig. 1).

2 Materials

2.1 Genome

2.1.1 Track 1 (Without a Publicly Available Genome Sequence)

1. Sequencing-grade genomic DNA (High Molecular Weight DNA; we recommend using Qiagen DNeasy kits).
2. High-memory server (Linux 64 bits, 256–512 GB RAM depending on genome size).
3. Sequence filtering tools: FASTX-Toolkit (http://hannonlab.cshl.edu/fastx_toolkit/) or Sickle (https://github.com/najoshi/sickle).

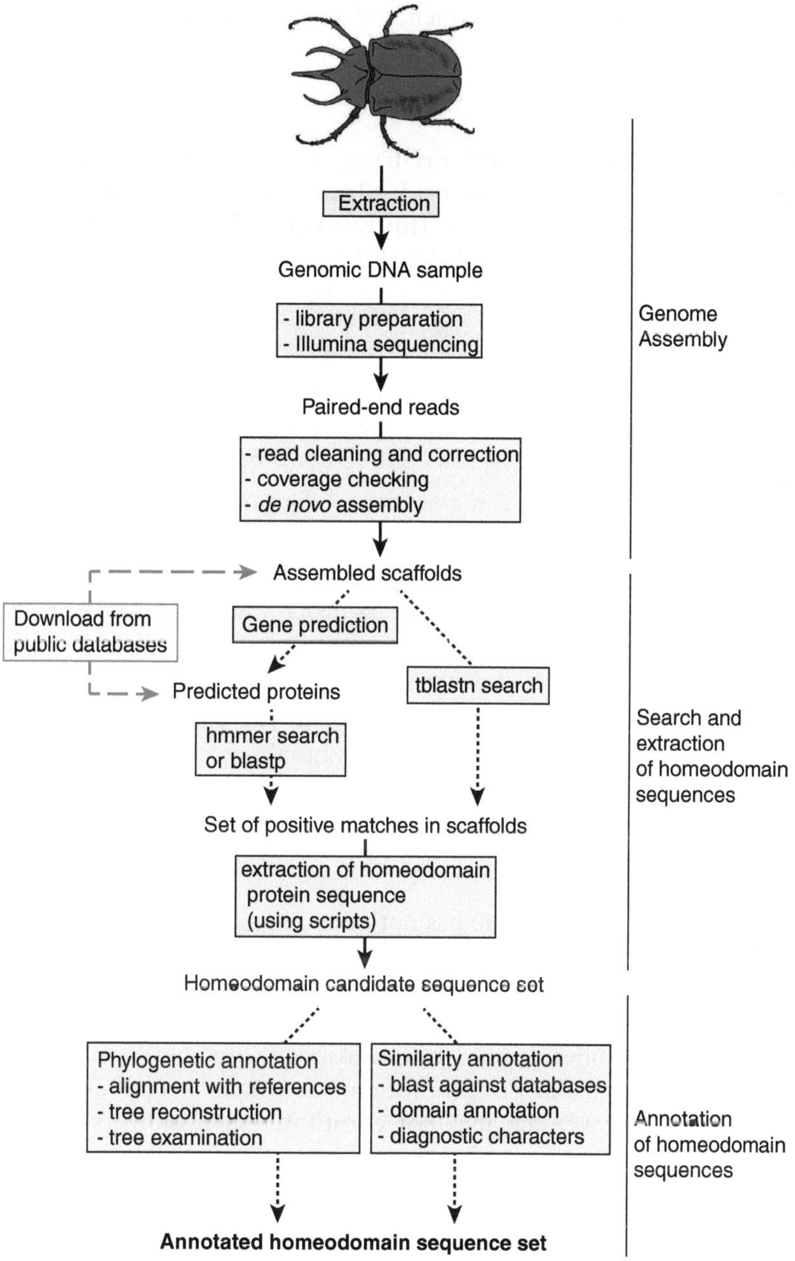

Fig. 1 Flowchart describing the main steps and alternatives in a protocol for extracting and classifying homeobox genes in metazoan genomes

4. Error correction tool: Musket (http://musket.sourceforge.net) [10].
5. *de Bruijn* assembler: Velvet (http://www.ebi.ac.uk/~zerbino/velvet/) [11].
6. Coverage estimation tool: JELLYFISH (http://www.cbcb.umd.edu/software/jellyfish/) [12].

2.1.2 Track 2 (with a Publicly Available Genome Sequence)

1. Genome sequence (FASTA file of scaffolds).
2. Gene models if available (FASTA file of proteins) and/or genome sequence translated in six frames.

2.2 Mining for Homeobox Sequences

Tools for similarity-based searching: the BLAST+ suite enables searching nucleotide databases (tblastn) or protein databases (program blastp) (ftp://ftp.ncbi.nlm.nih.gov/blast/executables/blast+/) [13]; HMMER is a sensitive protein similarity detection tool based on a probabilistic profile of a gene family (http://hmmer.janelia.org).

2.3 Alignment and Phylogeny Reconstruction

1. Custom programs enabling extraction of similarity domains from scaffolds/gene models: hmmer-parse.pl and parse-blast-hd.py (https://github.com/fmarletaz).
2. MAFFT is a sequence alignment tool (http://mafft.cbrc.jp/alignment/software/) [14].
3. RAxML is a computationally efficient maximum-likelihood tree inference software particularly well suited for large trees (https://github.com/stamatak/standard-RAxML) [15]. Version 7.2.6 can be downloaded from http://www.exelixis-lab.org/
4. Tree visualization program: TreeView (http://taxonomy.zoology.gla.ac.uk/rod/treeview.html).

3 Methods

If the genome has not been sequenced for your species of interest (Track 1), then Subheading 3.1 must be followed to determine and assemble a draft genome sequence, before Subheadings 3.2 and 3.3 are possible. If a genome sequence is already available, Subheading 3.1 can be skipped (Fig. 1). We stress, however, that availability of genome sequence data on a publicly accessible server or website does not constitute permission to use these data for extensive analysis since data are often deposited by researchers before their own analysis and publication. Prior permission should always be sought from the researchers who deposited genome data. In the following section, we will introduce a number of command lines, each preceded by a "$" sign. We attempt to provide an explanation for each of the parameters. However, we also encourage the users to read the documentation accompanying each computer program.

3.1 Genome Sequencing and Assembly

3.1.1 Genome Sequencing

The small size of homeobox sequences (~180 nucleotides, coding for a homeodomain of ~60 amino acids) makes it possible to detect and extract them from relatively small genomic contigs. The sequencing of a low coverage genome and its assembly in short contigs or scaffolds therefore constitutes a valuable alternative to a time-consuming and uncertain degenerate PCR screen. Using the current DNA sequencing platforms, a coverage of 15–30× is generally considered a minimum and 100× as an optimum to undertake a genome assembly [11, 16]. For detection of homeobox sequences, coverage towards the minimum range (15–30×) is suitable in most cases. For a metazoan genome, typically between 150 Mb (e.g. some nematodes and insects) and 3 Gb (e.g. mammals), this can be achieved by sequencing between 70 and 200 million typical Illumina 100 bp paired-end reads. This constitutes between 20 and 100 % of a single Illumina HiSeq2000 lane (*see* **Note 1**). The basic library type recommended is a fragment library for which 100 bp will be sequenced at each extremity of a fragment sized between 180 and 600 bp (the example below assumes 250 bp). During the assembly step, the paired-end information is deployed to build larger scaffolds.

3.1.2 Checking Coverage

1. We do not estimate genome size prior to sequencing, but use data analysis to estimate depth of coverage. The *k*-mer spectrum approach is suitable for this, and provides essential information about the depth of sequencing performed and the feasibility of assembly.
2. If the obtained coverage is too low (under 10×), it might be necessary to generate more sequence coverage; this can be done from the same library. For instance, the JELLYFISH program will count any occurrence of a string of K nucleotides (*k*-mer) in the genome data, which translates easily into genome coverage information (Fig. 2) using the formula $C_k = C \times (L - k + 1)/L$, where C_k is *k*-mer coverage, C is base coverage, L is read length, k is *k*-mer size.

```
$ jellyfish count -m 17 -t 8 -s 50G--both-strands -o yourgenome_17 reads_1.fq reads_2.fq
```

```
$ jellyfish histo -o yourgenome_17.hist.txt yourgenome_17_0
```

where -m indicates the *k*-mer size (17 in this example), -t the number of allocated processors (8 in this example), -s the allocated memory (50 Gb in this example), --both-strands indicates that *k*-mer should be counted on both strands of DNA sequence and -o specifies the output file.

3.1.3 Read Preparation

1. Before undertaking genome assembly itself, we recommend performing quality controls and filtering. After downloading read files in fastq format, fastQC allows one to check for read quality, possible PCR artifacts during library preparation and presence of remaining adapters or contaminants. Such undesirable sequence can then be removed using an adapter trimmer program, such as scythe. This adapter filtering is sometimes performed by a sequencing center before making the read files accessible.

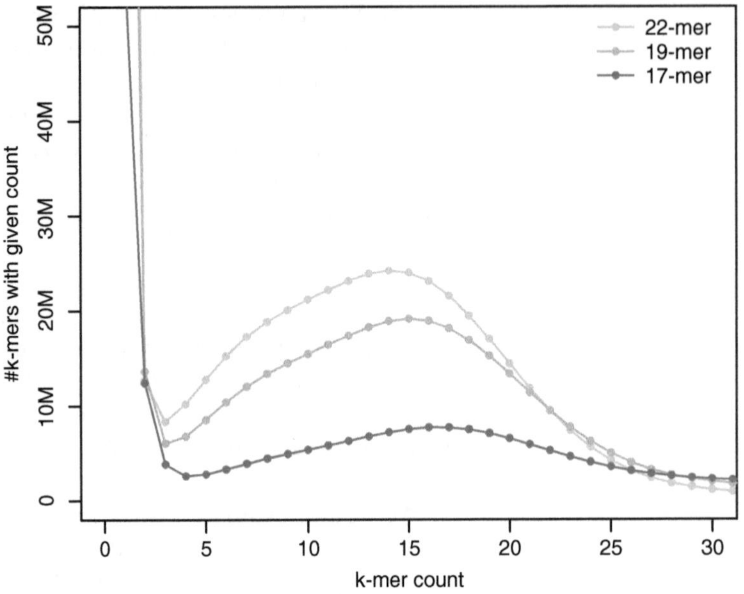

Fig. 2 Example *k*-mer coverage derived from 44 million paired-end reads (8.8 Gb) of DNA from a lepidopteran species. Higher *k*-mer size (17–22-mers) results in an increase of *k*-mer diversity and a slight decrease of the *k*-mer coverage. The high number of low frequency *k*-mers is attributable to sequencing errors. This graph can be used to provide an estimate of genome size and genome coverage. The observed *k*-mer coverage (approximately 15×) can be converted into base coverage (~depth of genome sequencing) using the formula $C_k = C^*(L - k + 1)/L$ where C_k is the *k*-mer coverage, C the base coverage, L the read length, and k the *k*-mer length. In this example, the base coverage is calculated to be 18.3×. This in turn translates to a genome size of 480 Mb (8,800 Mb sequence obtained divided by 18.3×). An alternative way is to directly use the *k*-mer coverage by calculating the total number of "good" *k*-mers likely represented at least three times (to account for sequencing errors), which gives 7.4 billion *k*-mers. Dividing this by the observed ~15× *k*-mer coverage indicates a ~490 Mb genome size

```
$ scythe -a adapter_file.fa -o trimmed_reads.
fq reads.fq
```

where -a points to the file containing the adapter sequences, -o names the output file and the final argument indicates the input read file.

2. After this initial step, any stretches of low quality bases still present in reads (for example, at the ends of reads) should be removed using a quality trimmer program, for instance Sickle. If the read length after trimming is below a minimal length, the read is discarded, but the now "orphan" sequence from the other end of the fragment is written to a separate file and used in assembly as a single read.

```
$ sickle pe -t sanger -f reads_1.fq -r reads_2.
fq -o reads_1.filt.fq -p reads_2.filt.fq -s
reads_single.filt.fq
```

where -t indicates the quality encoding scheme, -f and -r the input paired-read files and -o, -p, and -s the two paired and the single filtered read output files.

3. Finally, error correction of the reads allows removal of residual errors, thereby simplifying the subsequent assembly steps and decreasing RAM consumption. Most common error correction programs rely on *k*-mer spectrum to discriminate *bona fide* and error-prone *k*-mers (Fig. 2). We propose using the program Musket as an error correction tool as it is quite simple to use, choosing a *k*-mer size of 21 (-k), and 8 threads (-p):

```
$ musket -k 21 1000000000 -p 8 -omulti cor_
reads -inorder reads_1.filt.fq reads_2.filt.fq
reads_single.filt.fq
```

3.1.4 Genome Assembly

1. A wealth of guidelines and programs are now available to perform genome assembly using Illumina sequence data [16]. Assembly programs using the *de Bruijn* graph have been developed to handle efficiently the enormous amount of sequence produced by next-generation sequencing. Schematically, a *de Bruijn* graph assembler proceeds by splitting sequence data into *k*-mers (short sequences of length *k*), building an overlapping assembly graph by chaining successive *k*-mers together, and then recovering assembled contigs by traversing this graph in a way that minimizes the number of vertexes visited repeatedly [17]. We propose here using the Velvet package, which was one of the first *de Bruijn* assemblers released and in our experience performs remarkably well on small and medium sized datasets (*see* **Note 2**). Velvet uses large amounts of RAM but this can be partially overcome by use of trimmed and error-corrected reads as previously described. The most important parameter is the choice of a *k*-mer size, which represents a trade-off between selectivity and sensitivity. Low *k*-mer sizes ensure more data are included in assemblies but to the detriment of accuracy; long *k*-mer sizes ensure more accurate results but require higher genome sequence coverage (this relationship is not straightforward, and is affected by genome properties such as repeat content and polymorphism). Our recommended *k*-mer values are appropriate for the 15–30× genome sequence coverage suggested above.

2. The first program to run from the Velvet package is `velveth`, which generates hash tables for the sequence along a range of *k*-mers. We recommend exploring *k*-mer values ranging between 31 and 61 in steps 10 (*k*-mer values should always be odd numbers):

```
$ velveth63 yourgenome 31,61,10 -shortPaired
-separate-fastq  read_1.cor.filt.fq  read_1.
```

```
cor.filt.fq -short -fastq read_single.cor.
filt.fq
```

3. This command will generate several directories called `yourgenome_21, yourgenome_31 … yourgenome_61`. To launch the assembly itself (program `velvetg`), you must specify the fragment length of the sequenced library as this is important for the scaffolding step. In the command line below, this parameter is set to 250 nucleotides and its deviation to 50 but should be altered if a different library construction was used. Note that `velvetg` can use large amounts of RAM, typically 150–450 GB.

```
$ velvetg yourgenome_XX -ins_length 250 -ins_
length_sd 50  -cov_cutoff auto -exp_cov auto
```

4. The assembly produces a `contigs.fa` file which contains the scaffolds. The program also outputs useful statistics regarding the quality of the assembly and the fraction of reads used. The most important output parameters to examine are the N50 value and the total assembly size. N50 is a measure of the length of the assembled fragments (the larger the better), and indicates that half the assembly is represented in contigs at least as long as this value. The total assembly size should be close to the expected genome size, but may be lower if a genome has many repeated DNA sequences, as these may "collapse."
5. In most genome projects the step after genome assembly would be to run gene prediction across the entire assembly (*see* **Note 3**). If the goal is to extract homeobox sequences, prediction of all genes is not necessary. If the process generates an N50 above 1 kb and the assembly size seems reliable, you can start to use the assembly for mining homeobox genes.

3.2 Mining Genome Sequence for Homeobox Candidates

Due to their high level of sequence conservation in metazoans, homeobox genes can be detected by similarity searches using well-established tools such as BLAST or HMMER. BLAST has the advantage of versatility and can be applied directly to nucleotide sequences, which can be convenient for draft quality genomes. However, HMMER is more sensitive as it relies on a search profile built from multiple representatives of a gene family or class. We recommend using both approaches, and merging the results. In all cases, the "query" file will be known amino acid sequences of homeodomains from other species; searching using protein queries is far more sensitive than searching using nucleotide queries, even when searching a nucleotide database.

3.2.1 Search the Genome: BLAST

1. To perform BLAST searches, the first step is to generate a "query file" containing a diverse set of known homeodomains (amino acid sequences). We recommend making a query file

including the full diversity of homeodomains. The HomeoDB database (http://homeodb.cbi.pku.edu.cn; http://homeodb.zoo.ox.ac.uk) [7] provides curated and annotated sets of homeodomains for several species. The "download" command on the front page allows amino acid sequences of homeodomains from particular species to be rapidly downloaded. A larger range of species can be accessed via the Pfam databases (http://pfam.sanger.ac.uk/family/homeobox), although this does not include the full diversity of relevant sequences as it uses a more restricted definition of homeobox. To download complete sequences from Pfam (not just the homeodomain region) to be used in BLAST searches, click on the menu option "Species" then the tab "Tree", click on "Generate tree" and wait until it loads. Browse the tree for your taxonomic groups of interest (e.g. insects, vertebrates) and select them using tick boxes. Click on "Download sequences accessions" and after saving, use these to download the sequences as a FASTA file using "Swissprot Batch Download" (http://www.uniprot.org/batch/).

2. It is then necessary to build a BLAST database from your genome assembly in FASTA format. Different databases may be built depending on the type of genomic sequence data. In most cases, a nucleotide database will be built from the scaffolds, which will be searched using protein sequences downloaded from HomeoDB and/or SwissProt.

   To build the scaffolds database, use the following command:

```
$ makeblastdb -in contigs.fasta -dbtype 'nucl' -parse_seqids -out DB_scaffolds
```

   where `-parse_seqids` states that sequences should be indexed in the database for future retrieval.

3. A local BLAST search is then run on this scaffolds database, using the command:

```
$ tblastn -query query.fasta -db DB_scaffolds -out blast_output -outfmt 6
```

   where `query.fasta` is the file containing the amino acid sequences downloaded from HomeoDB or SwissProt, and `DB_scaffolds` is the BLAST database built from the scaffold DNA data. The flag `-outfmt 6` instructs BLAST to produce a tabular output; remove it if you prefer to see an output similar to traditional `BLAST` output.

4. If predicted proteins are available for your genome of interest, a protein database can be also built (for searching using blastp); the advantage is that homeobox sequences will not be split by introns, but we still recommend also building a nucleotide database in case some homeobox genes have been inadvertently missed from the predicted protein set.

To build a database from the predicted proteins:

```
$ makeblastdb -in predicted_proteins.fasta
-dbtype prot -out DB_proteins
```

To perform a BLAST search on the predicted proteins database, replace `tblastn` by `blastp` and the database name (`DB_proteins`).

```
$ blastp -query query.fasta -db DB_proteins
-outfmt 6 -out blastp_output
```

3.2.2 Search the Genome: HMMER

1. In the Pfam homeobox entry (http://pfam.sanger.ac.uk/family/homeobox) obtain the HMM profile for homeobox genes, which will be the query to use in a HMMER search: click "Curation & Model" then "Download." This profile will be used to search the predicted proteins of your genome or a file derived directly from genomic DNA translated in all six reading frames (e.g. using transeq, http://www.ebi.ac.uk/Tools/st/emboss_transeq/). The former may be missing genes if gene prediction is incomplete; the latter has the disadvantage that homeobox sequences may contain introns that are not taken into account in the HMMER profile.
2. Run:

```
$ hmmsearch hmmer_profile.hmm predicted_
proteins.fasta > hmmer_output
```

where `hmmer_profile.hmm` is the HMMER profile downloaded from Pfam in **step 1**, `predicted_proteins.fasta` is the genome data (predicted proteins or DNA translated in six frames) and `hmmer_output` is the output file.

3.2.3 Extract Homeobox Sequences

Homeodomains are short functional domains within larger proteins. To ease sequence characterization and alignment, we designed an approach to extract the homeodomain from the surrounding context. To perform this task, we provide two scripts specifically designed to handle the output of HMMER and tblastn, respectively.

1. The HMMER search uses just one sequence profile to search for all homeodomains, and so is devoid of redundancy in its report. The `hmm-parse.pl` script uses the Bioperl library that contains already written code for parsing the output of most common sequence analysis programs (e.g. BLAST, HMMER); it will parse the HMMER output and extract the homeodomain sequences from the corresponding fasta file. An option permits extension of the extracted region around the homology match (e.g. sequences flanking the homeobox) according to a specified "offset" parameter (the number 10 in the command line below specifies ten amino acids either side):

```
$ hmmer-parse.pl hmmer_output predicted_pro-
teins.fasta 10
```

2. In contrast to HMMER, the tblastn search uses a large number of query sequences, and each of these will detect overlapping sets of genes. In the tblastn report, therefore, each homeobox gene in the species of interest will be represented multiple times. The `parse-blast-hd.py` script attempts to filter this redundancy by reporting the "best" query for each matched homeodomain. It then extracts the nucleotide sequence and performs translation of the nucleotide sequences. The protein translation and in-frame nucleotide files are written in output files for subsequent analysis with relevant information in the header about the position in the scaffold and the name of the "best" query, plus a `prefix` which may be used to add a species tag. The script uses indexed blast databases obtained with `-parse_seqids` to retrieve scaffolds from large genomes and relies on the BioPython library for translation. Similar to Bioperl, Biopython contains code and modules generated by an active community of bioinformaticians to simplify handling sequence data and parsing files generated by common bioinformatics programs. The `querysize` parameter specifies the length of the blast query files to extend possibly incomplete blast alignments. We recommend setting this value to 60.

```
$ parse-blast-hd.py DB_scaffolds blast_out-
put prefix querysize
```

3. If predicted proteins were used, not genomic scaffolds, simple methods can be used to merge HMMER and BLAST results, removing redundant sequences. This can be done in either Unix (combining the commands `cut` to extract the column containing the hits and `diff` to compare the hits from BLAST in one file versus the hits in HMMR in another), or in Microsoft Excel (import the tabular outputs into different columns, order alphabetically and compare, or paste into a single column and use "Remove Duplicates"). These simple methods are not recommended if scaffolds are searched, since a scaffold may have multiple genes.

3.3 Classifying Homeobox Sequences

Correct classification of genes is necessary if genes are to be compared between species. We deploy a combination of approaches, combining BLAST searches against public databases, phylogenetic analysis, and scanning for companion domains or diagnostic characters.

3.3.1 BLAST Against GenBank and HomeoDB

1. Using the `-remote` option, it is possible to BLAST the non-redundant (nr) GenBank database without having to download it. For example, to BLAST the list of gene candidates

extracted from your genome of interest against GenBank, for a tabular output, use

```
$ blastp -query candidates.fasta -db nr
-remote -out outputfile -outfmt 6
```

2. It is also possible to use BLAST to compare the candidate genes against HomeoDB (http://homeodb.zoo.ox.ac.uk or http://homeodb.cbi.pku.edu.cn/) but only one by one. Alternatively, sequences from HomeoDB can be downloaded to make a local database against which a query database can be searched. Be aware that HomeoDB comprises homeodomain sequences for homeobox genes but does not contain the complete protein sequences.

3.3.2 Companion Domains

Some homeobox gene classes and families have other conserved domains in the protein (PRD, POU, CUT etc.) that can be of assistance in classification [8]. Companion domains can be found using Pfam batch (http://pfam.sanger.ac.uk/search), CDD (http://www.ncbi.nlm.nih.gov/Structure/cdd/wrpsb.cgi) or SMART (http://smart.embl-heidelberg.de/).

3.3.3 Alignment and Diagnostic Characters

Some homeobox gene classes or families have specific diagnostic characters that can help in correctly classifying genes. Before these can be checked, sequences must be aligned. Several alignment programs are available (several can be found at http://www.ebi.ac.uk/Tools/msa/). We recommend MAFFT which has a good balance between ease of use and accuracy (http://mafft.cbrc.jp/alignment/server/); the L-INS-I algorithm gives good results for homeobox genes. If the number of sequences is very high (>200 sequences), consider downloading MAFFT rather than using it online, to gain access to its multi-processor options (http://mafft.cbrc.jp/alignment/software/multithreading.html, only non-Windows version support multi-threading).

Example diagnostic characters are:

- ANTP class (some): Hexapeptide sequence I(Y/F)PWM(K/R) N-terminal to the homeodomain.
- NKL subclass of ANTP class (some): *tinman* motif TPFSVKDIL(S/N)L(P/E) N-terminal to the homeodomain.
- Gsc and Mix families (PRD Class): K at position 50 (not unique).
- HNF class: extra amino acids between helix 2 and 3.
- SINE class: K at position 50 (not unique).
- TALE class: three extra amino acids between helix 1 and helix 2 of homeodomain.
- Meis, Tgif families (TALE Class): Iso at position 50.
- Mkx family (TALE Class): Ala at position 50.

- Pbx family (TALE Class): Gly at position 50.
- Cmp family (CUT Class): ten extra amino acids between helix 1 and 2.
- Cux family (CUT Class): His at position 50.
- PROS/Prox: three extra amino acids between helix 2 and 3.

3.3.4 Phylogenetic Tree Reconstruction

1. Sequence alignment (Subheading 3.3.3) is also a necessary preliminary step to drawing a phylogenetic tree. The first decision to be made is which other species should be included in a tree alongside the sequences obtained from your genome of interest. We recommend alignment to homeodomains from least two reference species with well-annotated homeobox genes (Fig. 3). If a phylogenetic tree of all homeobox genes is attempted, the alignment should be restricted to the homeodomain itself (*see* **Note 4**). We also stress that phylogenetic

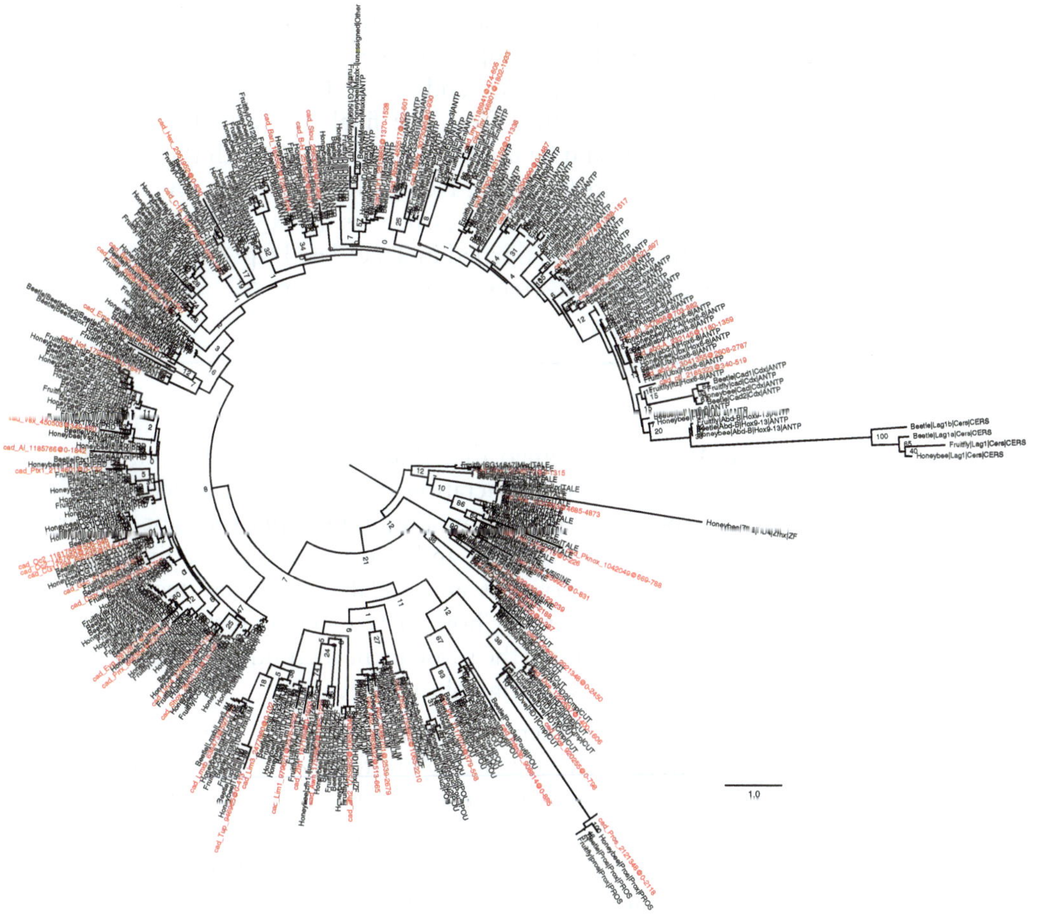

Fig. 3 Example of extraction of deduced homeodomain sequences from a trichopteran species using the approach described. Sequences in *red* are from the newly sequenced species, presented in a radial phylogenetic tree along with homeodomains from *Drosophila*, *Tribolium*, and *Apis*

trees of large gene datasets should be built from deduced amino acid sequence, not nucleotide sequence, as the ancient divergence times between different gene classes and families will have saturated silent nucleotide substitutions.

2. We propose use of maximum likelihood implemented in the RAxML program (*see* **Note 5**). To run a tree using the most complex evolutionary model (LG + Gamma + Invariants) type:

```
$ raxml -f a -x 12345 -d -# 1000 -m PROTGAMMALGI
-s input_file.phy -n output_file. out
```

The name of the binary file (`raxml`) may change depending on how the program was installed, or if the multithreaded version was compiled (the RAxML documentation provides further information). The flag `-f a` triggers the default phylogenetic analyses. The flag `-x <odd seed number>` is required to provide an odd seed number for the computer random number generator. The flag `-d` instructs RAxML to start the search from a randomly generated tree, instead of the default Maximum Parsimony tree. Parameter `-# 1000` sets the number of bootstrap replicates, `-m PROTGAMMALGI` sets the evolutionary model, and `-s` and `-n` define, respectively, the input and output files. The final tree with bootstrap support values should be found in a Newick file with the word "bipartitions" in its file name. Newick files represent phylogenetic trees as nested parentheses, such as ((tax_A,tax_B),(taxC,taxD)); such filescan be opened in TreeView http://taxonomy.zoology.gla.ac.uk/rod/treeview.html (*see* **Note 6**).

4 Notes

1. We have assumed that Illumina HiSeq2000 will be the platform used for generation of genome sequence data. In our experience, this technology represents the best compromise between read length and cost per Mb at the time of writing. Other technologies are available, and new technologies will certainly emerge. Choice of sequencing technology should not change radically the subsequent methods advised for extraction and analysis of homeobox sequences.
2. Alternative assembly programs, besides Velvet, include ABySS (http://www.bcgsc.ca/platform/bioinfo/software/abyss) and SOAPdenovo (http://soap.genomics.org.cn/soapdenovo.html) [18, 19]. All programs can perform well in certain situations, and we have recommended one that has worked well in our experience for this application.
3. If the genome assembly is of good quality enough, complete gene prediction could be envisaged across all genes. This is not necessary if only homeobox sequences are sought, but can be

helpful in predicting full gene sequences. We would recommend this only if 90 % of the genome sequence (N90) is in fragments longer than average gene size (typically 5 kb in invertebrate species and 15 kb in mammals). Accurate gene prediction can be a demanding process involving ab initio software and experimental data, including transcriptome sequencing both for direct annotation of genes and to "train" software in the codon usage bias of the organism under study [20]. If the species of interest is close to one for which some training has already been undertaken (such as insects), it can be worth trying a program like AUGUSTUS [21], even without transcriptome data from the species under study.

```
$ augustus --species your_species scaffolds.fa
```

4. Since the N-terminal and C-terminal limits of homeodomains are well defined, there is no need to check for nonalignable regions if the sequences comprise homeodomains only. However, if the sequences aligned include sequence outside the homeodomain (or indeed if these protocols are adapted for use on other types of genes), nonalignable regions should be removed prior to construction of a phylogenetic tree. Several programs are available for this, including GBlocks (http://molevol.cmima.csic.es/castresana/Gblocks_server.html) and TrimAL (http://trimal.cgenomics.org/) [22, 23]. In Gblocks, we recommend selecting "Options for a less stringent selection." Gblocks produces error messages if the sequences names contain non-alphanumeric characters.
5. There is a huge literature on phylogenetic reconstruction, with the two most popular approaches currently being Maximum Likelihood (ML) and Bayesian Inference (BI). Both methodologies can be implemented by an array of programs. For example, ML can be inferred in the popular RAxML or PhyML among others [24], while BI analyses can be performed in MrBayes and Phylobayes [25, 26]. We have recommended ML implemented in RAxML as it provides a good compromise between speed and accuracy, and works well for large numbers of short sequences.
6. TreeView is a highly versatile but easy-to-use tree visualization program. Alternatives include FigTree (http://tree.bio.ed.ac.uk/software/figtree/) and MEGA5 (http://www.megasoftware.net/) [27].

Acknowledgements

The authors' research is funded by the European Research Council under the European Union's Seventh Framework Programme (FP7/2007–2013)/ERC grant [268513]11.

References

1. Duboule D, Dollé P (1989) The structural and functional organization of the murine HOX gene family resembles that of Drosophila homeotic genes. EMBO J 8:1497
2. Graham A, Papalopulu N, Krumlauf R (1989) The murine and Drosophila homeobox gene complexes have common features of organization and expression. Cell 57:367–378
3. de Rosa R, Grenier JK, Andreeva T et al (1999) Hox genes in brachiopods and priapulids and protostome evolution. Nature 399:772–776
4. Garcia-Fernández J, Holland PWH (1994) Archetypal organization of the amphioxus Hox gene cluster. Nature 370:563–566
5. Amores A, Force A, Yan YL et al (1998) Zebrafish hox clusters and vertebrate genome evolution. Science 282:1711–1714
6. Pendleton JW, Nagai BK, Murtha MT et al (1993) Expansion of the Hox gene family and the evolution of chordates. Proc Natl Acad Sci U S A 90:6300–6304
7. Zhong Y-F, Holland PWH (2011) HomeoDB2: functional expansion of a comparative homeobox gene database for evolutionary developmental biology. Evol Dev 13:567–568
8. Bürglin TR (2011) Homeodomain subtypes and functional diversity. Subcell Biochem 52:95–122
9. Holland PWH, Booth HAF, Bruford EA (2007) Classification and nomenclature of all human homeobox genes. BMC Biol 5:47
10. Liu Y, Schröder J, Schmidt B (2013) Musket: a multistage k-mer spectrum-based error corrector for Illumina sequence data. Bioinformatics 29(3):308–315
11. Zerbino DR, Birney E (2008) Velvet: algorithms for de novo short read assembly using de Bruijn graphs. Genome Res 18:821–829
12. Marçais G, Kingsford C (2011) A fast, lock-free approach for efficient parallel counting of occurrences of k-mers. Bioinformatics 27(6): 764–770
13. Camacho C, Coulouris G, Avagyan V et al (2009) BLAST+: architecture and applications. BMC Bioinformatics 10:421
14. Katoh K, Standley DM (2013) MAFFT multiple sequence alignment software version 7: improvements in performance and usability. Mol Biol Evol 30:772–780
15. Stamatakis A (2006) RAxML-VI-HPC: maximum likelihood-based phylogenetic analyses with thousands of taxa and mixed models. Bioinformatics 22(21):2688–2690
16. Miller JR, Koren S, Sutton G (2010) Assembly algorithms for next-generation sequencing data. Genomics 95:315–327
17. Compeau PEC, Pevzner PA, Tesler G (2011) How to apply de Bruijn graphs to genome assembly. Nat Biotechnol 29:987–991
18. Li R, Zhu H, Ruan J et al (2010) De novo assembly of human genomes with massively parallel short read sequencing. Genome Res 20:265–272
19. Simpson JT, Wong K, Jackman SD et al (2009) ABySS: a parallel assembler for short read sequence data. Genome Res 19:1117–1123
20. Yandell M, Ence D (2012) A beginner's guide to eukaryotic genome annotation. Nat Rev Genet 13:329–342
21. Stanke M, Keller O, Gunduz I et al (2006) AUGUSTUS: ab initio prediction of alternative transcripts. Nucleic Acids Res 34(Web Server issue):W435–W439
22. Capella-Gutiérrez S, Silla-Martínez JM, Gabaldón T (2009) trimAl: a tool for automated alignment trimming in large-scale phylogenetic analyses. Bioinformatics 25:1972–1973
23. Castresana J (2000) Selection of conserved blocks from multiple alignments for their use in phylogenetic analysis. Mol Biol Evol 17(4):540–552
24. Guindon S, Gascuel O (2003) A simple, fast, and accurate algorithm to estimate large phylogenies by maximum likelihood. Syst Biol 52:696–704
25. Ronquist F, Teslenko M, van der Mark P et al (2012) MrBayes 3.2: efficient Bayesian phylogenetic inference and model choice across a large model space. Syst Biol 61:539–542
26. Lartillot N, Lepage T, Blanquart S (2009) PhyloBayes 3: a Bayesian software package for phylogenetic reconstruction and molecular dating. Bioinformatics 25:2286–2288
27. Tamura K, Peterson D, Peterson N et al (2011) MEGA5: molecular evolutionary genetics analysis using maximum likelihood, evolutionary distance, and maximum parsimony methods. Mol Biol Evol 28:2731–2739

Chapter 2

How to Study Hox Gene Expression and Function in Mammalian Oocytes and Early Embryos

Delphine Paul, Caroline Sauvegarde, René Rezsohazy, and Isabelle Donnay

Abstract

Mammalian oocytes and early embryos have unique characteristics and can only be obtained in small amounts. As a consequence, the techniques to be used to characterize gene expression and function have to be adapted. It is also important to keep in mind that differences exist between mammalian species. Here we describe a set of techniques useful to analyze gene expression in oocytes and early bovine embryos, including techniques to quantify maternal and embryonic transcripts by RT-qPCR, to evaluate the translation potential of maternal transcripts, to knock down HOX transcripts by injection of siRNA, and to localize HOX proteins by whole-mount immunofluorescence.

Key words Mammalian oocyte, Early embryo, RT-qPCR, RNA silencing, Whole-mount immunofluorescence

1 Introduction

Studying gene expression and protein function in oocytes and early mammalian embryos is always a challenge due to, first of all, the low quantity of available biological material, but also to the specificities of such material. The oocyte is a unique cell that accumulates and stores proteins, lipids, organelles as well as mRNAs to support its own maturation, but also fertilization and the first steps of embryonic development. Indeed, soon after the onset of nuclear maturation, transcription stops and completion of maturation, fertilization, and early embryo development up to the major onset of the embryonic genome is under the control of proteins and mRNAs stored in the oocyte (reviewed in ref. 1). Before oocyte maturation, stored maternal mRNAs are associated with specific proteins forming ribonucleoproteic particles. During maturation and early embryo development, protein synthesis is regulated at the translational level, namely through the regulation of cytoplasmic adenylation of the maternal transcripts. Indeed, a large part of the maternal

Yacine Graba and René Rezsohazy (eds.), *Hox Genes: Methods and Protocols*, Methods in Molecular Biology, vol. 1196, DOI 10.1007/978-1-4939-1242-1_2, mRNAs is deadenylated and some of them will be re-adenylated before being translated during oocyte maturation or during the first embryonic cleavages. This process is regulated through some mRNA-specific sequences, referred to as cytoplasmic polyadenylation elements (CPE) (reviewed in ref. 2).

The maternal to embryonic transition (MET) which refers to the onset of embryonic genes activation, occurs at specific embryonic stages depending on the species. For example, it occurs at the first cell cycle in the mouse, at the third cell cycle in the pig and the human or at the fourth cell cycle in the bovine. This transition is characterized by the degradation of the remaining maternal transcripts and by the major onset of the embryonic genome that takes the control of embryo development. But maternal proteins can persist in the embryo after the MET.

Considering the successive steps of oocyte maturation and early embryo development relying either on maternal determinants (mRNAs and proteins) or on zygotic gene expression, it is important to take into account the stage of embryo development (before or after the MET) when studying gene expression in oocytes and early embryos: before the MET, the very large majority of the transcripts and proteins are of maternal origin, part of the transcripts are deadenylated and most of them will not be translated into proteins. The way mRNA extraction and reverse transcription (RT) are performed thus determines the type of results obtained: for example, if oligo(dT) are used for RT, only the transcripts with a poly-A tail will be reverse transcribed, while the use of hexamers allows to reverse transcribe all the transcripts with the same efficiency [3]. Moreover, the possible presence of maternal proteins, sometimes long after the MET, has to be taken into account in functional studies, namely when studying KO embryos issued from heterozygous mothers.

From the zygote to the morula stage, all the blastomeres are undifferentiated (totipotent) but from the compact morula stage, two cell populations will progressively emerge: the inner cells that will give rise to the inner cell mass composed of embryonic stem cells (pluripotent), and the outer cells that will differentiate to form the trophectoderm. The presence of at least two cell populations with specific metabolism has thus to be taken into account when evaluating gene expression at the morula and blastocyst stages.

Besides the specific characteristics of the oocyte and early embryo, the scarcity of the material prevents or restricts the use of several techniques. For example, about a 1,000 embryos might be necessary to detect by Western blotting a protein present in small amount, like a transcription factor. It is why the favorite technique used to analyze gene expression in early mammalian embryos is still RT-PCR or quantitative RT-PCR (RT-qPCR), despite the poor correlation often observed between mRNA levels and the corresponding protein as explained before. Protein studies at those stages are mostly based on immunofluorescence. But this technique is not quantitative and raises the question of the specificity of the observed staining. This is particularly true for the study of HOX proteins as specific antibodies are notably difficult to obtain for those proteins.

Even if the main steps of oogenesis and embryo development are similar between mammalian species, differences can be observed. The kinetics of oogenesis and early embryogenesis, including the onset of embryonic genome activity or timing of implantation, may differ. Also, the control of oocyte maturation or of the first cell lineages differentiation may involve species-specific components [4–6]. Our main model is the bovine as oocytes can be easily collected and embryos produced in vitro from abattoir ovaries. Moreover, the kinetics of maturation, fertilization, and early development is much slower than in the mouse and closer to the human species. This allows easily studying each step and transition separately. However, considering early development specificities, we have to keep in mind that it can be difficult and even misleading to generalize data obtained from one species to all the mammals. In this chapter, we will first describe the techniques used to quantify maternal and embryonic transcripts in bovine oocytes and early embryos by RT-qPCR and how it is possible to evaluate the ratio between adenylated and deadenylated transcripts corresponding to a specific gene. Then a protocol of knock down by injection of siRNA in bovine oocytes and zygotes will be described and discussed. Finally a protocol for whole-mount immunofluorescence staining to localize HOX proteins will be proposed.

2 Materials

2.1 RNA Extraction

1. *DEPC* (Diethyl Pyrocarbonate): Mix 100 μl of DEPC and 100 μl of water and leave it overnight at room temperature. Autoclave for 20 min. Store the aliquots at −20 °C.
2. *PBS–PVP* (polyvinylpyrrolidone): 1 mg/ml of PVP in PBS. Mix well and autoclave. Store at 4 °C.
3. *Glycoblue* (Ambion): store the aliquots at −20 °C (*see* **Note 1**).
4. *Luciferase control RNA 1 mg/ml* (Promega): luciferase mRNA is diluted to 50 ng/μl, aliquoted by 6 μl and stored at −80 °C.
5. *Tripure isolation reagent* (Roche): aliquots are stored at 4 °C and protected from light.
6. *Chloroform* (*see* **Note 2**).
7. *Isopropanol.*

2.2 DNase Treatment

1. *DNase RQ1* 1 U/μl (Promega).
2. *DNase RQ1 Buffer* (Promega).
3. *DNase RQ1 Stop buffer* (Promega).

2.3 Reverse Transcription

1. *Hexamers*: hexanucleotide mix 10× concentrated (Roche). Add 10 μl hexanucleotide mix 10× to 490 μl water. Aliquot by 40 μl and store at −20 °C.
2. *Poly(dT)*: Primer poly$(dT)_{15}$ for cDNA synthesis (Roche).
3. *Expand RT* 50 U/μl and its buffer 5× (Roche).
4. *DTT (100 mM)*.
5. *Deoxynucleoside Triphosphate PCR grade 100 mM*: Prepare 10 mM dNTP by mixing 100 μl of dATP, 100 μl of dGTP, 100 μl of dTTP, 100 μl of dCTP and 600 μl of H_2O. Aliquot by 50 μl and store at −20 °C.
6. *RNase out* 40 U/μl (Invitrogen).

2.4 Quantitative Polymerase Chain Reaction (qPCR)

1. *Primers*: if possible, primers are designed to hybridize on distinct exons (*see* **Note 3**). The qPCR primers were designed using the Primer Express® software (Applied Biosystems) based on NCBI database sequences. Their specificity was checked in silico by blasting each primer against the bovine transcriptome. Their efficiency should be tested on three different samples that are diluted for example 5× and 25×. Several primer concentrations can be tested (100–1,000 nM). The primer concentration giving repeatedly the efficiency closest to 100 % will be chosen, it does not need to be the same for each pair of primers.
2. *SYBR green*: keep at −20 °C, away from light.
3. *qPCR 96-well plate and adhesive films.*
4. *qPCR block.*

2.5 RNA Silencing

1. *Microinjector* (FemtoJet, Eppendorf).
2. *Micromanipulator equipped with a heated plate.*
3. *siRNA*: siRNA kits directed against bovine *HOX* sequences are commercially available.
4. *Holding pipets*: pipets can be purchased or handmade. The outer diameter of the holding pipet for a bovine oocyte or zygote has to be approximately 100 μm and the inner diameter approximately 30 μm as recommended in [7] (*see* **Note 4**).
5. *Injection needles* (Femtotip, Eppendorf).
6. *Loading tips*: microloader 20 μl (Eppendorf).
7. *PVP.*

2.6 Oocytes and Embryos Collection, Fixation, and Permeabilization

1. *PBS-PVP*: 1 mg/ml of PVP in PBS. Mix well and autoclave. Store at 4 °C.
2. *PBS-Tween-20* (PBS-T): 0.5 % Tween-20 in PBS (*see* **Note 5**).
3. *PBS-paraformaldehyde* (PFA): 2 % PFA. Work under an extractor hood. Weigh 0.2 g PFA and transfer it into a beaker

containing 10 ml demineralised H_2O. The mix is heated on a magnetic stirrer for 30–60 min at 60 °C. Once all the PFA is dissolved, filter the solution through a 0.22 μm filter with a syringe. Store at 4 °C (*see* **Note 6**).

4. *PBS-T-Triton*: 0.5 % Triton-X-100. Add 50 μl Triton-X-100 to 10 ml PBS-T (0.5 %) in a beaker and mix on a magnetic stirrer (*see* **Note 7**).

2.7 Immunofluorescence

1. *Blocking buffer*: add 10 % normal serum to PBS-T. Use serum from the same species in which the secondary antibody to be used was produced (*see* **Note 8**). Store at 4 °C until use.
2. *Antibody dilution buffer*: 1 % BSA in PBS-T. Mix well on a magnetic stirrer. Store at 4 °C until use (*see* **Note 9**).
3. *Dako Pen* (Dako).
4. *Vectashield® with DAPI* (Vector laboratories).
5. *Lab-Tek® chamber slide* (Nunc™).

3 Methods

3.1 RNA Extraction

All manipulations are performed in RNase free conditions, with gloves. Samples are handled on ice and centrifuged at 4 °C to reduce RNA degradation (*see* **Note 10**).

The amount of samples extracted at once should be minimized to avoid long delays between manipulations. However, all samples that will be compared in the same qPCR should be extracted simultaneously.

1. Oocytes or embryos are grouped and washed three times in cold PBS-PVP. Groups of oocytes or embryos are transferred with a minimum amount of liquid into a 500 μl Eppendorf tube and immediately frozen in liquid nitrogen.
2. Oocytes or embryos are kept at −80 °C until extraction (*see* **Note 11**).
3. Samples are frozen and thawed three times in liquid nitrogen to mechanically break the zona pellucida (ZP) and cell membranes.
4. Dilute 10,000 times the luciferase control mRNA (from 50 ng/μl to 5 pg/μl) using four serial dilutions.
5. In each sample, as well as in a negative control tube, add 5 μl of glycoblue, 5 μl of diluted luciferase mRNA (*see* **Note 12**), 100 μl of Tripure isolation reagent.
6. Vortex for 15 s.
7. Add 20 μl of chloroform.
8. Vortex for 10 s.

9. Centrifuge for 10 min at 13,000 × *g* and 4 °C.
10. After centrifugation, two phases are obtained, the lower phase is pink because it is formed by the phenol and contains the DNA. The upper phase is clear, it contains the RNA. Proteins are found at the interface.
11. Take 50 μl of the upper phase and transfer in an Eppendorf tube containing 50 μl of cold isopropanol (*see* **Note 13**).
12. Mix gently.
13. Centrifuge for 20 min at 13,000 × *g* and 4 °C.
14. Remove the supernatant (*see* **Note 14**).
15. Add 100 μl of cold ethanol 70 %.
16. Centrifuge 5 min at 13,000 × *g* and 4 °C.
17. Remove the supernatant carefully; the pellet is usually poorly adhesive to the tube at this stage.
18. Dry the pellets in an air vacuum system for 10 min.
19. Store the samples at −80 °C until RT.

3.2 DNase Treatment

If qPCR primers are on the same exon, a DNase treatment step has to be included (*see* **Note 3**).

If no DNase treatment is necessary, go to Subheading 3.3, **item 3** and divide all reagents by 2.

1. Add to each sample 10 μl of the following mix:
   (a) 8 μl DEPC water.
   (b) 1 μl DNase RQ1 buffer.
   (c) 1 μl DNase RQ1.
2. Incubate for 30 min at 37 °C.
3. Add to each sample 1 μl STOP DNase RQ1.
4. Incubate for 10 min at 65 °C.
5. Take 1 μl from each sample and put it in a new Eppendorf tube. This will be the RT negative control (RT^-). Those RT^- controls will not be reverse transcribed but they will undergo the PCR reaction.

3.3 Reverse Transcription

1. Add to each sample 2 μl hexamers (*see* **Note 15**). If no DNase treatment was performed, add a premix of 5 μl of water and 1 μl hexamers and continue the RT protocol here below with half the indicated amount for each reagent.
2. Incubate for 10 min at 65 °C.
3. Centrifuge for 5 min at 13,000 × *g* and 4 °C.
4. Add to each sample 11.2 μl of the following premix:
   (a) 4.8 μl Expand RT Buffer 5×.
   (b) 2.4 μl DTT (100 mM).

   (c) 2.4 μl DNTP (10 mM).
   (d) 0.8 μl RNase OUT.
   (e) 0.8 μl Expand RT.
5. Incubate for 1 h at 42 °C.
6. Centrifuge 5 min at 13,000 ×*g*.
7. Add 20 μl of water.
8. Store at −80 °C or proceed with qPCR (*see* **Note 16**).

3.4 Quantitative Polymerase Chain Reaction (qPCR)

1. Design the qPCR 96-well plate (*see* **Note 17**): the samples are tested in duplicates or triplicates. A RT⁻ control has to be included for each sample if the primers are not trans-exon.
2. Fill the wells with 18 μl of the qPCR premix (*see* **Note 18**):
   (a) 10 μl SYBR green.
   (b) 7.6 μl water.
   (c) 0.2 μl forward primer.
   (d) 0.2 μl reverse primer.
   (e) Add 2 μl of each sample by well in a separate room.
3. Seal the plate with a plastic film and centrifuge at 13,000 ×*g* for 5 min. qPCR plates can be kept at 4 °C for a few hours before starting the PCR run.
4. Run the PCR.
5. Analyze qPCR results with the ΔΔCt method (*see* **Note 19**), we usually exclude the results when there is less than 5 Ct between the RT and the RT⁻ control for a given sample:
   (a) Calculate for each stage the ΔCt = Ct gene of interest − Ct reference.
   (b) Calculate the ΔΔCt by comparing each stage to a stage of reference.
   (c) ΔΔCt = ΔCt stage x − ΔCt stage of reference.

3.5 Poly(dT): Hexamers Comparison

In order to follow the polyadenylation status of a particular mRNA, RT will be performed on the same sample in parallel with hexamers (allowing the RT of all mRNAs, whatever their poly-A tail) and with polydT (allowing the RT of polyadenylated transcripts).

1. After the DNase treatment, split each sample in two:
   (a) 5 μl sample 1 + 1 μl polyd(T).
   (b) 5 μl sample 2 + 1 μl hexamers.
2. Heat for 10 min at 65 °C.
3. Centrifuge for 5 min at 13,000 ×*g* and 4 °C.
4. Proceed to RT as in Subheading 3.3, dividing the amount of each reagent by 2 (*see* **Notes 20–22**).

3.6 Injection of Zygotes

Loss of function studies can be used to investigate the role of HOX proteins in the oocyte and early embryo. In bovine oocytes and embryos, injecting silencing RNA (siRNA) at the zygote or the oocyte stage is a good alternative to traditional knockouts (*see* **Note 23**).

Zygotes are injected using a microinjector associated to a micromanipulator.

1. Load the siRNA or the scramble preparation in the needle (*see* **Note 24**).
2. Fix the needle to the microinjector. Set the parameters of the microinjector: 120 hPa for 2 s.
3. Place a drop of rinsing media on the lid of a petri dish, close to the side. Place the drop as far from the needle as possible.
4. Zygotes are freed from their cumulus cells by vortexing.
5. Place 60 zygotes into the drop, place the petri dish so that the zygotes appear at the top of the visual field.
6. Place the holding pipet and the injection needle roughly in front of each other.
7. Bring the holding pipet down until it touches the bottom of the dish, which is noticeable because it moves forwards a little bit.
8. Aspirate a zygote with the holding pipet and bring it in the middle of the field.
9. Set the focus so that the external part of the zona pellucida is on focus.
10. Lower the needle very carefully. It cannot touch the dish otherwise it will break.
11. The tip of the needle has to be on focus together with the exterior of the zona pellucida.
12. Move the needle into the zygote, until a slight resistance is perceived. Sometimes, the displacement of the cytoplasm is visible. Bring the needle a little bit backward to move the oocyte membrane close to the zona pellucida and avoid injecting in between.
13. Inject approximately 7 pl (120 hPa, 2 s) (*see* **Note 25**).
14. Remove the needle and free the zygote at the bottom of the field so that it does not get mixed up with the noninjected zygotes.
15. Flush the needle every ten embryos (using the flush option of the microinjector).
16. Culture the embryos until the desired stage.
17. The efficiency of the knock down can be checked by RT-qPCR 24 h post-injection (*see* **Notes 26–30**).

3.7 Injection of Oocytes

Immature oocytes can also be injected with siRNA to inactivate mRNA before fertilization. In that case, injection is difficult because the cumulus cells surrounding the oocyte cannot be removed without impairing the maturation and fertilization processes. The cumulus cells tend to stick to the holding pipet and the needle, obstructing them.

1. Reduce the size of the cumulus by pipetting until only a few complete layers of cumulus cells remain attached to the oocyte.
2. Place 20 oocytes in a drop of rinsing media containing 10 mg/ml PVP to avoid sticking.
3. Inject with an injecting pressure of 70 hPa for 4 s (*see* **Note 31**).
4. Injected oocytes can be matured and then fertilized.
5. The efficiency of the knock down can be checked 24 h later by RTqPCR.

3.8 Hox Protein Detection by Immunofluorescence

3.8.1 Oocytes and Embryos Collection, Fixation, and Permeabilization

At each step, manipulate embryos with a 0–10 μl micropipette.

1. Collect immature/mature oocytes and embryos at the desired stage.
2. Rinse them three times in PBS-PVP (*see* **Notes 32** and **33**).
3. Transfer oocytes/embryos into a well of a plate containing PFA (2 %) (*see* **Notes 34** and **35**). Work under an extractor hood.
4. Incubate the oocytes/embryos for 20 min at room temperature in an extractor hood (*see* **Note 36**).
5. Rinse them 3× 5 min in a well of a plate containing PBS-T (*see* **Notes 37** and **38**).
6. Permeabilization step: transfer oocytes/embryos into a well of a plate containing PBS-T-Triton. Let them incubate for 60 min at room temperature (*see* **Note 39**).
7. Rinse them 3× 5 min in a well of a plate containing PBS-T (*see* **Note 40**).

3.8.2 Immunofluorescence Staining of Oocytes and Embryos

At each step, manipulate embryos with a 0–10 μl micropipette.

Controls (positive and negative) have to be run in each assay.

Negative controls: Immunostaining without primary antibody and/or without secondary antibody for each stage as well as on a sample known not to contain the protein (when possible) (*see* **Note 41**).

Positive control: Immunostaining of a sample known to contain the protein (when possible).

1. Draw a circle of about 1 cm of diameter with a Dako Pen on a microscope slide (*see* **Note 42**).

2. Add 40 µl blocking buffer inside the circle.
3. Add oocytes/embryos and incubate them in the blocking buffer for 60 min at room temperature (*see* **Note 43**).
4. Draw a Dako Pen circle on another microscope slide and fill it in with 40 µl primary antibody diluted in the antibody dilution buffer (*see* **Notes 44–46**).
5. Transfer and incubate oocytes/embryos in the primary antibody solution O/N at 4 °C in a humidified box (*see* **Note 47**).
6. Rinse them 3× 5 min in PBS-T on an agitator plate (*see* **Note 48**).
7. In semi-darkness, put 40 µl secondary antibody diluted in the antibody dilution buffer in a Dako Pen circle drawn on a microscope slide (*see* **Note 49**).
8. Transfer and incubate oocytes/embryos in the secondary antibody solution 60 min at room temperature in a dark humidified box.
9. Rinse them 3× 5 min in 1 ml PBS-T on an agitator plate (*see* **Note 48**).
10. Incubate them in Vectashield® with DAPI (put the Vectashield® in a small Dako Pen circle on a microscope slide) for 60 min at 4 °C in a dark moistened box.
11. For an observation with:
    - An epifluorescent microscope, mount the oocytes/embryos on a microscope slide in Vectashield® with DAPI (*see* **Note 50**). Engrave a circle on a microscope slide thanks to a point of diamond (*see* **Note 51**). Place oocytes/embryos at the center of the circle in a 2 µl Vectashield® with DAPI droplet. Cover it with a cover slip and fix it with nail polish.
    - A confocal microscope, place oocytes/embryos on a Lab-Tek® chamber slide. Place one oocyte/embryo in each chamber in a 2 µl Vectashield® with DAPI droplet (*see* **Note 52**).
12. Store the slides/Lab-Tek® at 4 °C until observation (*see* **Note 53**).

4 Notes

1. Glycoblue is tinted glycogen which facilitates the formation and visualization of the pellet.
2. Tripure and chloroform are toxic by inhalation. Samples should be manipulated with caution under an extractor hood until the pellet is resuspended in water. Chloroform is highly volatile and cannot be stored in aliquots for a long period of time.

3. Although the reason for this is unclear, it appeared that DNase treatment can decrease the yield of RNA extraction or RT. Therefore, when working on lowly expressed genes such as *HOX* genes and starting from minute amounts of biological material, it might be better to avoid DNase treatment. Therefore, to discriminate between cDNA resulting from the RT and genomic DNA that could contaminate RNA preparations, the PCR primers have to be designed on two different exons, if possible.
4. The holding pipet can be used for several successive injections as long as it is properly rinsed between each injection. Before starting, check that the holding pipet is not blocked up by aspirating washing medium.
5. Tween-20 is a viscous solution. To pipet it, cut the end of tips slantwise. It is also important to aspirate and to evacuate the liquid slowly to obtain the correct volume.
6. PFA is toxic and should be manipulated and dissolved in an extractor hood. Filtering the solution is useful to remove the last PFA grains that are not dissolved. Remove the PFA solution from the heating magnetic stirrer once the PFA is dissolved. Heating too long could denature it. It is more suitable to prepare a fresh solution for each experiment or to store concentrated aliquots (PFA 20 %) at −20 °C.
7. Triton-X-100 is a viscous solution. To pipet it, cut the end of tips slantwise. It is also important to aspirate and to evacuate the liquid slowly to obtain the correct volume.
8. We found it is better to prepare it fresh each time.
9. We found it is better to prepare it fresh each time.
10. When working with small amounts of biological material, special care should be taken to avoid any risk of contamination by DNA. Surfaces should be cleaned with NaOH before starting reagents preparation or samples manipulation. Reagents should not be prepared after manipulating biological material containing DNA or RNA. All solutions are prepared with DEPC-water or water of equivalent quality (RNAse free).
11. Samples can be kept at −80 °C for days. Nevertheless they should be extracted as soon as possible because RNA is unstable and can be degraded within weeks.
12. The luciferase mRNA is an exogenous mRNA added to each sample to normalize the qPCR results with respect to variations in extraction efficiency, or in RT and qPCR reactions.
13. Sometimes, it is difficult to take 50 μl because the tip gets really close to the phenol phase. In that case, do not hesitate to take less than 50 μl. It is better to take a lesser amount of the RNA

containing phase but free of phenol which could interfere with enzymatic reactions.

14. Although the phenol and chloroform should have been removed at that stage, we suggest the sample to be manipulated under the extractor hood until the end of the extraction procedure, in case trace of phenol or chloroform remains.
15. It is always a better choice to use hexamers instead of poly(dT) to prime the RT reaction when working with oocytes or early embryos before the MET. Indeed, while RT using polyd(T) is more efficient if the region amplified by PCR is close to the 3′ end, it obviously introduces a bias towards the polyadenylated transcripts. Hexamers on the other hand, allow all mRNAs to be reverse transcribed with the same efficiency regardless of their polyadenylation status. Alternatively, when working with a limited number of genes, specific primers can be used for the RT step.
16. cDNA can be kept at −20 °C for a few weeks. However, even at −20 °C cDNA can be unstable to some extent which might become a problem to amplify low copy numbers. Therefore, qPCR should be performed within weeks.
17. All genes or samples to be compared should be loaded on the same plate. If one plate is not enough to test all genes and samples, normalization can still be made between genes that have been tested on different plates if, for each gene, all samples are tested on the same plate.
18. The PCR premix should be prepared in a separate room where no biological material is being manipulated. The PCR premix manipulation area can be equipped with an UV light that will damage possible nucleic acid contaminant if turned on 30 min before starting the manipulation. Premix has to be prepared on ice.
19. To analyze qPCR results, one must chose a reference: in our studies two kinds of reference genes were used. First, the luciferase mRNA, added prior extraction, can be used as an external reference. This will normalize for differences due to variation in pipetting or enzymatic reaction efficiency. However, it is recommended to use several "housekeeping" genes or internal references because, beside the variations due to technical reasons, normalizing with internal references also take into account variations in the starting biological material for example due to cell number, RNA decay, …. Variation in the amount of starting material always occurs when comparing embryos at various stages of development. Which genes to use and how many have to be included should be assessed for each experimental design using the geNorm software [5]. When comparing gene expression in bovine oocytes and embryos until the

blastocyst stage, three reference genes have been used [8, 9]: Gapdh, Ywhaz and H2a.

20. The qPCR reverse primer should be designed close to the poly(A) tail.
21. The mRNA encoding the reference genes are also controlled by polyadenylation. Therefore, when investigating the modifications of the poly(A) tail, it is better to use an external reference such as luciferase mRNA.
22. qPCR results will be normalized as described above. The expression profiles obtained with the two methods are compared. If at a particular stage, the amount of mRNA reverse transcribed with poly(dT) decreases massively while the amount of mRNA reverse transcribed with hexamers is stable, it indicates that a massive deadenylation process has occurred for this transcript. Oppositely, if the expression profiles are similar, it indicates that no massive deadenylation occurred and that the mRNA is susceptible to be translated into proteins.
23. Since there is no (or basal level) transcription before the MET, plasmids encoding shRNA, which have to be transcribed cannot be used. Instead, siRNAs have to be injected into oocytes or embryos individually.
24. Needles are extremely fragile; never touch anything with the tip or it will break. To take off the protecting cap, hold the needle vertically, tip down, and loose the cap until it falls.
25. The needle will be progressively damaged by the injections, increasing the injected volume. If the cytoplasm gets lighter at the site of injection, it probably means that too much solution is injected. Pressure of injection can be progressively reduced.
26. To take into account the effect of the injection procedure, the mRNA levels of the gene targeted by the siRNAs are compared between the embryos injected with the siRNAs and the embryos injected with a negative control siRNA (scrambled). When inactivating HOX mRNA, it is important to check in parallel mRNA levels of other HOX genes, in particular the genes from the same paralog group, sharing the highest degree of similarity with the gene of interest.
27. If the efficiencies of different siRNAs are in the same range, the use of a siRNA mixture instead of individual siRNA decreases the risk of side effect. The effects of HOX silencing should be compared to a negative control siRNA targeting no mRNA (scrambled).
28. In most cases, inactivation of the mRNA will not be complete. A partial reduction of the amount of mRNA can however lead to a substantial decrease in protein level and induce a phenotype. The fact that distinct functions fulfilled by a given Hox

protein can distinctly rely on different amounts of the protein has been well demonstrated for Hoxa2 [10]. Unfortunately it is virtually impossible to correlate the level of mRNA with the amount of protein, which also depends on its stability.

29. Knocking down a gene in early embryos or oocytes using a silencing strategy is complicated by the possible presence of the maternal protein stored in the oocyte, which might persist long enough, even after the MET, to prevent any decrease in protein levels despite a substantial decrease in mRNA levels.
30. Another drawback of silencing strategies in early embryos is the limited stability of the injected siRNA. Indeed, for unknown reasons, it appears that sometimes siRNA are still active after the MET while sometimes not (maybe because of degradation, which could occur at the same time as for maternal RNA). Therefore, when using a silencing strategy in early embryos, the impact of mRNA knock down on protein expression should ideally be checked before proceeding to phenotypical analysis.
31. The cumulus cells make it difficult to visualize the needle penetration into the oocyte. It needs some practice to monitor the needle penetrating the zona pellucida. Injecting a fluorescent dye might help training for the injection procedure.
32. Adding PVP prevents oocytes/embryos to stick to the tips or to the bottom of the dish.
33. Proceed with PBS-PVP droplets; it will be easier to find and to gather the oocytes/embryos.
34. At each step, the more the well surface in which you handle the embryos is small, the easier it will be to find them back.
35. The choice of the fixative agent is very important [11]. This choice is dictated by the intra- or extracellular localization of the protein of interest. Two main classes of fixative agents are usually used: the additive ones such as glutaraldehyde or formaldehyde, and the coagulative ones such as alcohols and acetone. The first group induces the formation of cross-links between proteins and the second group leads to cell dehydration and protein precipitation. Forming cross-links may hamper antibodies penetration in oocytes/embryos and elimination of the unfixed antibodies, generating a certain level of background staining. These cross-links may also mask the protein epitopes. Additive fixatives are nonetheless the best fixative agents to highlight an intracellular protein. However, this kind of fixation requires a permeabilization step. For membrane-bound proteins, acetone fixation is the best choice while both additive and coagulative fixatives can be used for extracellular proteins.
36. Increasing PFA concentration or incubation time should be limited to avoid excessive cross-link formation between proteins.

37. Rinsing the oocytes/embryos at that step is very important to eliminate PFA.
38. At that step, oocytes/embryos can be stored in PBS-T (0.5 %) for 2 weeks at 4 °C in a plate sealed with parafilm.
39. An appropriate permeabilization step is important after the use of additive fixatives. Moreover, oocytes and embryos are spherical (from 130 to 200 μm of diameter) and surrounded by a glycoprotein membrane, the zona pellucida. Permeabilization should be strong and long enough to allow the antibodies reaching the inner cells of the embryo or the center of the oocyte.
40. After the permeabilization step, an antigen unmasking procedure can be applied to improve the signal ([12–15]; Beaujean N., personal communication). This treatment breaks the protein cross-links formed by additive fixatives and thus releases hidden antigenic sites.
    - Heat for 15 min at 90 °C on a heating block an Eppendorf tube filled with citrate buffer (10 mM Na citrate in H_2O—pH 6).
    - Rinse oocytes/embryos 3× 5 min in PBS after the permeabilization step in a well of a plate.
    - Transfer heated citrate buffer in a well of a plate.
    - Place oocytes/embryos into the well and put the plate into an oven at 80 °C (or on a heating plate) for 10 min.
    - Rinse them 3× 5 min in a well of a plate containing PBS.
    - Proceed with the blocking step, Subheading 3.8.2, **steps 1–3**.
41. The negative control without the primary antibody is important to control the specific binding of the secondary antibody while the negative control without the secondary antibody allows the detection of samples auto-fluorescence.
42. This avoids wasting solutions and allows finding out more easily oocytes/embryos.
43. There are no "strict rules" for the blocking reagent used. Depending on the antibody and the sample, a blocking step with 10 % BSA (PBS-T (0.5 %) 10 % BSA) or with 10 % milk (PBS-T (0.5 %) 10 % milk) could improve the signal-to-background ratio.
44. The appropriate antibody dilution to use has to be determined by the investigators for each antibody. Refer to the antibody datasheet to select the starting point dilution and establish a dilution range. Often, you can use a higher dilution than the one provided by the supplier.

45. Antibody specificity has to be checked by the investigators. The main problem encountered with immunofluorescence assay to localize Hox proteins is the specificity of the available antibodies. Indeed Hox proteins share a high degree of sequence similarity, mainly inside paralog groups. Moreover, studying species other than human and mouse might add to the difficulty. Another crucial problem is that Hox proteins are probably (at least are they predicted to be so) highly disordered, this means low specificity in the antigen/antibody association. Here are different strategies that could be set up to predict or assay antibody specificity.
    - *Bioinformatic analysis of the immunogenic sequence.* It is important to check if the immunogenic sequence used to produce the antibody shares some similarity with other sequences, especially within the proteins of the same paralog group. This can be simply evaluated by BLAST (http://blast.ncbi.nlm.nih.gov/Blast.cgi).
    - *Knockout animals.* The best way to check the specificity of an antibody is to perform immunohistochemistry on knockout animals for the gene/protein to be detected, wild type animals being used as positive control.
    - *Overexpression of Hox proteins in cells.* When knockout animals are not available, such as for the bovine, a good solution to check antibody specificity is to overexpress the Hox protein of interest in a cell line. In addition, to verify that the antibody does not recognize another protein of its paralog group, all Hox proteins of this paralog group should be overexpressed one by one. Antibody specificity can be then checked by Western blot and immunofluorescence. Proteins to be expressed in cell lines can also be fused to a tag (a flag epitope, for instance) to confirm protein expression and Western blot or immunofluorescence results (appropriate size of the proteins, intracellular distribution, …). Cells transfected with an empty vector are used as negative control.
46. While preparing the antibody dilution, keep the antibody vial on ice. If the antibody of interest has to be kept at −20 °C, sampling the antibody into smaller aliquots will avoid the antibody to undergo multiple temperature shifts.
47. Depending on the antibody, primary antibody incubation time could vary from 60 min to O/N. Incubation at 4 °C increases the specific antibody binding to its antigen.
48. Washing oocytes/embryos after primary/secondary antibody incubation is essential to discard the unbound antibodies. Shaking could facilitate antibodies' passage through the zona pellucida and so could facilitate elimination of unbound antibodies and reduce background signal.

49. The appropriate dilution for the secondary antibody has to be determined by the investigators to obtain an optimal signal in combination with the primary antibody while having no signal in the negative controls.
50. Using of polymerizing mounting medium, such as Prolong® Gold Antifade mounting medium (Life Technologies), is not recommended for oocytes/embryos observation, especially for the blastocyst stage. Indeed, blastocysts contain a blastocoel cavity filled with fluid. By polymerizing, such mounting medium will compress blastocysts and it will make difficult to distinguish the inner cell mass cells from the trophectoderm cells.
51. The engraved circle will help finding the oocytes/embryos and to bring them in focus.
52. Putting oocytes/embryos on Lab-Tek® chamber slide makes possible their three-dimensional observations. Pay attention to prepare a very small droplet (2 μl) of Vectashield® with DAPI otherwise the oocytes/embryos will move during the confocal acquisition.
53. Samples can be observed until a few days after staining if conserved correctly.

Acknowledgements

We gratefully acknowledge the team of the professor Schellander from the University of Bonn and in particular Franca Rings and David Tesfaye for the training in microinjection and the fruitful discussions about RNAi. We gratefully acknowledge Karen Goossens from the Ghent University, Belgium and Rozenn Dalbiès-Tran from the INRA, Nouzilly, France for the fruitful discussions about qPCR normalization. The authors also deeply thank Nathalie Beaujean from INRA, France, Françoise Gofflot from UCL, Belgium, and Bernard Knoops from UCL, Belgium for their precious advices about the immunofluorescence. We also acknowledge Philippe Bombaerts, Raphael Chiarelli, Nathan Nguyen, Wendy Sonnet, Laure Bridoux, and Emmanuelle Ghys for their assistance in embryo production and Marie-Anne Mauclet for her help with administrative procedures.

References

1. Li L, Zheng P, Dean J (2010) Maternal control of early mouse development. Development 137:859–870
2. Evsikov AV, Graber JH, Brockman JM et al (2006) Cracking the egg: molecular dynamics and evolutionary aspects of the transition from the fully grown oocyte to embryo. Genes Dev 20:2713–2727
3. Paul D, Bridoux L, Rezsöhazy R et al (2011) HOX genes are expressed in bovine and mouse oocytes and early embryos. Mol Reprod Dev 78:436–449

4. Kuijk EW, van Tol LT, Oei CH et al (2008) Differences in early lineage segregation between mammals. Dev Dyn 237:918–927
5. Bilodeau-Goeseels S (2011) Cows are not mice: the role of cyclic AMP, phosphodiesterases, and adenosine monophosphate-activated protein kinase in the maintenance of meiotic arrest in bovines oocytes. Mol Reprod Dev 78:734–743
6. Rossant J (2011) Developmental biology: a mouse is not a cow. Nature 471:457–458
7. Favetta LA, Madan P, Mastromonaco GF et al (2007) The oxidative stress adaptor p66Shc is required for permanent embryo arrest in vitro. BMC Dev Biol 7:132
8. Vandesompele J, De Preter K, Pattyn F et al (2002) Accurate normalization of real-time quantitative RT-PCR data by geometric averaging of multiple internal control genes. Genome Biol 3:7
9. Goossens K, Van Poucke M, Van Soom A et al (2005) Selection of reference genes for quantitative real-time PCR in bovine preimplantation embryos. BMC Dev Biol 5:7
10. Ohnemus S, Bobola N, Kanzler B et al (2001) Different levels of Hoxa2 are required for particular developmental processes. Mech Dev 108:135–147
11. Goossens K, Vandaele L, Wydooghe E et al (2011) The importance of adequate fixation for immunofluorescent staining of bovine embryos. Reprod Domest Anim 46: 1098–1103
12. Khan DR, Dube D, Gall L et al (2012) Expression of pluripotency master regulators during two key developmental transitions: EGA and early lineage specification in the bovine embryo. PloS One 7:e34110
13. Shi SR, Cote R, Taylor C (2001) Antigen retrieval techniques: current perspectives. J Mol Histol 49:931–937
14. Yamashita S (2007) Heat-induced antigen retrieval: mechanisms and application to histochemistry. J Histochem Cytochem 41: 141–200
15. Montero C (2003) The antigen–antibody reaction in immunohistochemistry. J Histochem Cytochem 51:1–4

Chapter 3

Genetic Lineage Tracing Analysis of Anterior Hox Expressing Cells

Brigitte Laforest, Nicolas Bertrand, and Stéphane Zaffran

Abstract

Cell lineage studies have been widely used in developmental biology to establish which cells, and how many cells, in the early embryo will give rise to a specific structure and its derivatives. Several methods have been developed to label progenitor cells in the early embryo. Here, we describe the genetic tracing approach that relies on the use of the recombinase to genetically and permanently label progenitor cells as well as their progeny through specific activation of a conditional reporter gene, the *ROSA26* reporter mouse. Labeling of progenitor cells is spatially controlled by the use of a tissue-specific promoter driving *Cre*, the *Hoxb1*$^{IRES\text{-}Cre/+}$ and the *Hoxa1-enhIII-cre*. ROSA26R mice and *Hoxb1*$^{IRES\text{-}Cre/+}$ or *Hoxa1-enhIII-cre* mice are crossed together to generate embryos at different stages of development. Embryos are collected and dissected at a specific stage of development and fixed in paraformaldehyde. To follow *Hoxb1*$^{+}$ and *Hoxa1*$^{+}$ progeny, X-gal staining is performed to detect β-galactosidase activity in embryos or developing organ such as the heart. Finally, X-gal-positive cells are observed on whole-mount embryos or dissected organ to determine the lineage contribution of *Hoxa1* and *Hoxb1* during development.

Key words Genetic lineage tracing, Knock-in, Transgene, Cre recombinase, Hoxa1, Hoxb1, Rosa26 reporter

1 Introduction

Morphogenesis is the result of different types of cell behavior, including cell proliferation and differentiation, which are regulated temporally and ordered spatially as the embryo develops. Cell labeling studies provide knowledge about the relationships between cells within a structure and give insight into developmental processes such as cell movements. Hence, lineage analysis has emerged as a powerful tool to follow the fate of a cell or a group of cells during development in order to determine to which structure or substructure they will give rise to. Several methods have been developed over the years to specifically label cells, each coming with its merits and limitations. Lipophilic carbocyanine dies, such as DiI for example, are used to trace groups of cells and examine

Yacine Graba and René Rezsohazy (eds.), *Hox Genes: Methods and Protocols*, Methods in Molecular Biology, vol. 1196, DOI 10.1007/978-1-4939-1242-1_3, © Springer Science+Business Media New York 2014

cell movement, whereas microinjection of a fluorescent dye allows single cell labeling but is more challenging [1]. Genetic lineage tracing, a cell-tracing technique used for lineage analysis, consists in following the fate (contribution and differentiation) of a group of progenitor cells and its descendants during embryonic development [1]. The classical approach in this context is based on the use of a reporter gene that can be permanently and irreversibly activated in a chosen progenitor cell population at any stage of development. The cre-lox site-specific recombination system, coupled to the ROSAβgeo26 reporter (R26R) line [2–4], has been used extensively over the past decade to achieve conditional and irreversible reporter gene activation by taking advantage of the use of a tissue-specific promoter.

The Cre-lox recombination system is based on the capacity of the CRE (*C*auses *RE*combination) recombinase from bacteriophage P1 to use the *loxP* site for site-specific recombination [5–7]. *LoxP* site is a 34 base pair motif comprising two inverted repeats of 13 base pairs separated by an 8 base pair spacer [7]. The recombination event between two *loxP* sites in direct orientation leads to the deletion of the sequence comprised between these two sites.

The R26R mouse line has been developed taking into account the knowledge of the ROSA26 locus in which the expression of the β-galactosidase (β-gal) reporter gene is ubiquitous and starts at pre-implantation stages [2–4]. Indeed, to generate the R26R line, the initial ROSA26 locus was targeted by homologous recombination with an engineered construct containing the *lacZ* gene downstream of a cassette containing a splice acceptor site followed by the *neo* gene flanked by two *loxP* sites in the same orientation and a triple polyadenylation signal. Hence, this cassette does not allow *lacZ* expression prior to recombination between the two *loxP* sites catalyzed by the Cre recombinase, which is expressed under the desired expression pattern.

For the purpose of this chapter, we will demonstrate the genetic lineage tracing approach by giving an illustrative case study. Thus, for anterior Hox genetic lineage tracing analysis, R26R mice were crossed to the $\mathit{Hoxb1}^{IRES\text{-}Cre/+}$ and *Hoxa1-enhIII-cre* mice. These two *Cre* lines are used for this study to follow the fate of Hoxb1 and Hoxa1 expressing cardiac progenitor cells during embryogenesis and get a better understanding of the structures they contribute. Pregnant females were sacrificed by cervical dislocation and embryos were dissected at different stages of embryonic development (E8.5–E16.5) and fixed in paraformaldehyde. For fetuses between E12.5–E16.5, the heart was removed from the thoracic cavity and also fixed in paraformaldehyde. X-gal staining was performed on whole embryos or hearts to detect β-gal activity. Embryos and hearts were post-fixed in paraformaldehyde and observed to visualize X-gal-labeled cells. Embryos and hearts were then embedded in OCT and sectioned using a cryostat. Sections were analyzed to visualize more specifically cells that contribute to the *Hoxb1* and *Hoxa1* lineage in the embryo as development proceeds.

2 Materials

2.1 Mouse Lines

Mouse handling and experimentation was done in accordance with national and institutional guidelines. All protocols were approved by the national animal care ethical committee (protocol 32-08102012).

1. The $Hoxb1^{IRES\text{-}Cre/+}$ line is used to follow the fate of Hoxb1 expressing cells during embryogenesis. The $Hoxb1^{IRES\text{-}Cre/+}$ mice were generated by targeting the Hoxb1 locus via homologous recombination as previously described [8]. Briefly, an engineered vector was designed containing an internal ribosomal entry site (IRES) driving expression of the *Cre* recombinase (*IRES-Cre* cassette) and a removable *neomycin* (*neo*) selection gene (flanked by *FRT* sequences) (Fig. 1a). This cassette was inserted into the 3′ untranslated region of the Hoxb1 locus, more precisely just downstream of the translation stop codon and upstream of the polyadenylation site so the CRE protein could be expressed following the Hoxb1 expression but without interfering with its activity.
2. The *Hoxa1-enhIII-Cre* line we are using is a classical transgenic line generated by Li and Lufkin [9]. The transgene integrates the *Hoxa1* enhancer-III previously identified [10]. This enhancer is located 3′ of the *Hoxa1* polyadenylation signal, is 0.5 kb in length and contains a functional retinoic acid responsive element (RARE) of the DR5 type, in combination with a TATA box promoter upstream of the *Cre* gene followed by an IRES-alkaline phosphatase cassette (Fig. 1b).

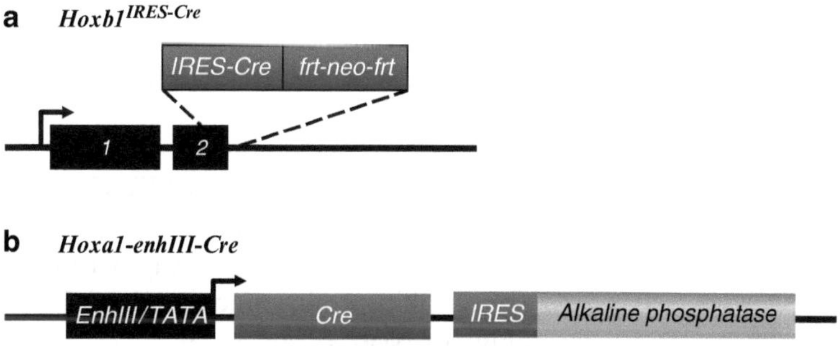

Fig. 1 Schematic representation of the targeting construct and transgene used for generation of $Hoxb1^{IRES\text{-}Cre}$ and *Hoxa1-enhIII-Cre* mice respectively

2.2 Dissection of Embryos or Hearts

1. Dissection tools: scissors, forceps.
2. Phosphate-buffered saline (PBS) 1×.
3. Plates: 6-well or 24-well plates.

2.3 Para-formaldehyde (PFA)

Dissolve 20 g of paraformaldehyde in 450 ml of PBS 1× (*see* **Note 1**). The powder will not immediately dissolve in solution. Raise the pH by adding 200 μl of a 10 M NaOH solution (20 μl NaOH/50 ml PBS 1×) to help dissolution. Shake the bottle vigorously and place on a hot plate/stirrer set at 65 °C. Allow the solution to heat while stirring, and when ready, it will change from cloudy to clear. Be careful that the solution does not boil. When the PFA has completely dissolved, turn off the heat, continue stirring and allow to cool down. Adjust the volume to 500 ml with PBS 1× and adjust the pH between 7.3 and 7.4 with concentrated HCl (*see* **Note 2**). Filter the PFA and prepare 10 mL aliquots that are stored at −20 °C until further use (*see* **Note 3**).

2.4 X-Gal Staining

1. Phosphate-buffered saline (PBS) 10× (Lonza).
2. 200 mM Potassium ferricyanide (K FerriCN): Dissolve 3.293 g in 50 ml of deionized water. Keep 10 ml aliquots at −20 °C.
3. 200 mM Potassium ferrocyanide (K ferroCN): Dissolve 4.224 g in 50 ml of water. Keep 10 ml aliquots at −20 °C.
4. Magnesium Chloride ($MgCl_2$) 1 M: Dissolve 4.76 g in 50 ml of deionized water. Keep at room temperature.
5. X-gal 40 mg/ml: Dissolve 1 g in 25 ml of deionized water. Keep aliquots of 500 μl in the dark at −20 °C.
6. Nonidet P-40 (NP-40): 20 % solution in water. Keep in the dark.

2.5 Freezing Embryos or Hearts for Cryosectioning

1. Dry ice.
2. Embedding cryomold (Leica).
3. Optimal cutting temperature (OCT) medium.
4. Sucrose 15 %: Dissolve 7.5 g of sucrose in 50 ml of PBX 1×. Keep the solution at 4 °C.

3 Methods

3.1 Crossing Mice

Hoxb1$^{IRES\text{-}Cre/+}$ or *Hoxa1-enhIII-Cre* males are mated with R26R females to generate embryos at the desired stage for analysis (*see* **Note 4**). The morning, a vaginal plug is observed. Embryos are staged taking embryonic day (E) 0.5 as the morning of the vaginal plug (*see* **Note 5**).

3.2 Collecting Embryos and Samples

Embryos or hearts are dissected in PBS 1× under a stereomicroscope (Leica M80). The pregnant female is sacrificed by cervical dislocation. The uterus is then dissected by pulling up on the uterus with forceps. Remove embryos from the uterus and collect them in a petri dish containing PBS 1×. Separate embryos from one another by cutting carefully in between the embryos with small scissors. Remove the muscular wall of the uterus on the embryo, the reichert's membrane as well as the visceral yolk sac with fine forceps. Transfer embryos to a 6-well or 24-well plate containing PBS 1×. For heart dissection, cut the head of the embryo, open the thoracic cavity, and remove the heart carefully.

3.3 X-Gal Staining

1. To visualize β-gal activity, embryos or hearts are fixed for 20 min in 4 % paraformaldehyde at 4 °C directly following dissection (*see* **Note 6**).
2. Rinse the embryos twice with PBS 1× after fixation is complete and add the X-gal solution mix. The X-gal solution is prepared by mixing 1 ml of PBS 10×, 200 μl of K FerriCN 200 mM, 200 μl of K FerroCN 200 mM, 20 μl of 1 M $MgCl_2$, 100 μl of X-gal and 10 μl of 20 % NP40, adjust the volume to 10 ml with deionized water.
3. Incubate embryos in X-gal staining solution overnight at 37 °C in the dark (*see* **Note 7**).
4. The following day, wash embryos or hearts with PBS 1× and post-fix in 4 % PFA overnight at 4 °C.
5. Whole embryos or hearts are photographed using a macroscope (Zeiss Axiozoom) (*see* Figs. 2 and 3).

Using this technique, we could detect positive *Hoxb1*$^+$ progeny in the rhombocephalon, with strong expression in rhombomere 4, the second pharyngeal arch, the migrating neural crest cells, the trunk, tail, and limb buds at E9.5 and E10.5 of whole embryos (Fig. 2a, b, d, e) [11]. X-gal staining of whole hearts revealed β-gal positive cells in the outflow tract at E9.5 and E10.5 (Fig. 2c, e), in the left and right atrium as well as in the pulmonary trunk at later stages of development (Fig. 2g–i). The *Hoxa1* lineage was observed in the rhombocephalon, the migrating neural crest cells, the apical ectodermal ridge of the limb buds and distal part of the OFT at E9.5 and E10.5 (Fig. 3a–d) [11]. In whole hearts, *Hoxa1*$^+$ progeny could be observed in the distal OFT at E10.5 (Fig. 3e), the myocardium at the base of the pulmonary trunk (Fig. 3f–h) and in some cells of the right and left atrium.

3.4 Sample Treatment

1. For cryosectioning, embryos (or hearts) are first immersed in a 15 % sucrose/PBS solution (*see* **Note 8**).
2. Label the OCT cryomold on the side.
3. Once the samples have fallen to the bottom of the tube, transfer the embryos (or hearts) in the cryomold containing the OCT medium (*see* **Note 9**).

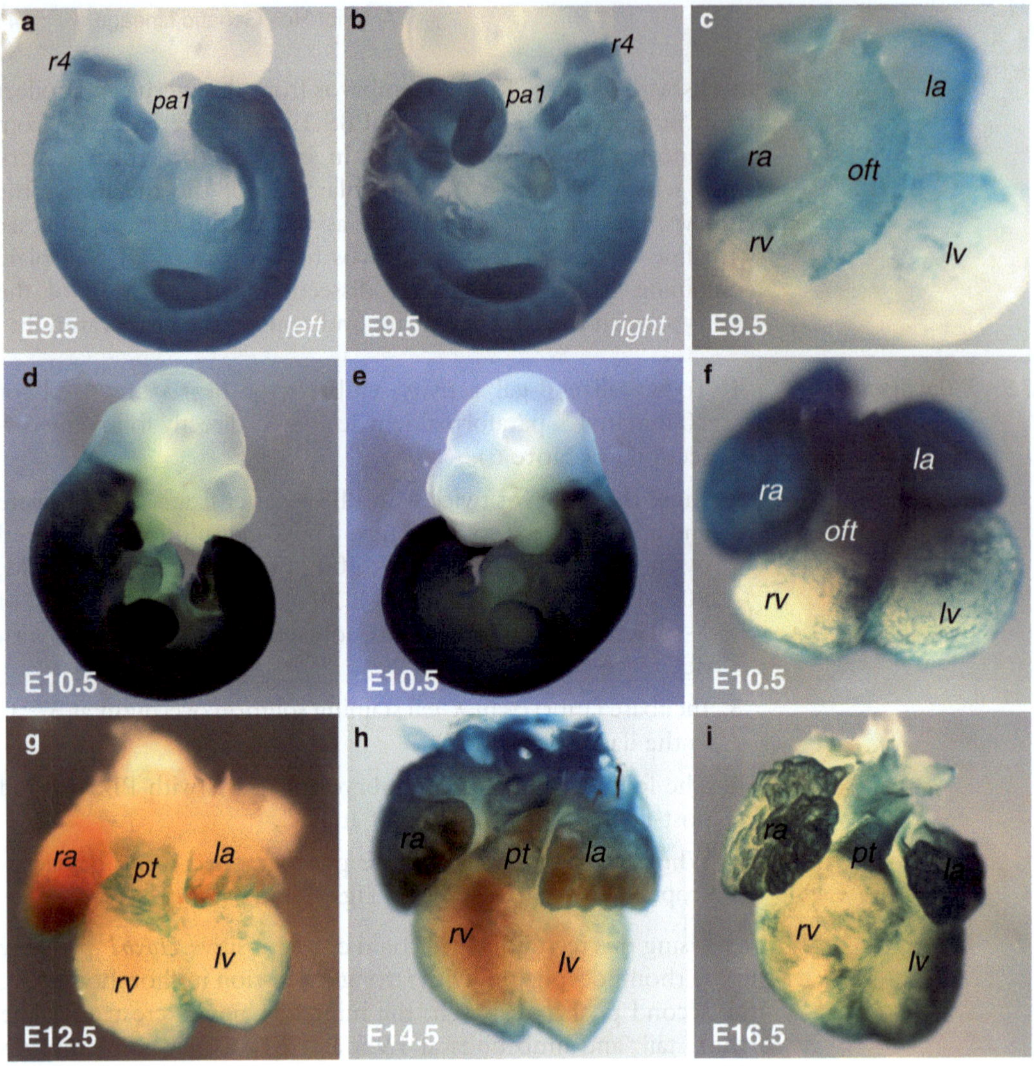

Fig. 2 Genetic lineage contribution of the *Hoxb1*+ progeny. (**a**–**i**) Hoxb1 lineage visualized by X-gal staining of *Hoxb1*$^{IRES\text{-}Cre}$; R26R embryos or hearts. (**a**, **b**, **d**, and **e**) X-gal staining of *Hoxb1*$^{IRES\text{-}Cre}$; R26R embryos at E9.5 and E10.5. β-galactosidase (β-gal) is detected in the region of the rhombocephalon, being strong in rhombomere 4, and also in the second pharyngeal, the migrating cardiac neural crest cells, the trunk, the tail, and the limb buds. (**c**, **f**, and **g**–**i**) Ventral view of X-gal-stained heart at E9.5, E10.5, E12.5, E14.5, and E16.5. β-gal activity is detected in the inferior part of the cardiac outflow tract (OFT) (**c** and **f**), right and left atria and the epicardium. At later stages, X-gal-labeled cells are detected at the myocardial base of the pulmonary trunk and the epicardium (**g**–**i**). *la* left atrium, *lv* left ventricle, *oft* outflow tract, *pa1* pharyngeal arch 1, *pt* pulmonary trunk, *r4* rhombomere 4, *ra* right atrium, *rv* right ventricle

Fig. 3 (continued) ectodermal ridge of the limb buds and the distal part of the outflow tract (OFT). Ventral view of X-gal-stained heart at E10.5 (**e**), E12.5 (**f**), E14.5 (**g**), and E16.5 (**h**). X-gal-positive cells are detected in the distal region of the OFT (**e**), at subpulmonary myocardium (**f**–**h**) as well as in a small number of left and right atrial cells. *la* left atrium, *lv* left ventricle, *oft* outflow tract, *pt* pulmonary trunk, *r4* rhombomere 4, *ra* right atrium, *rv* right ventricle

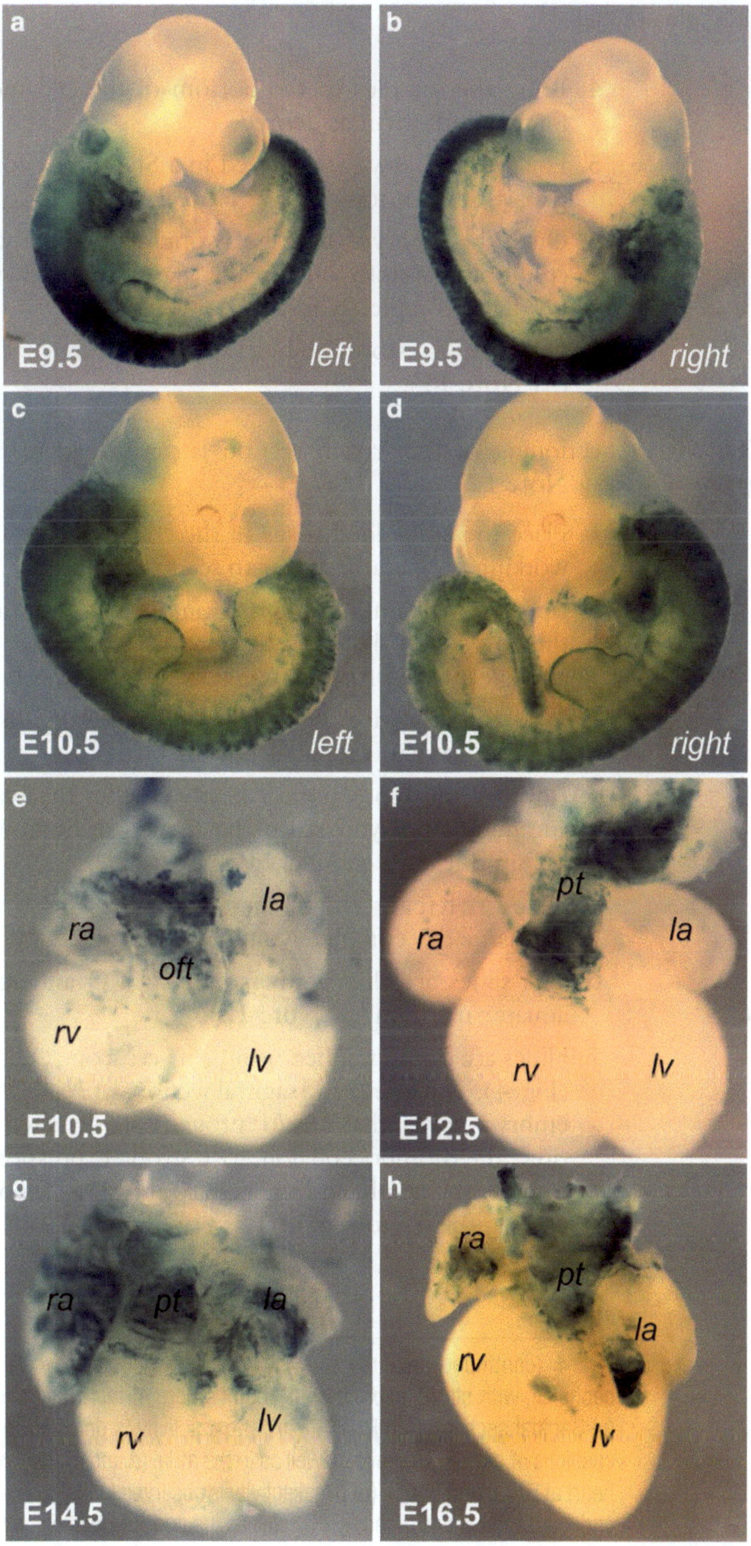

Fig. 3 Genetic lineage contribution of the *Hoxa1*⁺ progeny. Hoxa1 lineage visualized by X-gal staining of *Hoxa1-enhIII-Cre*;R26R embryos or hearts. (**a–e**) X-gal staining of *Hoxa1-enhIII-cre*;R26R embryos at E9.5 and E10.5. β-gal activity is detected in the rhombocephalon, in migrating neural crest cells, the trunk, the tail, the apical

4. Bring the samples to the bottom of the cryomold with fine forceps and orient as desired.
5. For freezing, dry ice is placed in a Styrofoam container and a glass placed on top of it (*see* **Note 10**).
6. Once the embryos have been oriented properly, transfer the cryomold carefully on top the glass plate and wait until the OCT has completely frozen.
7. Once frozen, store the blocks at −20 °C in a sealed container (*see* **Note 11**).

3.5 Sectioning and Image Capture

Sectioning of the frozen section is performed using a cryostat (*see* **Note 12**).

1. Place the embedded tissue in the chamber to warm up to the working temperature prior to sectioning.
2. Apply a small amount of OCT on the disk and quickly transfer the embedded tissue on it; place back into the chamber and allow to freeze properly.
3. Place the disk with the attached block on the disk holder and position so that it is parallel to the blade.
4. Advance the block until it nearly reaches the blade and start trimming until you reach the tissue. Sections were cut at a 16 μm thickness and picked up on a slide.
5. Slides are left at room temperature for about 1 h in order to let the tissue dry completely on the slide.
6. The slides are then immersed into H_2O and mounted with aquatex medium (*see* **Note 13**).
7. Slides are photographed using a macroscope (Zeiss axiozoom) (Fig. 4). Transverse and sagittal sections of *Hoxb1*$^{IRES\text{-}Cre/+}$;R26R embryos reveal that *Hoxb1*$^{+}$ progeny are detected in the neural tube, the rhombocephalon, with strong expression in rhombomere 4, the trunk, and the tail at E9.5 (Fig. 4a, b). Transverse and sagittal sections of hearts between E11.5 and E16.5 shows

Fig. 4 (continued) embryo showing positive X-gal-stained cells in the rhombocephalon, with strong expression in rhombomere 4, the trunk, and the tail region. (**c**) Sagittal section of an E10.5 heart. β-gal activity is detected in the endocardial cushions of the OFT and avc as well as in the atria. (**d** and **e**) Transverse section of a heart at E11.5 and E16.5. (**d**) β-gal activity is observed in the outflow tract (OFT), the endocardial cushions of the OFT as well as the endocardium lining the cushions, the left atrium, and the epicardium. (**e**) X-gal-positive cells are detected in the mitral and tricuspid valves, the atria, the epicardium, and some cells in the left ventricle. *a* atria, *avc* atrio-ventricular canal, *ep* epicardium, *ivs* interventricular septum, *la* left atrium, *lv* left ventricle, *mv* mitral valve, *nt* neural tube, *oft* outflow tract, *r4* rhombomere 4, *rv* right ventricle, *tv* tricuspid valve, *v* ventricle

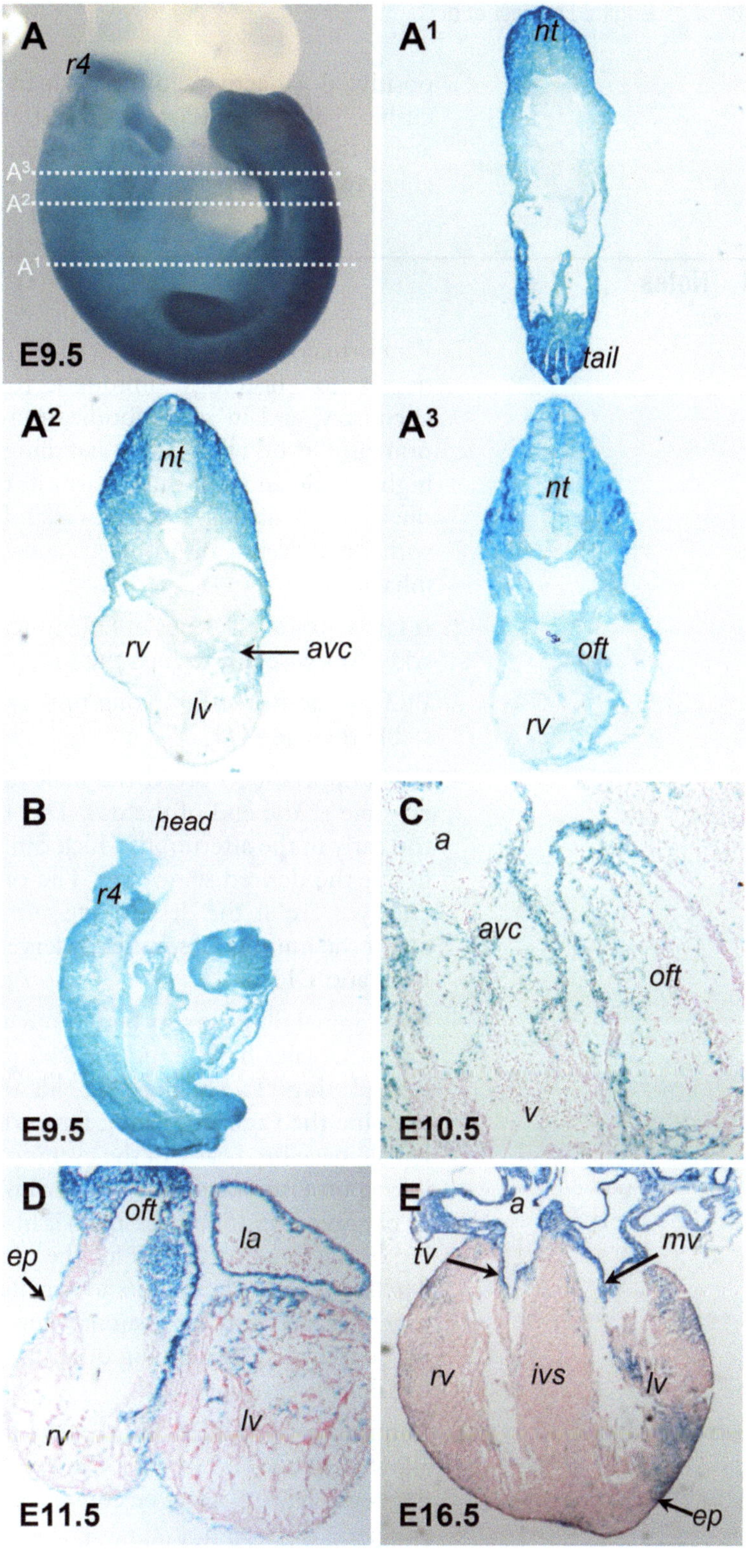

Fig. 4 Cryosections showing genetic lineage contribution of the *Hoxb1*⁺ progeny. (**a**–**e**) Sections of Hoxb1 lineage visualized by X-gal staining of $Hoxb1^{IRES\text{-}Cre}$;R26R embryos or hearts. (**a** and **a**^{1}–**a**^{3}) Transverse sections of the embryo at E9.5. β-gal is detected in the neural tube, the endoderm, and the tail. (**b**) Sagittal section of an E9.5

positive β-gal activity in mesenchymal cells of the endocardial cushions in the outflow tract and atrio-ventricular canal, the atria, the epicardium, and the mitral and tricuspid valves (Fig. 4c–e).

4 Notes

1. Paraformaldehyde is a strong fixative used to fix and preserve tissues for histology, immunocytochemistry/immunohistochemistry, and in situ hybridization protocols. Extreme caution should be taken when handling PFA as it is volatile and highly toxic and will cause harm if it comes into contact with the body or inhaled. Wear gloves and a lab-coat when working with PFA. Weigh PFA on a balance under a hood to prevent inhalation.
2. It takes only a few drops of HCl to reach the desired pH. Thus, add drop-wise not to miss the pH.
3. PFA can be stored for a long time at −20 °C, where it is more stable than at 4 °C.
4. It is important to breed the male and female mice as late as possible at the end of the day. This helps to prevent breeding too early in the afternoon, which can result in embryos that are not at the desired stage/age. The only way to assure that the embryos are at the desired age and properly developed is to count the number of somites observed on the embryo between E8.0 and E11.5.
5. The vaginal plug persists between 8 and 24 h following breeding so females have to be checked early in the morning for a vaginal plug. To see the plug, lift the female by the tail and examine the vaginal opening for a white mass. If the male and female have bred early in the evening, the mass may not be visible upon visual examination. In this case, a metal instrument or plugger (*Fine Scientific Tools*) is inserted in the vaginal opening of the female to verify for the presence of a plug. If a plug is felt, separate the male and female. Otherwise, leave them together and verify for a vaginal plug every morning. However, the presence a vaginal plug does not necessarily mean that the female is pregnant.
6. Aliquots of 4 % PFA are thawed in a water bath and then allowed to cool down on ice just before use. Once an aliquot has been thawed, it can be conserved at 4 °C for about 1 week.
7. If the color develops rapidly, X-gal staining can be left at room temperature overnight in the dark. It is also possible to incubate over the weekend.

8. Sucrose is used as a cryo-protector. Make sure that the samples have fallen to the bottom of the tube before embedding in OCT.
9. The embryo is directly transferred from the sucrose solution into the OCT. Carefully pipet away as much of the sucrose as possible from the embryo. It is also recommended to roll the embryo in the OCT medium before positioning to remove excess sucrose. If there is a little bit of sucrose left around the tissue before embedding, the tissue will not freeze properly and will not be as stable in the OCT medium. This will result in blocks that are hard to cut.
10. This method is simple but the tissue does not freeze as fast as if it was completely immersed in a freezing medium. Isopentane on dry ice pellets can also be used as another technique. Once oriented, the mold is immersed in this liquid medium for freezing.
11. It is important to store the blocks in a container or in plastic wrap as they tend to dry, which renders the sectioning very difficult. The blocks may also be kept at −80 °C for a long period of time.
12. The optimal temperature for sectioning at the cryostat is between −15 and −20 °C. If the temperature goes too high, the block will not stay frozen and if it becomes too cold, the block will become like powder.
13. Light eosin counterstain can be done at that stage to visualize the tissues on the section.

Acknowledgement

The authors would like to thank Pr. Mario Capecchi and Pr. Tom Lufkin for the *Hoxb1*$^{IRES\text{-}Cre}$ and *Hoxa1-enhIII-Cre* lines respectively. This work was supported by the "Association Française contre les Myopathies" (NMH Decrypt Project). Brigitte Laforest received postdoctoral fellowship from the "Fondation pour la Recherche Médicale".

References

1. Buckingham ME, Meilhac SM (2011) Tracing cells for tracking cell lineage and clonal behavior. Dev Cell 21:394–409
2. Soriano P (1999) Generalized lacZ expression with the ROSA26 Cre reporter strain. Nat Genet 21:70–71
3. Zambrowicz BP, Imamoto A, Fiering S et al (1997) Disruption of overlapping transcripts in the ROSA beta geo 26 gene trap strain leads to widespread expression of beta-galactosidase in mouse embryos and hematopoietic cells. Proc Natl Acad Sci U S A 94:3789–3794
4. Friedrich G, Soriano P (1991) Promoter traps in embryonic stem cells: a genetic screen to identify and mutate developmental genes in mice. Genes Dev 5:1513–1523
5. Sauer B, Henderson N (1988) Site-specific DNA recombination in mammalian cells by

the Cre recombinase of bacteriophage P1. Proc Natl Acad Sci U S A 85:5166–5170
6. Sauer B, Henderson N (1989) Cre-stimulated recombination at loxP-containing DNA sequences placed into the mammalian genome. Nucleic Acids Res 17:147–161
7. Hoess RH, Ziese M, Sternberg N (1982) P1 site-specific recombination: nucleotide sequence of the recombining sites. Proc Natl Acad Sci U S A 79:3398–3402
8. Arenkiel BR, Gaufo GO, Capecchi MR (2003) Hoxb1 neural crest preferentially form glia of the PNS. Dev Dyn 227:379–386
9. Li X, Lufkin T (2000) Cre recombinase expression in the floorplate, notochord and gut epithelium in transgenic embryos driven by the Hoxa-1 enhancer III. Genesis 26:121–122
10. Frasch M, Chen X, Lufkin T (1995) Evolutionary-conserved enhancers direct region-specific expression of the murine Hoxa-1 and Hoxa-2 loci in both mice and Drosophila. Development 121:957–974
11. Bertrand N, Roux M, Ryckebusch L et al (2011) Hox genes define distinct progenitor sub-domains within the second heart field. Dev Biol 353:266–274

Chapter 4

A Genetic Strategy to Obtain P-Gal4 Elements in the *Drosophila* Hox Genes

Luis de Navas, David Foronda, Delia del Saz, and Ernesto Sánchez-Herrero

Abstract

The *Drosophila* Gal4/UAS system allows the expression of any gene of interest in restricted domains. We devised a genetic strategy, based on the P-element replacement and UAS-y^{+} techniques, to generate Gal4 lines inserted in Hox genes of *Drosophila* that are, at the same time, mutant for the resident genes. This makes possible to express different wild-type or mutant Hox proteins in the precise domains of Hox gene expression, and thus to test the functional value of these proteins in mutant rescue experiments.

Key words *Drosophila*, Gal4/UAS system, P-element, *Ultrabithorax*, *Abdominal-B*

1 Introduction

Hox genes are transcribed in precise domains of expression during development [1–3], and changes in these restricted domains result in strong pattern modifications, the homeotic transformations. In the mouse, loss-of-function mutations in a Hox gene frequently result in partial changes in pattern, and a complete homeotic transformation is only observed when all genes of a particular Hox paralog group from the different Hox complexes are eliminated [4]. In fact, several cases of redundancy within Hox genes have been described in the mouse [4–7]. In *Drosophila*, by contrast, only one Hox cluster (split in two complexes) is present, suggesting unique functions for each Hox gene. However, some cases of very similar function between different Hox genes have also been reported in the fly. For example, the Hox genes *Ultrabithorax* (*Ubx*) and *abdominal-A* (*abd-A*) can equally promote the development of gonads [8] or halteres [9], and different Hox genes can promote the development of the tritocerebrum [10]. Nevertheless, extensive studies on Hox redundancy in different tissues and developmental stages have not been carried out.

Yacine Graba and René Rezsohazy (eds.), *Hox Genes: Methods and Protocols*, Methods in Molecular Biology, vol. 1196, DOI 10.1007/978-1-4939-1242-1_4, © Springer Science+Business Media New York 2014

Different functional domains of the Hox proteins have been conserved in evolution, notably the DNA-binding domain, the homeodomain [11]. Particular domains in the Hox proteins serve to correlate Hox products from different species [12], and the analysis of proteins mutated in these domains is useful to assign them a specific function. Precise mutations in specific domains of the Hox proteins are rare, but they can be generated in vitro and introduced into the fly to study their function. However, the precise replacement of a wild-type gene by its mutant variant, required to study the protein in its proper developmental context, is laborious in *Drosophila* [13], although one such replacement in the *Ubx* Hox gene has been achieved [14]. An alternative approach to get around this problem is to isolate lines with P-Gal4 elements inserted in Hox genes that direct the Gal4 protein in the same pattern of expression as the endogenous gene and which are, at the same time, mutant for the gene. By introducing the mutant Hox cDNA under the control of UAS sequences, one can functionally replace the wild-type protein by the altered one and test its activity in a specific Hox gene mutant background.

This method has been used for several *Drosophila* genes. As an example, studies carried out about the gene *apterous* have enabled to assess the importance of different domains for the function of the Apterous protein [15, 16]. In these and other studies, the key element in the method was the isolation of the appropriate lines with P-Gal4 insertions in the gene of interest. A way to isolate such lines for the Hox genes that we describe here is based on two previous techniques: the first one is the P-replacement strategy, whereby one P-element can be replaced by a different one present in the same genome [17, 18]; the second one is the "yellow" method of identifying P-Gal4 insertions [19].

The P-element to be replaced is typically a P-lacZ insertion, although other P-elements may be used. Several P-lacZ elements have been isolated in *Drosophila* Hox genes, particularly in the Bithorax complex (BX-C) [20–28], and these transposons can be targeted for P-element replacement. The identification of the new insertion relies on the use of the "yellow method" [19]. In this technique, a line with a P-Gal4 insertion is mobilized with a specific P-element transposase and crossed to a line with an *UAS-y^+* construct in a *yellow* mutant background. The original P-Gal4 insertion which is mobilized does not drive any pattern of expression in the adult cuticle (or a localized one, which does not interfere with the identification of the new one). By contrast, the new line with a P-Gal4 insertion (replacing the P-lacZ one) can be selected by the *yellow$^+$* (brownish) color observed in particular domains of the cuticle of the adult fly (where the Gal4 protein is expressed), while the rest of the cuticle bears a *yellow* color. The new *yellow$^+$* pattern corresponds to the expression domain of the Hox gene where the P-Gal4 insertion has been inserted.

We have used this method to isolate P-Gal4 insertions in the *Ubx* and *Abdominal-B* (*Abd-B*) genes of the BX-C [29] and identified the replacement by the *yellow*⁺ color of the halteres (in the case of the *Ubx* insertions) or the female seventh abdominal segment and genitalia (in *Abd-B* P-Gal4 insertions). A P-Gal4 element inserted in the third BX-C gene, *abdominal-A* (*abd-A*) was also isolated using the same approach [30]. Although we specifically designed the method to obtain P-Gal4 insertions in the BX-C, it may be used not only to get such insertions in other Hox genes but also in any gene with restricted expression during development.

2 Materials

1. A stock with a "recipient" P-element insertion in a Hox gene in a *yellow* mutant background (*see* **Note 1**).
2. A stock with a "donor" P-Gal4 element, also in a *yellow* mutant background.
3. A source of P-element transposase (*see* **Note 2**).
4. A stock with a *UAS-y*⁺ insertion, in a *yellow* mutant background [19].

3 Methods

1. Cross en masse the recipient P-element stock (P-lacZ or other) with the stock carrying the donor P-Gal4 to make a stock with both the P-elements, the donor and the recipient; the latter inserted in a Hox gene. Both can be recombined on the third chromosome (where the Hox complexes are) but it is better if they are combined with appropriate balancers on two different chromosomes. The P-Gal4 line we normally use is on the X and does not direct any pattern of expression on the adult cuticle. It is important that both stocks carry a *yellow* mutation, which affects the color of the cuticle, since the recognition of the insertion is based in the *y*⁺ rescue in the adult cuticle (*see* **Notes 3** and **4**) (Fig. 1).
2. Cross en masse males from a stock carrying the donor and the recipient P-elements to a stock carrying the Δ2-3 P-element transposase, both in a *yellow* mutant background (*see* **Note 5**). It is in the germ line of these dysgenic males, carrying the three elements (donor and recipient Gal4 insertions and transposase), where the P-Gal4 may substitute for the Hox-resident P-element (*see* **Note 6**). The efficiency of replacement is increased if the P-recipient element is over a deficiency for the resident gene or over a balancer chromosome (*see* **Note 7**) (Fig. 1).

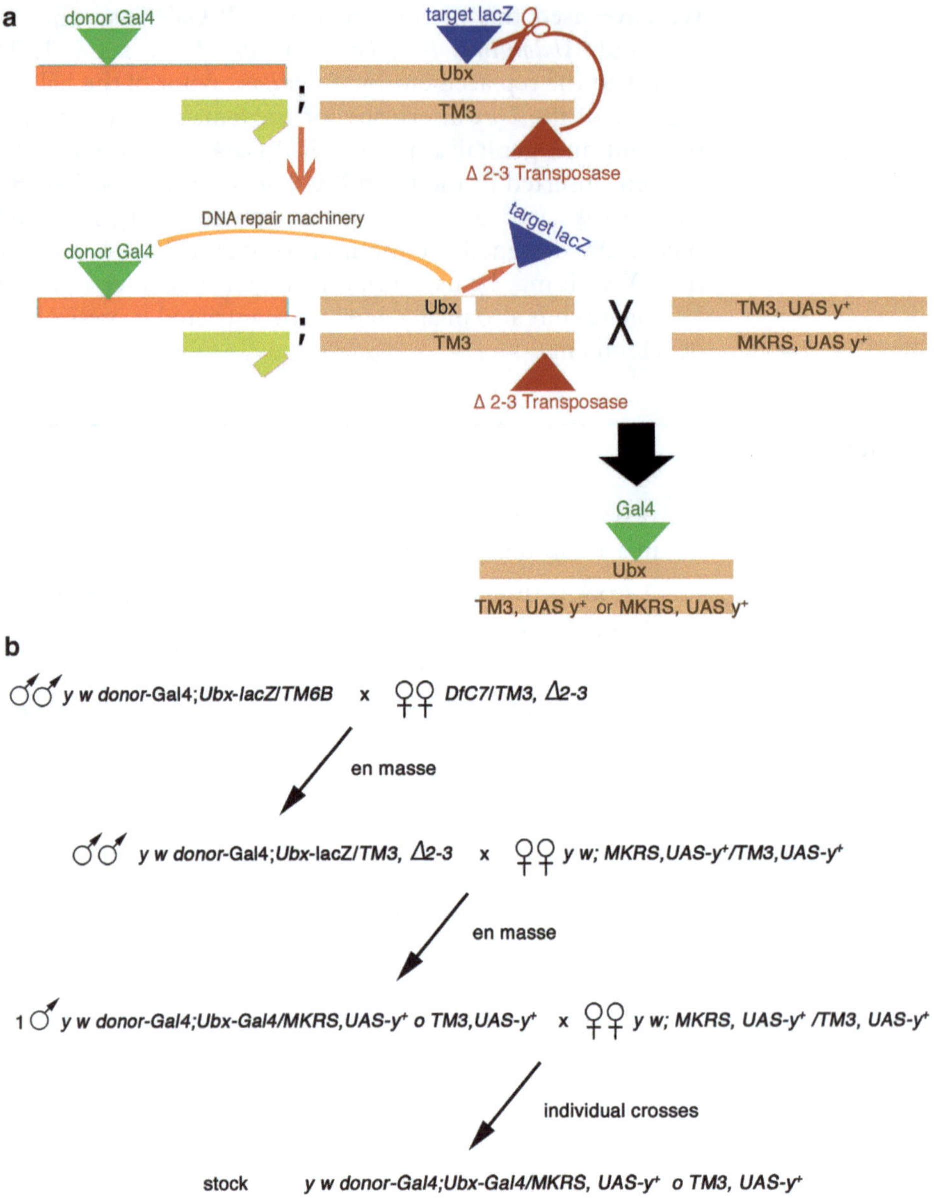

Fig. 1 Schemes of crosses needed to obtain the P-Gal4 insertion in a Hox gene (insertion in the *Ubx* gene is shown as an example). (**a**) Representation of the replacement of a P-lacZ transposon inserted in the *Ubx* gene by a P-Gal4 mobile element located on the X chromosome. The X, Y, and third chromosomes are represented by *orange*, *ocre*, and *light brown colors*, respectively. (**b**) Crosses needed to get the P-element replacement in the *Ubx* gene

3. Cross en masse the dysgenic males carrying the donor and "Hox gene inserted" P-elements as well as the Δ2-3 transposase to a UAS-y^+ stock (*see* **Note 8**). A convenient stock to use in these crosses is *y w*; TM3, UAS-y^+/MKRS, UAS-y^+ since it carries the UAS-y^+ insertion in balancers [19, 29].

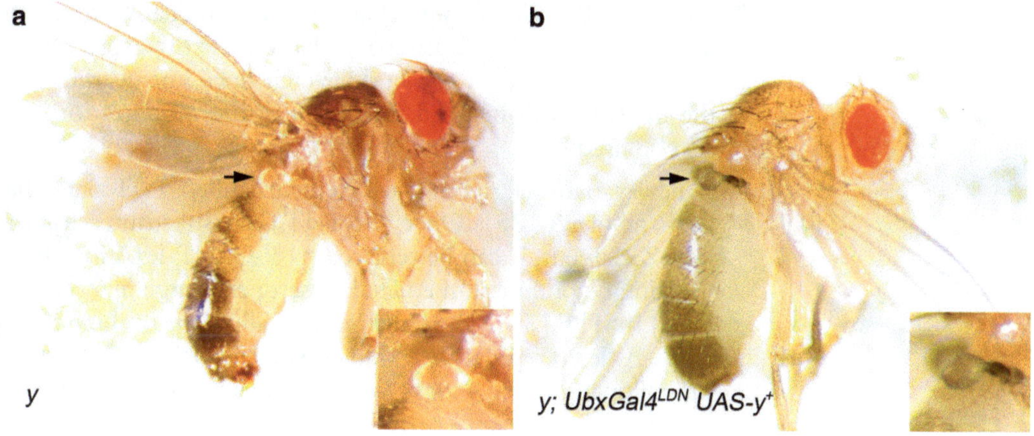

Fig. 2 Detection of an *Ubx*-Gal4 insertion. (**a**) A *yellow* (*y*) fly. An *arrow* indicates the haltere and a *close-up* is shown below. (**b**) A *y*; *UbxGal4*^{LDN} *UAS-y*$^{+}$ fly, showing the dark aspect of the haltere (*arrow*). A *close-up* is also shown below

4. Select individual flies by the y^+ color in part of the cuticle. A patch of y^+ observed in the cuticle, mimicking the expression of the endogenous Hox gene, suggests the existence of a P-Gal4 insertion in the Hox locus (Figs. 1 and 2) (*see* **Notes 9** and **10**).
5. Check the putative new P-Gal4 insertion by crossing it again with a UAS-y^+ in a *yellow* mutant background.
6. Balance the chromosome with the putative P-Gal4 insertion and check the expression pattern driven by the Gal4 line during development by crossing it with a UAS-GFP or UAS-lacZ stocks. Make a stock with the new insertion. It can be subsequently mapped by inverse PCR.

4 Notes

1. There are several P-elements inserted in Hox genes of *Drosophila*, almost all of them being P-lacZ elements. P-elements carry different markers used to recognize the transformed flies, most commonly the mini-*white* gene, the *rosy* gene in some of the initial insertions (but still quite used), or other. A method to recognize replacement of P-lacZ by P-Gal4 elements has been designed based on these different eye markers [18]. For our experiments it does not matter which marker they carry.
2. Normally, the Δ*2-3* P-transposase inserted at chromosomal position 99B is used because it is very stable [31]. There is a balancer chromosome carrying this insertion, *TM3*, Δ*2-3* [32], very useful to reduce pairing and facilitate the interchange of P-elements [18, 33, 34] (*see* **Note** 7).

3. We used a P-Gal4 "donor" that does not direct any expression in the adult cuticle. An insertion that directs the expression in a restricted region of the body, which does not overlap with the one that will be used to recognize the mobilization, can also be used.
4. A new variant of the P-Gal4 element, called P-IT.Gal4, has been reported [35]. Once the P-IT.Gal4 substitutes for the P-lacZ element, it can be replaced by P-Q or P-lexA insertions by using recombinase-mediated cassette exchange in a method called InSITE [35]. The new elements allow to combine the Q or lex-A/lexO regulatory modules with the Gal4/UAS system, thus allowing many genetic combinations to express or inactivate genes in adjacent or overlapping groups of cells [36, 37].
5. Since the dysgenic flies we use (those carrying the P-lacZ and P-Gal4 elements plus the P-element transposase (Fig. 1)) are males, and the *yellow* marker and donor element are on the X chromosome, the stock carrying the *TM3, Δ2-3* P-transposase does not need to be mutant for *yellow* if females carrying the two P-element insertions, and mutant for *yellow*, are used in the cross; when crossed to the stock with the *TM3, Δ2-3* chromosome, all the male progeny will be mutant for this gene (Fig. 1).
6. Apart from the correct replacement of one P-element by the other there could be cases of more than one insertion of the mobilized transposons. The original P-Gal4 element may also remain at their original place. Crosses with appropriate markers or balancers should be done to identify and eliminate these supernumerary P-elements.
7. The double-strand gap produced when a P-element is mobilized is filled with sequences from the sister chromatid or homologous chromosome (targeted gene conversion) [33, 34]. In the case of the replacement of one P-element by another, the homology of P-element sequences enables this repair by using the donor P-element as a template, thus obtaining the replacement of a P-element by the P-Gal4 [18, 29]. The P-element mobilization may leave a gap that is repaired by using sequences from the homologous chromosome as a template [33]. If the homolog harbors a deficiency for the gene where the P-element is inserted, the repair mechanism would favor the use of other homologous sequences present elsewhere in the genome, in our case the P-element ends, common to P-lacZ (or other) and P-Gal4 elements, and which may remain after the excision of the P-lacZ element [33, 38]. Therefore, a deficiency in the homologous chromosome for the locus where the recipient P-element is inserted would increase the rate of this substitution [33, 34]. However, many insertions in Hox genes are lethal or produce adults that are not very

healthy or fertile. As an alternative, a balancer chromosome, which disrupts pairing, may be used, since this would make difficult the replacement by the sequences in the homolog [33, 34]. That is why it is convenient to use the *TM3*, *Δ2-3* balancer. If the resident P-element is located on another chromosome, appropriate deficiencies or balancers may be used (as to the X chromosome, the use of males is preferred, since the X chromosome has no homolog).

8. Males that are dysgenic undergo P-element excisions also in the somatic line. This can be recognized by mosaicism of the *white* gene carried in the P-Gal4 element. Eyes with part of the ommatidia being white or different shades of red and orange should be seen.

9. Hox genes are normally expressed in the adult cuticle, so the insertion of a P-Gal4 line in one of these genes should be recognized by the y^+ color in the adult cuticle. For those genes whose expression is limited or difficult to identify, a UAS-GFP stock can be used instead of a UAS-y^+ one. The insertion is then recognized by looking for a GFP pattern that reproduces the pattern of the gene at the larval or pupal stages. This method can be used for any gene with defined expression during development, not only Hox genes, and requires that the donor P-Gal4 line does not drive GFP expression in a pattern that can overlap with the desired P-Gal4 one. We used this modification to obtain a P-Gal4 insertion in the *headcase* gene [29].

10. The P-Gal4 element carries the mini-*white* gene as a marker to identify the transformants. However, Hox genes are not expressed in the eye primordium of the eye-antennal disc and these genes are under active repression by *Polycomb*-group genes. Therefore, it would be expected that P-Gal4 insertions into Hox loci have the *white* gene within the transposon repressed [21]. In fact, the flies carrying some of the P-Gal4 insertions already isolated in the BX-C have white eyes ([29]; unpublished observations). However, others show a weak or variable expression of white, their eyes ranging from pale yellow to orange (*see* ref. 29 for a discussion of this point).

Acknowledgements

Work in the laboratory is being supported by a grant from the Spanish Ministerio de Economía y Competitividad (BFU2011-26075) and an institutional grant from the Fundación Ramón Areces. Delia del Saz is being supported by an FPI fellowship from the Spanish Ministerio de Economía y Competitividad.

References

1. Maeda RK, Karch F (2006) The ABC of the BX-C: the bithorax complex explained. Development 133:1413–1422
2. Kaufman TC, Seeger MA, Olsen G (1990) Molecular and genetic organization of the antennapedia gene complex of *Drosophila melanogaster*. Adv Genet 27:309–362
3. Alexander T, Nolte C, Krumlauf R (2009) Hox genes and segmentation of the hindbrain and axial skeleton. Annu Rev Cell Dev Biol 25:431–456
4. Wellik DM, Capecchi MR (2003) Hox10 and Hox11 genes are required to globally pattern the mammalian skeleton. Science 301:363–367
5. Greer JM, Puetz J, Thomas KR et al (2000) Maintenance of functional equivalence during paralogous Hox gene evolution. Nature 403: 661–665
6. Zhao Y, Potter SS (2001) Functional specificity of the *Hoxa13* homeobox. Development 128:3197–3207
7. Zhao Y, Potter SS (2002) Functional comparison of the *Hoxa4*, *Hoxa10* and *Hoxa11* homeoboxes. Dev Biol 244:21–36
8. Greig S, Akam M (1995) The role of homeotic genes in the specification of the *Drosophila* gonad. Curr Biol 5:1057–1062
9. Casares F, Calleja M, Sánchez-Herrero E (1996) Functional similarity in appendage specification by the Ultrabithorax and abdominal-A *Drosophila* Hox genes. EMBO J 15:3934–3942
10. Hirth F, Loop T, Egger B et al (2001) Functional equivalence of Hox gene products in the specification of the tritocerebrum during embryonic brain development of *Drosophila*. Development 128:4781–4788
11. Gehring WJ, Affolter M, Bürglin T (1994) Homeodomain proteins. Annu Rev Biochem 63:487–526
12. Merabet S, Hudry B, Saadaoui M et al (2009) Classification of sequence signatures: a guide to Hox protein function. Bioessays 31: 500–511
13. Rong YS, Golic KG (2000) Gene targeting by homologous recombination in *Drosophila*. Science 288:2013–2018
14. Hittinger CT, Stern DL, Carroll SB (2005) Pleiotropic functions of a conserved insect-specific Hox peptide motif. Development 132: 5261–5270
15. O'Keefe DD, Thor S, Thomas JB (1998) Function and specificity of LIM domains in *Drosophila* nervous system and wing development. Development 125:3915–3923
16. Rincón-Limas DE, Lu CH, Canal I et al (2000) The level of DLDB/CHIP controls the activity of the LIM homeodomain protein Apterous: evidence for a functional tetramer complex in vivo. EMBO J 19:2602–2614
17. Engels WR (1996) P-elements in Drosophila. In: Saedler H, Gierl A (eds) Transposable elements. Springer, Berlin, pp 103–123
18. Sepp KJ, Auld VJ (1999) Conversion of lacZ enhancer trap lines to GAL4 lines using targeted transposition in *Drosophila melanogaster*. Genetics 151:1093–1101
19. Calleja M, Moreno E, Pelaz S et al (1996) Visualization of gene expression in living adult *Drosophila*. Science 274:252–255
20. Engström Y, Schneuwly S, Gehring WJ (1992) Spatial and temporal expression of an *Antennapedia*/lacZ gene construct integrated into the endogenous *Antennapedia* gene of *Drosophila melanogaster*. Roux's Arch Dev Biol 201:65–80
21. Galloni M, Gyurkovics H, Schedl P et al (1993) The bluetail transposon: evidence for independent cis-regulatory domains and domain boundaries in the bithorax complex. EMBO J 12:1087–1097
22. McCall K, O'Connor MB, Bender W (1994) Enhancer traps in the *Drosophila* bithorax complex mark parasegmental domains. Genetics 138:389–399
23. Casares F, Bender W, Merriam J et al (1997) Interactions of *Drosophila Ultrabithorax* regulatory regions with native and foreign promoters. Genetics 145:123–137
24. Zhou J, Levine M (1999) A novel cis-regulatory element, the PTS, mediates an anti-insulator activity in the *Drosophila* embryo. Cell 99:567–575
25. Barges S, Mihaly J, Galloni M et al (2000) The Fab-8 boundary defines the distal limit of the bithorax complex *iab-7* domain and insulates iab-7 from initiation elements and a PRE in the adjacent iab-8 domain. Development 127:779–790
26. Bender W, Hudson A (2000) P element homing to the *Drosophila* bithorax complex. Development 127:3981–3992
27. Fitzgerald DP, Bender W (2001) *Polycomb* group repression reduces DNA accessibility. Mol Cell Biol 21:6585–6597
28. Estrada B, Casares F, Busturia A et al (2002) Genetic and molecular characterization of a novel *iab-8* regulatory domain in the *Abdominal-B* gene of *Drosophila melanogaster*. Development 129:5195–5204

29. de Navas LF, Foronda D, Suzanne M et al (2006) A simple and efficient method to identify replacements of P-lacZ by P-Gal4 lines allows obtaining Gal4 insertions in the bithorax complex of *Drosophila*. Mech Dev 123: 860–867
30. Hudry B, Viala S, Graba Y et al (2011) Visualization of protein interactions in living *Drosophila* embryos by the bimolecular fluorescence complementation assay. BMC Dev Biol 9:5
31. Robertson HM, Preston CR, Phillis RW et al (1988) A stable genomic source of P element transposase in *Drosophila melanogaster*. Genetics 118:461–470
32. Reuter G, Hoffmann G, Dorn R et al (1993) Construction and characterization of a TM3 balancer carrying P[(ry^{+}) Δ2-3] as a stable transposase source. Dros Info Serv 72: 78–79
33. Engels WR, Johnson-Schlitz DM, Eggleston WB et al (1990) High-frequency P element loss in Drosophila is homolog dependent. Cell 62:515–525
34. Gloor GB, Nassif NA, Johnson-Schlitz DM et al (1991) Targeted gene replacement in Drosophila via P element-induced gap repair. Science 253:1110–1117
35. Gohl DM, Silies MA, Gao XJ et al (2011) A versatile in vivo system for directed dissection of gene expression patterns. Nat Methods 8:231–237
36. Potter CJ, Tasic B, Russler EV et al (2010) The Q system: a repressible binary system for transgene expression, lineage tracing, and mosaic analysis. Cell 141:536–548
37. Yagi R, Mayer F, Basler K (2010) Refined LexA transactivators and their use in combination with the *Drosophila* Gal4 system. Proc Natl Acad Sci U S A 107:6166–61671
38. Nassif N, Penney J, Pal S et al (1994) Efficient copying of nonhomologous sequences from ectopic sites via P-element-induced gap repair. Mol Cell Biol 14:1613 1625

Chapter 5

Hox Complex Analysis Through BAC Recombineering

Mark Parrish, Youngwook Ahn, Christof Nolte, Bony De Kumar, and Robb Krumlauf

Abstract

BAC transgenesis in mice has proved to be useful in exploring the regulatory mechanisms and functions of the *Hox* complexes. The large constructs used may include most of the relevant components of the *cis-regulatory* landscape. Manipulations can be accomplished without compromising the integrity of the endogenous complex which reduces the likelihood of producing confounding phenotypic abnormalities. The development of recombineering tools has been critical in providing the means necessary to make many types of precise and varied manipulations of these large constructs. Here, we will discuss the methodologies necessary to manipulate *Hox* complex BACs, generation of transgenic animals bearing these constructs and the utilization of these resources to address fundamental aspects of *Hox* biology.

Key words BAC, Recombineering, *Hox*, Transgenic, CRE Recombinase, LoxP

1 Introduction

Regulatory and functional analyses of the *Hox* complexes has often been limited in the past by the size of the regions involved and the need to make a series of precise modifications. While plasmid-based transgenesis and manipulation of the endogenous complexes were feasible, each methodology had its own pitfalls. Plasmids often failed to reproduce the full complement of *Hox* expression patterns because regulatory modules are spread over a long range [1], and manipulations of the endogenous complexes could disrupt the structure of the complex itself or compromise normal development, making data interpretation difficult [2]. BAC transgenesis, as a method to explore the expression and regulation of the *Hox* complexes has, to a large degree, alleviated both of these problems. BACs are large enough to contain most or all of the relevant gene regulatory sequences, even if they are substantially distant from the target gene or at unknown locations. Manipulations are performed on a wild-type background, minimizing phenotypic anomalies that might result from alterations in the timing or patterns of

Yacine Graba and René Rezsohazy (eds.), *Hox Genes: Methods and Protocols*, Methods in Molecular Biology, vol. 1196, DOI 10.1007/978-1-4939-1242-1_5, © Springer Science+Business Media New York 2014

endogenous *Hox* expression. BAC transgenesis, in combination with recombineering tools that allow extensive manipulation of the BACs [3], has facilitated exploration of many complicated features of *Hox* complex regulation and function.

Given the instability and sequence complexity in large constructs, standard molecular cloning tools are not amenable for use on BACs. Recombineering (*Recombi*nation-mediated genetic engi*neering*) [4], instead, has emerged as one of the principle ways to manipulate these large constructs. A system developed utilizing the bacteriophage ʎ "red" proteins has become the standard system used in our lab and all methods described herein are based on this system. There are three "red" proteins Exo, Beta, and Gam. Exo processes dsDNA templates via 5′–3′ exonuclease activity producing a 3′ ssDNA overhang to facilitate Beta binding. Beta associates with ssDNA and protects it from degradation. Gam further prevents DNA degradation by inhibiting the nucleases RecBCD and SbcCD (reviewed in) [5]. A defective ʎ prophage containing these three genes under control of the temperature sensitive ʎ repressor *cI857* was introduced into the *E. coli* genome (strain DY380/SW102). CRE or FLP recombinase was then also added under L-arabinose control (EL350/SW106 and EL250/SW105 strains respectively). Recombination can then be stimulated by incubating these strains for 15 min at 42° without undue effects on viability. LoxP or FRT bounded selection cassettes can then be removed in EL350/SW106 and E250/SW105 by growth with LB + L-arabinose [6, 7].

With these basic tools a wide variety of manipulations of BACs are possible. In combination with transgenic technologies these BACs provide a platform for analyzing many different aspects of *Hox* regulation or function during development or in adult animals. In the following sections we will describe techniques to make serial modifications to BACs, seamless insertions/deletions, how to isolate portions of a BAC, and how to produce transgenic mice using these constructs (procedural flow chart Fig. 1). We also detail aspects of multiplex transgene expression analysis, the use of insulators in BAC construction, and identification of Hox binding sites through Chip-Seq analysis of transgenic expression of epitope-tagged Hox proteins.

2 Materials

2.1 Vector Construction Materials

1. DNA oligos (IDT, *see* **Note 1**).
2. Thermocycler.
3. T4 DNA Ligase Kit (NEB cat#M0202M).
4. *E. coli* competent cells.
5. Restriction enzymes (various).
6. Electrophoresis Power Pack.

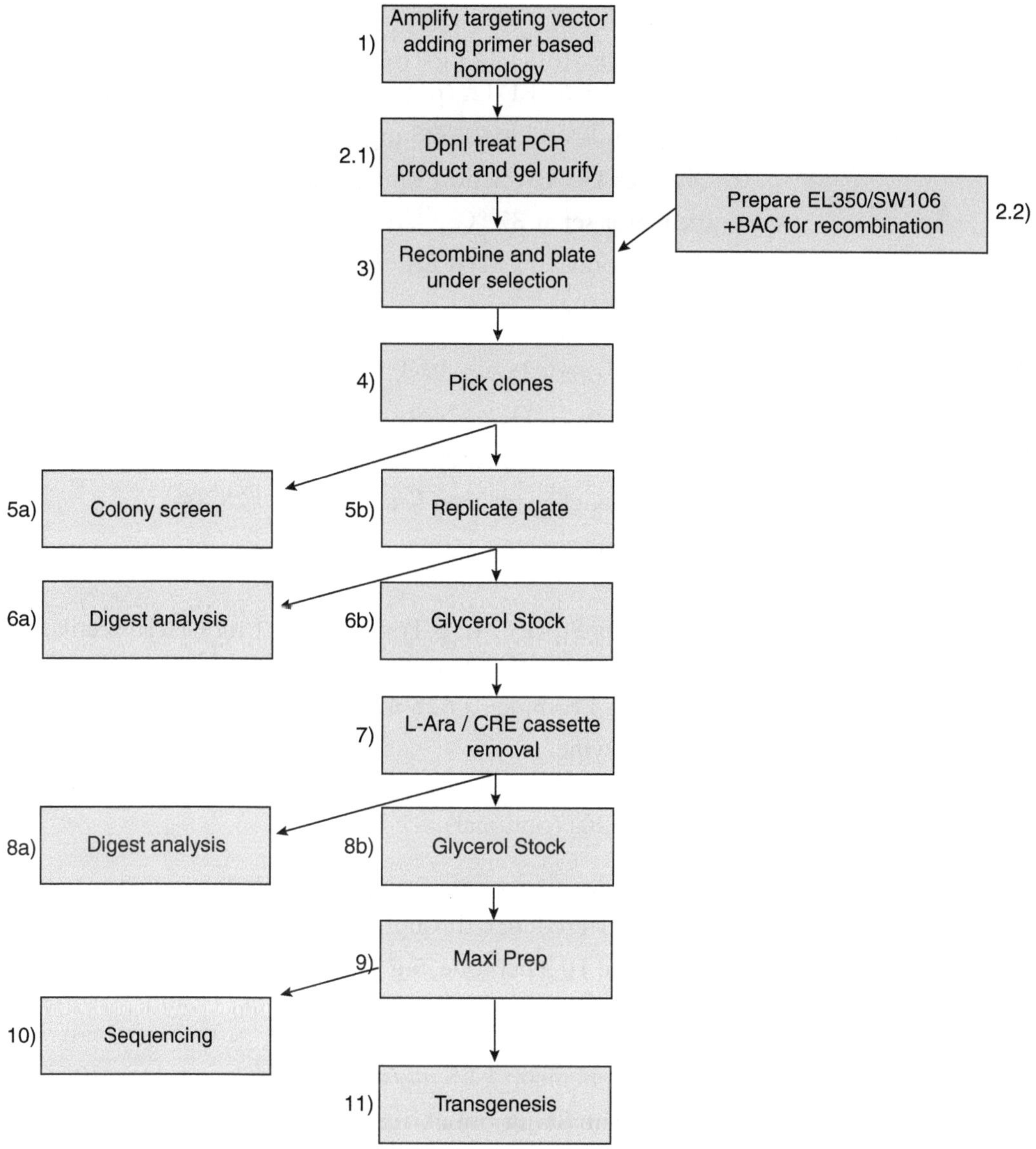

Fig. 1 Flowchart for processes involved in generating transgenic animals from recombinant BAC clones. "a" and "b" step parts occur concurrently. The process begins with the production of the targeting PCR product and preparation of the BAC bearing *E. coli* strains to make them electro-competent and induce the temperature regulated recombination machinery (1, 2.1, and 2.2). Afterward, the targeting PCR product is introduced into the BAC bearing *E. coli* strain through electroporation, recombination occurs, and transformants are plated under selection for about 24 h until the clones are large enough to analyze. The first clone screening step is a colony PCR (5a) and all clones are simultaneously stabbed onto a replicate plate for later retrieval (5b). Positive clones detected in the colony PCR are then mini-prepped overnight. A portion of this culture is saved as a glycerol stock (6b) and the remainder is used to isolate BAC DNA and analyze by digestion and gel electrophoresis (6a). Once the correct targeting is evidenced in the digest profile without any additional rearrangements, the selection cassette is removed by treating a culture grown from the glycerol stock with L-Arabinose and plated without the antibiotic used in step 3 (7). Clones are then mini-prepped, glycerol stocked, and subjected to a second digest analysis to verify cassette release and no additional rearrangements. After these steps a maxi should be made and the sequence double checked through critical regions to verify there are no point mutations (10). At this point the modified BAC is ready to be used for production of transgenic animals

7. Agarose.
8. TAE: (50× Stock) 242 g Tris-Base, 57.1 ml Glacial acetic acid, 100 ml, 0.5 M EDTA (pH 8.0), and ddH_2O to 1,000 ml.
9. Ethidium Bromide (0.625 mg/ml).
10. Qiaquick Gel Extraction Kit (Quiagen #28704).
11. Incubator set at 37 °C.
12. Nucleobond Xtra Maxi Kit or Qiagen Maxi Kit.

2.2 Targeting PCR Materials

1. Thermocycler.
2. Electrophoresis Power Pack.
3. DNA oligos (IDT, *see* **Note 1**).
4. Platinum Pfx polymerase kit (Invitrogen).
5. Qiaquick Gel Extraction Kit (Quiagen #28704).
6. Spectrophotometer.
7. Agarose.
8. TAE: (50× Stock) 242 g Tris-Base, 57.1 ml Glacial acetic acid, 100 ml, 0.5 M EDTA (pH 8.0), and ddH_2O to 1,000 ml.
9. Ethidium Bromide (0.625 mg/ml).
10. DpnI enzyme.
11. Topo cloning kit (Invitrogen Zero Blunt Topo cloning kit #K2800-20) (optional).

2.3 Recombineering Materials

1. Bacterial strains EL350 or SW106 (SW106 available via MTA from NCI Frederick through website).
2. LB media: 10 g peptone, 5 g yeast extract, 10 g NaCl + ddH_2O. pH to 7.0, adjust with ddH_2O to 1 L and sterilize by autoclaving.
3. Chloramphenicol: 12.5 μg/ml in ethanol stored at −20 °C.
4. Kanamycin: 25 μg/ml in dH_2O. Stored at −20 °C.
5. Multitron Incubator shaker.
6. Heated/shaking water bath.
7. Avanti J20I high-speed centrifuge.
8. 50 ml tube inserts for JA14 Centrifuge.
9. Table-top microcentrifuge.
10. Electroporator (BioRad Gene Pulser Xcell).
11. 0.1 cm cuvette.
12. SOC media: 5 g Bacto Yeast Extract, 20 g Bacto Tryptone, 0.5 g Sodium Chloride, 10 ml 250 mM Potassium Chloride. Autoclaved. 18 ml/L 20 % glucose added when cool.
13. LB plates with appropriate antibiotic.
14. Incubator set at 32 °C.

2.4 Colony PCR Materials

1. Thermocycler.
2. Owl/Thermoscientific electrophoresis apparatus.
3. 2× ReddyMix PCR premix.
4. DNA oligos (IDT, *see* **Note 1**).
5. Ethidium Bromide (0.625 mg/ml).
6. Agarose.
7. TAE: (50× Stock) 242 g Tris-Base, 57.1 ml Glacial acetic acid, 100 ml, 0.5 M EDTA (pH 8.0), and ddH_2O to 1,000 ml.
8. Mineral Oil.
9. LB-agar plates + appropriate antibiotic.
10. Electrophoresis Power Pack.
11. Incubator set at 32 °C.

2.5 Digest Screen Materials

1. LB media: 10 g peptone, 5 g yeast extract, 10 g NaCl + ddH_2O. pH to 7.0, adjust with ddH_2O to 1 L and sterilize by autoclaving.
2. Chloramphenicol: 12.5 μg/ml in ethanol stored at −20 °C.
3. Kanamycin: 25 μg/ml in dH_2O. Stored at −20 °C.
4. Glycerol.
5. Incubator set at 32 °C.
6. Heated water bath (55 °C).
7. Table-top microcentrifuge.
8. P1 (Resuspension): 50 mM Tris pH 8, 10 mM EDTA, 100 μg/ml RNase A. For 500 ml add 25 ml of 1 M Tris pH 8, 10 ml of 0.5 M EDTA, and 5 ml of 10 mg/ml RNase A (RNase stock should be stored at −20 °C).
9. P2 (Lysis): 200 mM NaOH, 1 % SDS. For 500 ml add 4 g of NaOH pellets and 50 ml of 10 % SDS solution.
10. P3 (Neutralization): 3 M Potassium Acetate pH 5.5. For 500 ml add 147.21 g of potassium acetate to 350 ml dH_2O + 100 ml pure glacial acetic acid. pH to 5.5 and increase volume to 500 ml with dH_2O.
11. Isopropanol.
12. 70 % Ethanol: Absolute alcohol (700 ml) + dH_2O (300 ml).
13. TAE: (50× Stock) 242 g Tris-Base, 57.1 ml Glacial acetic acid, 100 ml, 0.5 M EDTA (pH 8.0), and ddH_2O to 1,000 ml.
14. Heated shaking block.
15. Typhoon 8600Imaging System.
16. Nucleobond Xtra Maxi Kit or Qiagen Maxi Kit.

2.6 CRE Recombinase Induction Materials

1. LB media: 10 g peptone, 5 g yeast extract, 10 g NaCl + ddH_2O. pH to 7.0, adjust with ddH_2O to 1 L and sterilize by autoclaving.
2. Chloramphenicol: 12.5 μg/ml in ethanol stored at −20 °C.
3. Kanamycin: 25 μg/ml in dH_2O. Stored at −20 °C.
4. Multitron Incubator shaker.
5. L-Arabinose. 1,000× stock 1 g/ml in water stored at −20 °C sterile filtered.
6. LB-agar plates with appropriate antibiotics.
7. Glycerol.
8. Incubator set at 32 °C.

2.7 Recombineering Capture Materials

1. DNA oligos (IDT, *see* **Note 1**).
2. Annealing buffer (10 mM Tris pH 8.0, 50 mM NaCl, 1 mM EDTA).
3. Heat block.
4. Thermocycler.
5. Topo cloning kit (Promega (pGEMT-Easy #A3600) or Invitrogen (Topo-TA #K450001).
6. Taq DNA polymerase kit.
7. Qiaquick Gel Extraction Kit (Quiagen #28704).
8. Spectrophotometer.
9. Incubator set at 37 °C.
10. Electroporator (BioRad Gene Pulser Xcell).
11. 0.1 cm cuvette.
12. SOC media: 5 g Bacto Yeast Extract, 20 g Bacto Tryptone, 0.5 g Sodium Chloride, 10 ml 250 mM Potassium Chloride. Autoclaved. 18 ml/L 20 % glucose added when cool.
13. LB-agar plates with appropriate antibiotics.
14. Incubator set at 32 °C.
15. Nucleobond Xtra Maxi Kit or Qiagen Maxi Kit.

2.8 End Trimming, Homing Endonuclease Site/Insulator Insertion Materials

1. Materials from Subheadings 2.2–2.6.
2. Topo cloning kit (Promega (pGEMT-Easy #A3600) or Invitrogen (Topo-TA #K450001).
3. Taq DNA polymerase kit.

2.9 Seamless Insertions/Deletions (Two-Step Recombination with sacB-Kan Cassette) Materials

1. Materials from Subheadings 2.2–2.6.
2. Tetracycline 12.5 mg/ml in water (1,000× stock). Working conc. 12.5 μg/ml.
3. Sucrose.

2.10 BAC Transgenesis Materials

1. Qiagen Maxi Kit.
2. Avanti J20I high-speed centrifuge.
3. PI-Sce-I enzyme (NEB #R0696S).
4. Spectrophotometer.
5. 1,000× Polyamine: 30 mM Spermine (Sigma, tetrahydrochloride, #S-1141) 70 mM Spermidine (Sigma, trihydrocholoride, #S-2501). Dissolve in dH_2O, filter sterilize, and store at −20 °C.
6. BAC injection buffer: For 50 ml add 500 μl (1 M Tris–HCl pH 7.5), 10 μl (0.5 M EDTA pH 8.0), and 1 ml (5 M NaCl). Final concentration of 10 mM Tris, 0.1 mM EDTA, and 100 mM NaCl. Sterile filter. Can be stored at room temp. Add polyamine mix at the time of use.

2.11 Additional Materials/Equipment/ Software

1. mHoxb1 antibody (Abcam #ab24708)
2. Flag antibody (Sigma #F1804)

Equipment

1. Zeiss LSM 710 microscope
2. Leica M205 Stereomicroscope

Software

1. Vector NTI (Invitrogen)
2. Clc Main Workbench (CLC-Bio)
3. Imaris (Bitplane)
4. Image J (NIH)
5. Illustrator (Adobe)
6. Photoshop (Adobe)

3 Methods

3.1 Vector Construction

1. Order oligo containing Lox71 (Left side inverted repeat mutant) on the 5′ end and Lox66 (Right side inverted repeat mutant) on the 3′ end and include restriction enzyme sites in between as desired for insertion of a selection cassette (Fig. 2). Enzyme sites may be added 5′ and 3′ external to the Lox sites for convenient cloning. Alternatively, forward and reverse oligos may be designed to overlap in such a way as to produce over-hangs that are compatible with restriction enzyme sites contained in the backbone of the vector of choice (Fig. 2e, *see* **Note 2**).
2. Clone oligo via standard ligation into a vector of choice (Ampicillin-resistant vectors preferred).

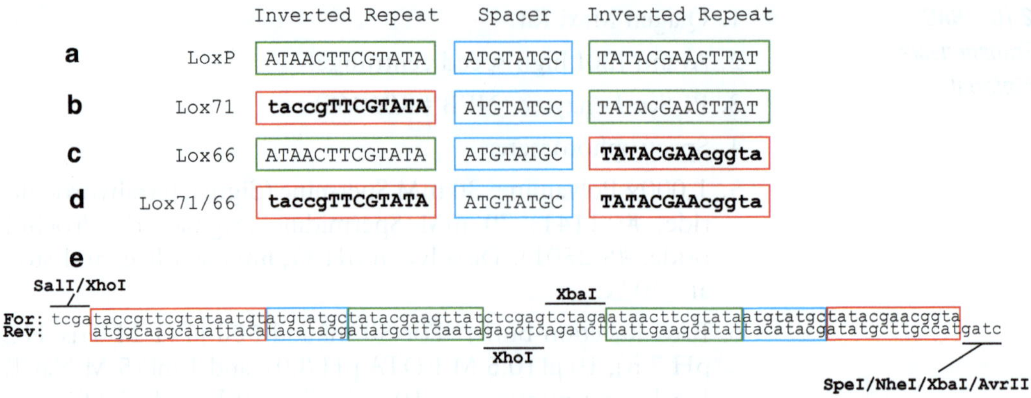

Fig. 2 LoxP sites. (**a**) depicts the wt LoxP target site. (**b**) and (**c**) show LoxP sites that have inverted repeat mutations 71 and 66. The composite 71–66 site (**d**) is the product of the reaction between a Lox71 and Lox66 when the spacers are in the same orientation and has almost negligible activity [3]. (**e**) Depicts a sample set of oligos that could be annealed and cloned into the pBluescriptII XhoI and XbaI sites via compatible overhangs (5′ SalI/XhoI compatible and 3′ SpeI/NheI/XbaI/AvrII compatible, original XhoI and XbaI sites in pBluescriptII are destroyed). Next a selection cassette could be inserted between the Lox71 and Lox66 sites via the internal XhoI/XbaI sites (Reverse oligo would need to be reversed before ordering)

3. Digest Lox71—Lox66 vector between the two lox sites, gel purify the vector and clone the desired antibiotic selection cassette via standard ligation. Kanamycin cassettes are recommended for several reasons (Fig. 3a, *see* **Note 3**).

3.2 Targeting PCR

Design primers that include 50 bp (base pairs) of 3′ or 5′ homology to the target region and a region for amplification of the targeting cassette (Fig. 3, *see* **Note 1**).

1. Amplify the cassette via PCR using the primers. Reaction composition:

   Template: 1 μl (50 ng)

   Water: 27.5 μl

   Buffer (Pfx buffer): 10 μl

   Enhancer (Pfx enhancer): 5 μl

   $MgSO_4$ (50 mM): 1 μl

   dNTPs (10 mM each of 4 nucleotides): 1.5 μl

   Primer A (10 μM): 1.5 μl

   Primer B (10 μM): 1.5 μl

   Plat. Pfx: *0.5*μl

   Total volume 50 μl

Cycling conditions:

1. 94 °C × 5 min.
2. 94 °C × 30 s.

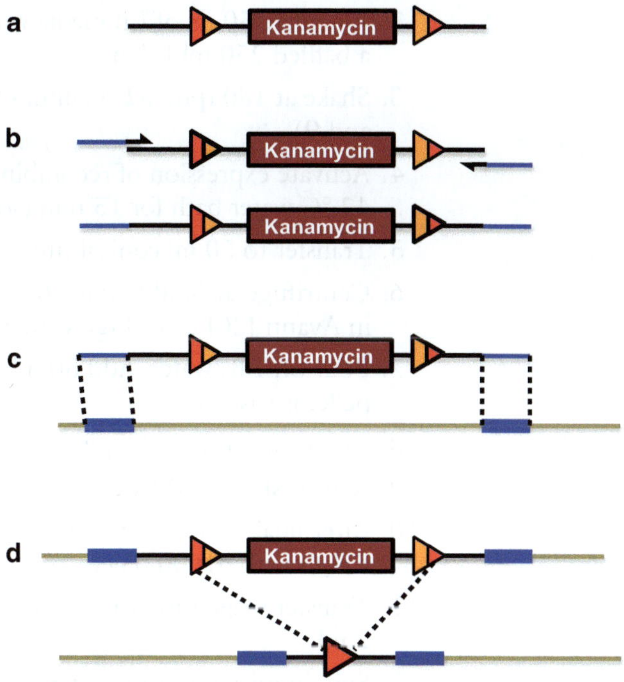

Fig. 3 Steps in the recombination reaction. (**a**) The initial targeting cassette contains a selectable marker flanked by LoxP sites and any additional insert, mutation, or modification that is desired in the final BAC. The loxP sites bear the Lox71 (5′ LoxP) and Lox66 (3′ LoxP) mutations (*red*). (**b**) Regions of homology to the target location (*blue*) are added onto the targeting cassette by PCR. (**c**) The targeting vector pairs with the target site in the BAC during recombination (*dashed lines*) via the regions of homology (*blue*) and replaces the intervening sequence. (**d**) The selectable marker is removed from the inserted targeting cassette by CRE-mediated excision. This produces a hybrid Lox71/Lox66 LoxP (*red*) site that is an unfavorable substrate for further CRE recombinase reactions

3. 58 °C × 30 s.
4. 68 °C × (time (min) = Length of product (bp)/1,000) (Repeat **steps 2–4** for 30 cycles).
5. 68 °C × 10 min.
6. 4 °C Hold.

Treat the PCR product with 1 μl DpnI for 1 h at 37 °C and gel purify using Qiaquick gel extraction kit (*see* **Note 4**).

Resuspend at final Qiagen elution step in ddH_2O. Quantitate and store at −20 °C (*see* **Notes 5** and **6**).

3.3 Recombineering

1. Inoculate BAC bearing bacterial strain EL350 or SW106 (for loxP-based selection cassettes) from glycerol stock in 5 ml of LB media and culture overnight shaking at 180 rpm and 32 °C (*see* **Note 7**).

2. Inoculate 50 ml of LB media with 1 ml of overnight culture in a baffled 250 ml Erlenmeyer flask.
3. Shake at 180 rpm, 32 °C until OD 600 is 0.4–0.6 (*see* **Notes 8** and **9**).
4. Activate expression of recombination machinery by shaking in 42 °C water bath for 15 min (*see* **Note 10**).
5. Transfer to 50 ml conical tube.
6. Centrifuge at 5,500 × *g* for 6 min (can be done in JA14 rotor in Avanti J20I centrifuge with 50 ml tube adapter).
7. Pour off LB/water. Add 50 ml ice cold dH_2O and shake until pellet is suspended.
8. Centrifuge at 5,500 × *g* for 6 min (*see* **Note 11**).
9. Repeat **steps 7**, **8** twice.
10. After final wash, pour off water and resuspend in 1 ml ice cold dH_2O with p1000 pipet.
11. Transfer suspension to 2 ml tube and fill with ice cold dH_2O to 2 ml.
12. Spin in microcentrifuge at 5,500 × *g* for 1 min at 4 °C.
13. Pour off water and resuspend pellet in 550 μl of ice cold dH_2O (*see* **Note 12**).
14. Add purified/Dpn1 treated targeting PCR fragment (1–4 μl) to 0.1 cm electroporation cuvette.
15. Add 50 μl of electrocompetent bacterial cells to cuvette.
16. Electroporate 1× at 1.8 kv, 25 μF, 200 Ω (*see* **Note 13**).
17. Add electroporation to 400 μl of SOC media and incubate shaking at 180 rpm and 32 °C for 2–4 h in 2 ml tube (*see* **Note 14**).
18. Plate on chloramphenicol 12.5 μg/ml/Kanamycin 25 μg/ml containing LB-agar plates overnight at 32 °C.
19. Colonies should be visible for picking/analysis the next afternoon.

3.4 Colony PCR

1. PCR primers should be designed to create a product that spans the joint where recombination has occurred. One primer location should be in the targeting PCR fragment and a second in the target BAC near the insertion site. PCR primers are generally designed with a Tm (melting temperature) near 60 °C (*see* **Note 15**) and PCR fragments should be planned to be fairly short (200–1,000 bp) (*see* **Note 16**).
2. A mastermix should be made using standard Taq (2× ReadyMix) + water + primers at a concentration of 0.3 μM

(3 μl/100 of a 10 μM stock). 10 μl/reaction in a 96 well PCR type plate is sufficient (*see* **Note 17**).

3. Using 200 μl pipet tips pick individual colonies from the electroporation plating. Make a stab culture of each on a 10 cm LB + antibiotic plate labeled with a 96 grid (12 × 8) and then place the pipet tip down in corresponding well of the PCR plate that contains the mastermix. Continue until finished or the plate is full. Afterward remove tips, add 30 μl of mineral oil and cycle, 95 °C ×5 min: 1 cycle, (95 °C × 30 s, 55–58 °C 30 s (*see* **Note 18**), 72 °C 10–60 s (*see* **Note 19**)):30 cycles, 72 °C 10 min: 1 cycle.
4. Run PCR samples in 1–2 % agarose TAE gel (*see* **Note 20**) + ethidium at 100 V for 20 min (*see* **Notes 21** and **22**, Fig. 4).
5. Incubate replicate plate overnight 32 °C then store at 4 °C until positive clones are identified.

3.5 Digest Screen

1. Inoculate clones from the replicate plate that were identified as positive in the colony PCR screen into 5 ml of LB + antibiotic media and grow shaking at 180 rpm/32 °C overnight.
2. Warm 100 % glycerol to 55 °C and add 200 μl to 2 ml, gasketed, cold storage tubes.
3. After glycerol has cooled to room temp, add 800 μl sample and vortex to make a 20 % glycerol/sample solution. Label tubes and store at −80 °C as glycerol stocks.
4. Concentrate remaining cells by centrifugation in a 2 ml tube (10k × *g*/30 s 2×).
5. Add 250 μl P1 and resuspend by vortexing (*see* **Note 23**).
6. Add 250 μl P2 and mix by inversion.
7. Add 250 μl P3 and mix by inversion.
8. Pellet debris by centrifugation (10k × *g*/10 min).
9. Prepare second set of tubes with 750 μl of isopropanol and label.
10. Transfer supernatant from debris tubes to second set of labeled, isopropanol containing tubes.
11. Precipitate DNA by centrifugation (10k × *g* × 15 min).
12. Wash DNA with 500 μl 70 % ethanol.
13. Centrifuge in microcentrifuge (10k × *g* × 5 min).
14. Aspirate alcohol and dry pellet (*see* **Note 24**).
15. Resuspend in 29 μl of buffer appropriate for enzyme digestion.
16. Incubate in heated shaking block @80 °C/500 rpm for 15 min.

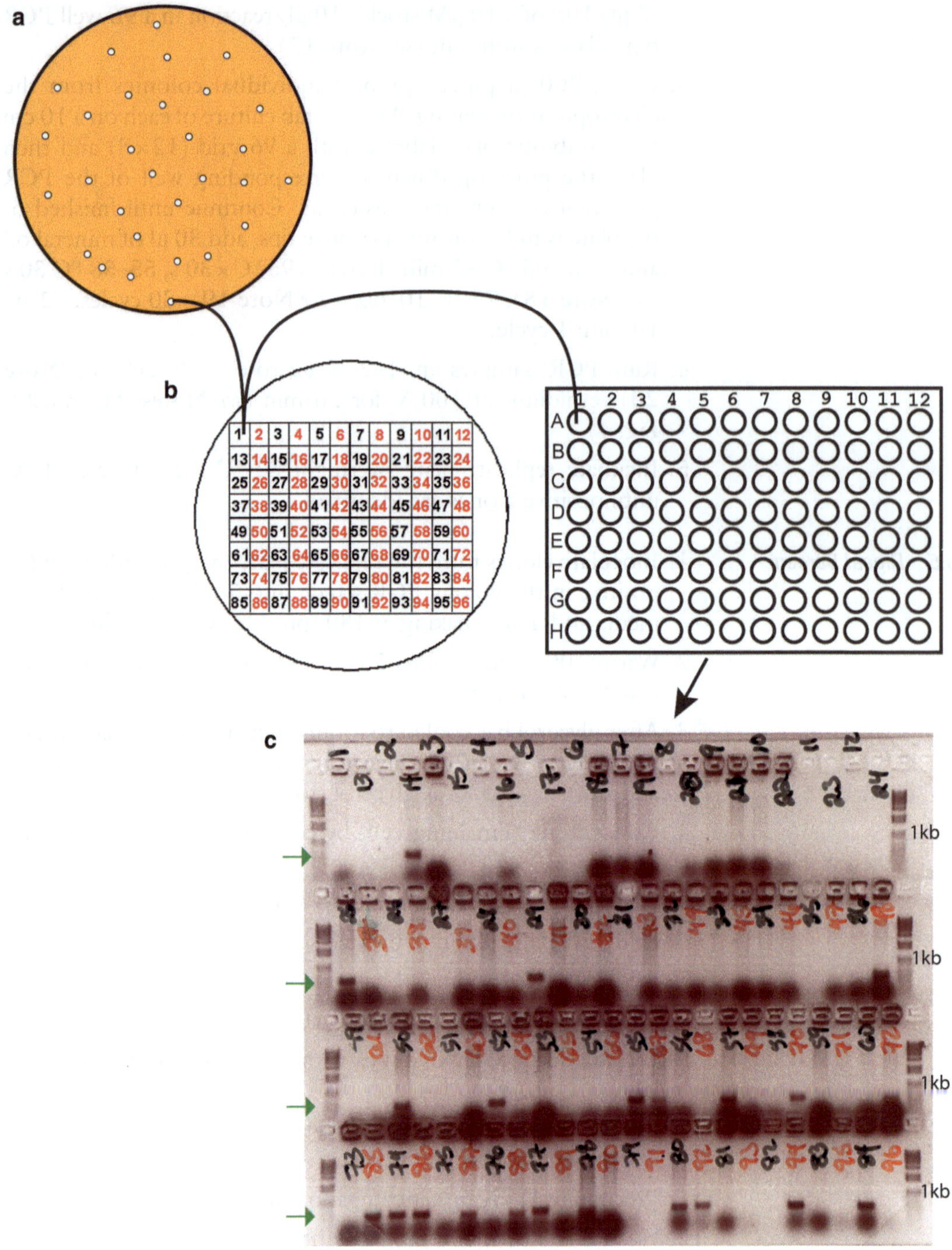

Fig. 4 Colony screening. (**a**) After recombination the reaction is plated at clonal density on LB + selective antibiotic plates. Colonies are picked from this plate and stabbed into a culture plate and then placed in the respective well of a 96 well plate containing amplification mix for colony screening (**b**). (**c**) After cycling, the reactions are loaded on a gel and electrophoresed to detect clones positive for recombination

17. Add 1 μl of restriction enzyme and digest at appropriate temp for 1–4 h (*see* **Note 25**).
18. Run overnight on large gel (10 in/25 cm) 1 % agarose/TAE gel + ethidium at 50 V overnight (18 h).
19. Analyze band pattern on gel documentation apparatus or phosphor imager (Fig. 5, *see* **Notes 26** and **27**).

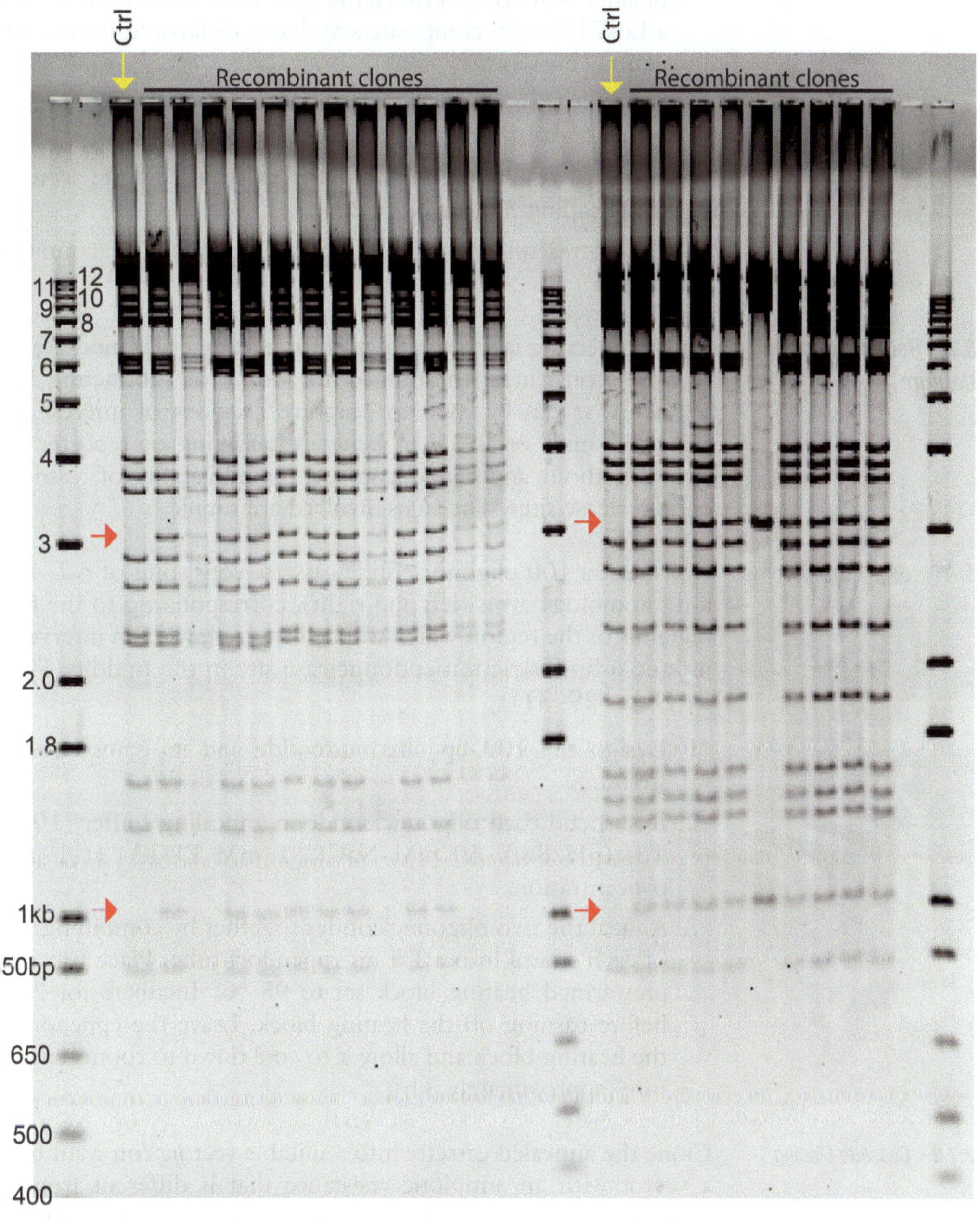

Fig. 5 Digest analysis. NheI digests of the BAC before ("Ctrl" *Yellow Arrow*) and after (*Recombinant clones*) recombination. Diagnostic bands present after recombination indicated at left sides with *red arrow*. 1 kb + ladder included at *left*, *right*, and *center*. Size markers indicated at *left* in bp (400–850) or kb (1–12)

3.6 CRE Recombinase Induction

1. Streak LB-agar plates + appropriate antibiotics from glycerol stocks of recombined BAC clones and incubate overnight at 32 °C.
2. Inoculate individual colonies in LB, grow to log phase and add 0.1 % L-arabinose and grow for 1 h shaking at 180 rpm/32 °C. CRE expression will be induced and recombination will occur between the Lox71 and Lox66 sites creating a Lox71-Lox66 composite site that is disfavored for recombination (Fig. 3).
3. Plate reaction on LB + chloramphenicol only selection and incubate overnight at 32 °C.
4. Analyze resultant colonies by enzyme digest screening (Subheading 3.6, *see* **Note 28**).
5. Confirm sequence of BAC over inserts before continuing (Fig. 6).

3.7 Recombineering Capture

Recombineering is a powerful tool for making exact modifications of DNA constructs. In addition to adding or subtracting DNA segments, it can be used to "capture" regions of interest from either plasmids or BACs to generate targeting or reporter constructs without any dependence on the availability of restriction endonuclease sites. The steps involved are simple.

3.7.1 Oligo Construction

Construct a 100 oligonucleotide cassette consisting of two 47 bp long homology arms (left and right), corresponding to the flanking ends of the region which will be captured, with an intervening unique 6 bp restriction endonuclease site in the middle (Fig. 7, *see* **Notes 29–31**).

1. Design the 100 bp oligonucleotide and its complementary strand.
2. Resuspend each oligonucleotide in annealing buffer (10 mM Tris (pH 8.0), 50 mM NaCL, 1 mM EDTA) at 1 μg/μl concentration.
3. Anneal the two oligonucleotides together by combining 10 μl of each strand into a 1.5 ml eppendorf tube. Place tubes in a prewarmed heating block set to 95 °C. Incubate for 2 min before turning off the heating block. Leave the eppendorf in the heating block and allow it to cool down to room temperature (approximately 3 h).

3.7.2 Cassette Cloning

Clone the annealed cassette into a suitable vector. You want to use a vector with an antibiotic resistance that is different from the plasmid or BAC from which you will "capture" the fragment. The easiest method to subclone the annealed cassette into a vector is to use a TA-cloning kit (from either Promega or Invitrogen):

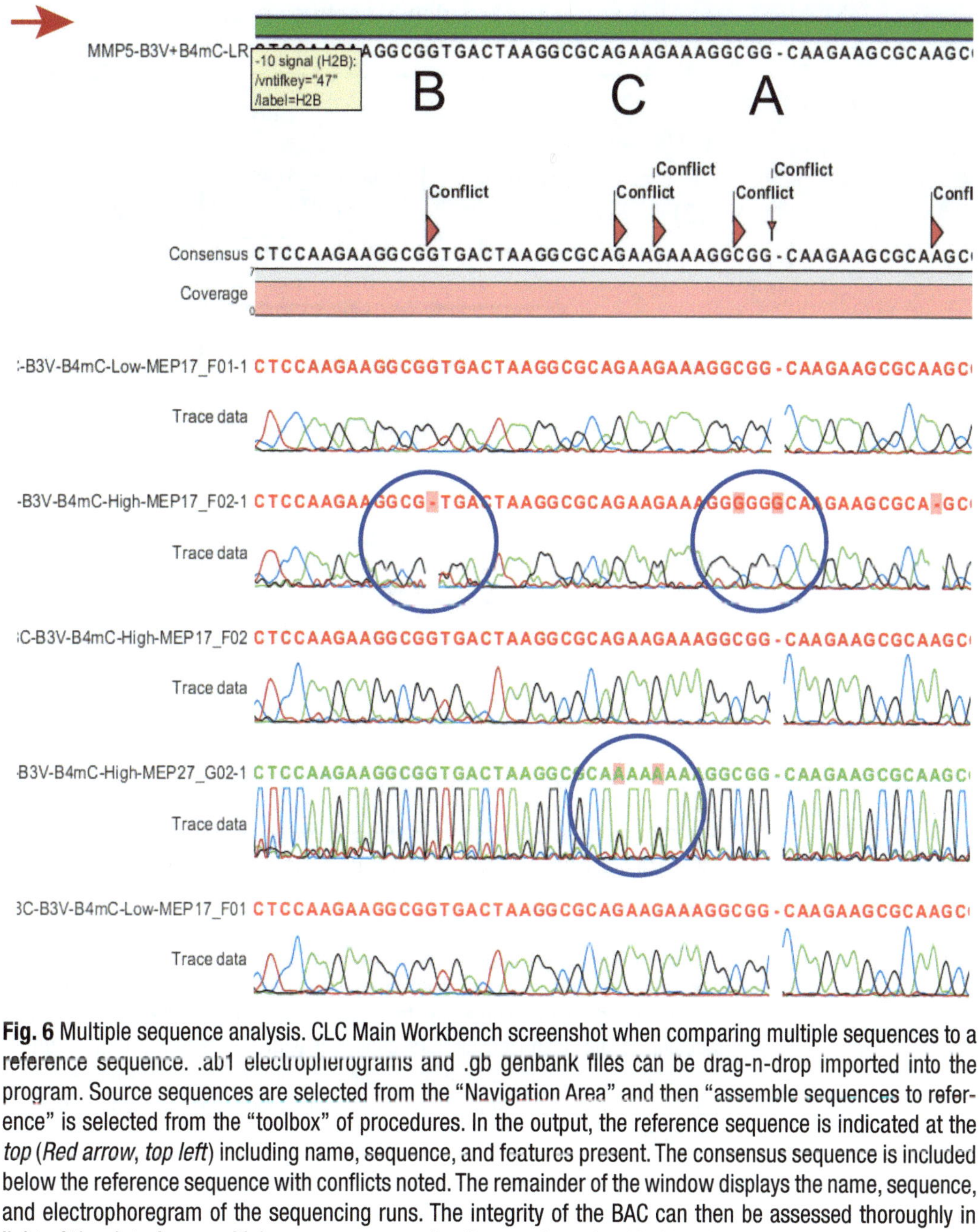

Fig. 6 Multiple sequence analysis. CLC Main Workbench screenshot when comparing multiple sequences to a reference sequence. .ab1 electropherograms and .gb genbank files can be drag-n-drop imported into the program. Source sequences are selected from the "Navigation Area" and then "assemble sequences to reference" is selected from the "toolbox" of procedures. In the output, the reference sequence is indicated at the *top* (*Red arrow, top left*) including name, sequence, and features present. The consensus sequence is included below the reference sequence with conflicts noted. The remainder of the window displays the name, sequence, and electrophoregram of the sequencing runs. The integrity of the BAC can then be assessed thoroughly in light of the data from multiple sequence runs simultaneously and guided by the sequencing quality at each point as indicated in the electrophoregrams. (**a**) indicates an instance of an insertion, (**b**) indicates a deletion, and (**c**) indicates a sequencing miss-read (*blue circles*). All these errors may be ignored in this case because they are not evidenced in other over-lapping reads and some appear to be related to the quality of the sequencing read at that position

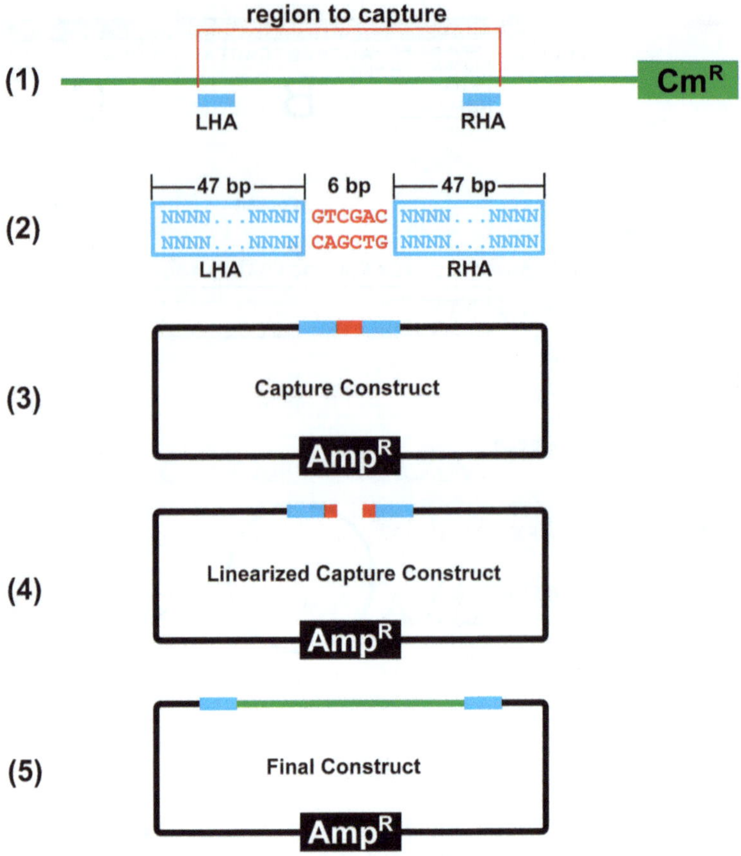

Fig. 7 Recombineering capture methodological steps. (1) Region to capture is identified and homology arms selected upstream and downstream (*blue*). (2) An oligo is generated containing 47 bp left and right homology arms (*blue*) separated by a unique restriction site (*red*). (3) This oligo is topo cloned into a vector. (4) Vector bearing the oligo is linearized at the unique restriction site separating the two homology arms. (5) After co-electroporation of the linearized capture construct and the BAC containing the region to be captured, the result of recombination is the linearized capture construct with the captured region (*green*) between the two homology arms (*blue*). LHA and RHA indicate left and right homology arms. Cm^R and Amp^R represent chloramphenicol and ampicillin resistance

1. To the annealed oligonucleotide cassette, add 2.5 μl of 10× Taq buffer, 1 μl of 10 mM dNTP mix, 0.5 μl of dH_2O and 1 μl of Taq-polymerase (do not use a high-fidelity Taq).
2. Incubate the reaction for 10 min at 72 °C, before cooling on ice for 2 min.
3. Follow the specific manufacturer's instructions for cloning a duplex DNA fragment containing A-overhangs into their vector.

3.7.3 Vector Linearization

Once the capture vector has been identified (i.e., by sequencing) and generated, linearize the vector using the unique restriction endonuclease site within the cassette.

1. Make sure to optimize the restriction digestion so that it proceeds as far as possible to completion.
2. Band purify the linearized plasmid. We usually use the gel extraction kit from Qiagen.
3. Resuspend the band purified linearized plasmid in dH_2O and determine its concentration via spectrophotometry.
4. Prepare the recombineering bacterial strain that contains the donor DNA for electroporation according to the instructions described earlier in this chapter (Subheading 3.3).
5. Combine 200 ng of the linearized, gel purified capture vector with 50 μl of electroporation readied bacteria and electroporate in a 1 mm gap cuvette using standard electroporation conditions.
6. Following 1 h recovery, plate out all of the electroporated bacteria on LB-agar plates containing an antibiotic that is selective for the capture construct.
7. The next day screen colonies either by colony PCR (Subheading 3.4) or mini-prep analysis (Subheading 3.5) (*see* **Note 32**).

3.8 Recombination-Mediated End Trimming, Homing Endonuclease Site/Insulator Insertion: Reduction of Transgenic Array Effects via Insulators and Homing Endonucleases

Hox genomic DNA fragments in a BAC clone often need to be trimmed off at its ends. This can be done with a targeting vector with two homology arms, one for the vector backbone of BAC clones and the other for the *Hox* region. During this process, unnecessary sequences in the pBACe3.6 backbone vector such as the partial *Tet* cassette, the *sacBII* promoter and the *sacBII* coding region can be removed by using appropriate homology arms. This is essential to remove these sequences to reduce unwanted recombinants if *Tet* or the *sacB-Kan* cassette is to be used for selection afterward as described in the following section (Fig. 8).

During pronuclear injection, circular BACs can integrate into the genome after random breakage resulting in potential end effects. Some vectors used for BAC library construction including pBACe3.6 contain a target site for homing endonuclease such as I-SceI and PI-SceI, which can be used to linearize BACs to minimize end effects. If not present, these sites can also be introduced into BACs during the trimming process.

It has been reported that insulators can reduce positional effect of integration sites and transgenic array effects due to multi-copy integration. Therefore, we introduced a copy of the 1.2 kb chicken β-globin insulator [8] into one end of *Hox* genomic fragments.

The following describes two rounds of recombineering to modify a mouse BAC, RP23-196F5, containing the *Hoxb* cluster [9]. The first round removes the *sacB* promoter and a LoxP511

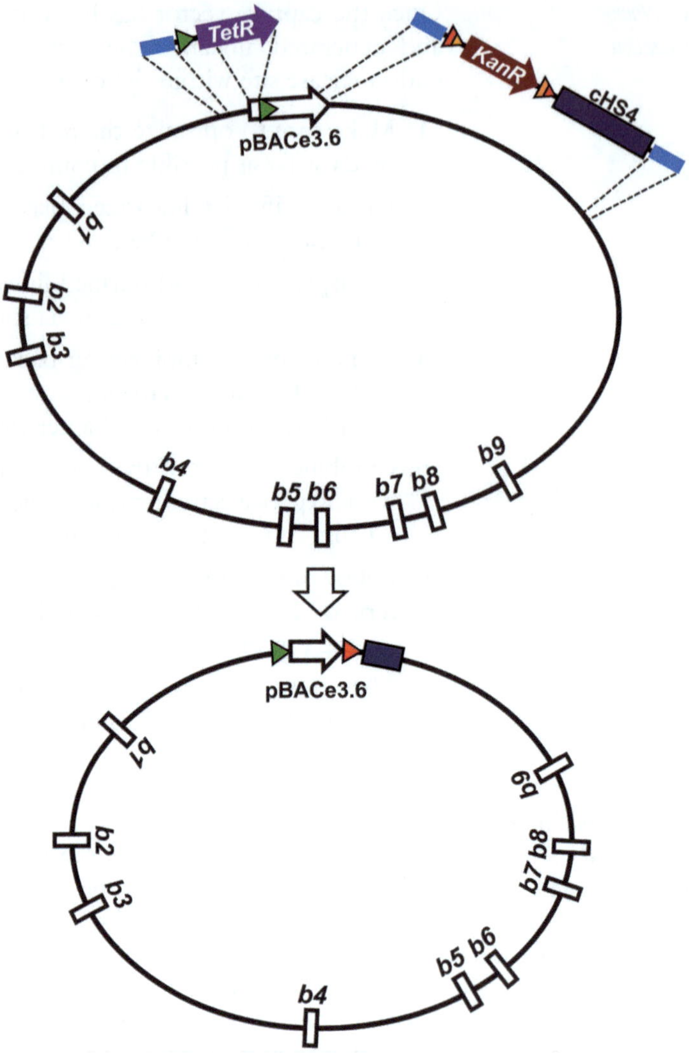

Fig. 8 Recombination-mediated end trimming and insertion of an insulator. The original mouse Hoxb BAC clone, RP23-196 F5, was modified by two rounds of recombineering. Each recombination was followed by Cre-mediated removal of the selection cassettes. The recombineering constructs are shown above the original BAC (*top*). As a result, the 5′ end of the *Hoxb* cluster was trimmed off and the 1.2 kb chicken β-*globin* insulator, cHS4, was inserted (*bottom*). *Green triangle*: wild-type LoxP site; *Red-orange triangle*: variant LoxP site

site and inserts the 1.2 kb chicken β-globin insulator at the 5′ end of the *Hoxb* cluster. The second round removes the partial *Tet* cassette and the *sacBII* coding region.

1. Amplify the 1.2 kb chicken β-globin insulator named cHS4 by PCR using chicken genomic DNA as template, clone it into pGEMTeasy vector (Promega) and confirm the sequence of the inserted DNA.

2. Clone the Kan cassette flanked by a LoxP variant site upstream of cHS4 to make pGEMTeasy-Kan-cHS4.
3. Design two primers to amplify the Kan-cHS4 fragment. At their 5′ end, forward primer contains 5′ homology arm identical to the region around PI-SceI site and reverse primer contains 3′ homology arm identical to the 5′ end of the *Hoxb* cluster.
4. PCR amplify the Kan-cHS4 with the homology arms with the above primers using pGEMTeasy-Kan-cHS4 as template, clone it into pGEMTeasy vector and confirm the sequence of the inserted DNA.
5. Digest with appropriate restriction enzymes and gel extract the Kan-cHS4 with the homology arms.
6. Perform recombineering (Subheading 3.3), selected clones on LB-Chl-Kan plates and identify recombinants with the Kan-cHS4 at the correct position (Subheadings 3.4 and 3.5).
7. Remove the Kan cassette by inducing Cre-mediated recombination (Subheading 3.6) and verify by PCR (Subheading 3.4) and sequencing.
8. Clone the Tet cassette into pGEMTeasy vector and confirm the sequence of the inserted DNA.
9. Design two primers to amplify a homology arm identical to 3′ end of the *Hoxb* cluster. The reverse primer contains a LoxP site at its 5′ end (This LoxP site should be in the same orientation as the other LoxP site in the original BAC after recombineering with the amplified fragment at **step 12**). Amplify the fragment, clone it into pGEMTeasy vector and confirm the sequence of the inserted DNA.
10. Clone the homology arm with a LoxP site downstream of the Tet cassette in a reverse orientation so that LoxP site is positioned between the Tet cassette and the homology arm.
11. Digest with appropriate restriction enzymes and gel extract the Tet cassette with the homology arm.
12. Perform recombineering, select clones on LB-Chl-Tet plates and identify recombinants with the Tet cassette at the correct position (Subheadings 3.3–3.5).
13. Remove the Tet cassette by inducing Cre-mediated recombination and verify by PCR and sequencing (Subheadings 3.4 and 3.6).

3.9 Seamless Insertions/Deletions (Two-Step Recombination with sacB-Kan Cassette)

Cre-mediated removal of a selection cassette after recombineering described above can be conveniently used to insert or delete any DNA sequence in BAC. However, a leftover LoxP site in the modified BAC can be problematic if it alters gene expression potentially by interfering with function of *cis*-regulatory elements nearby. The two-step selection and counter-selection method was

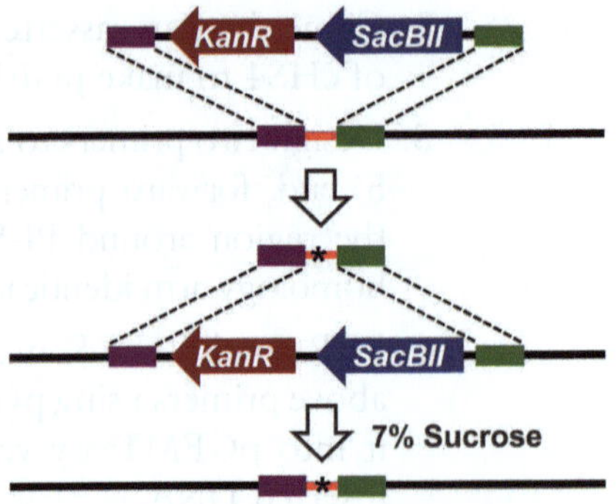

Fig. 9 Strategy used to introduce seamless insertions/deletions into BAC. First, the *SacB-KanR* cassette is introduced into a target site (*red line*) with homology arms (*purple* and *green boxes*) flanking the site. Second, recombineering is performed in the presence of sucrose for selection against *SacBII* with a construct identical to the target site except a desired sequence modification (*asterisk*)

developed by Muyrers et al. in order to introduce point mutations into BAC without leaving any other changes utilizing the *sacB-Kan* cassette [10].

The *sacBII* gene from *Bacillus amyloliquefaciens* encodes an enzyme that converts sucrose into polyfructose, which is toxic to *E. coli* and hence can be used for negative selection [11]. The first targeting construct contains the *sacB-Kan* fusion gene flanked by 45–50 bp homology arms, which are flanking sequences of the target site in the *Hox* BAC. The second targeting construct is identical to the wild-type sequence except the point mutations located in the middle (Fig. 9).

1. Clone the sacB cassette with its own promoter upstream of the Kan cassette to make pGEMTeasy-sacB-Kan.
2. Design two pairs of primers for homology arms with appropriate cohesive ends when annealed.
3. Clone the first annealed DNA upstream of sacB and then the second annealed DNA downstream of Kan (PCR amplification of sacB-Kan with homology arms is very difficult likely due to the secondary structure within the *sacB* promoter).
4. Perform recombineering, select clones on LB-Chl-Kan plates and identify recombinants with the *sacB-Kan* at the correct position (Subheadings 3.3–3.5).
5. Generate the second targeting construct by site-directed mutagenesis of 0.3 kb DNA fragment covering the target site.
6. Digest with appropriate restriction enzymes and gel extract the DNA fragment with desired mutations.
7. Perform recombineering and select clones on LB-Chl-7 % Sucrose plates (Subheading 3.3).

8. Re-streak clones on LB-Chl-7 % sucrose and LB-Chl-Kan plates and screen ~500 clones by PCR (Subheading 3.4).
9. Select PCR-positive clones which do not grow on LB-Chl-Kan plates and analyze by maxi-prep/digestion and sequencing (Subheading 3.5).

3.10 BAC Transgenesis

1. Inoculate 2 ml LB media containing the appropriate antibiotics with a fresh colony of bacteria and grow for 8 h at 32 °C with shaking at 200–250 rpm.
2. Seed 200 ml LB media with 0.2 ml of the above culture and grow overnight at 32 °C with shaking at 200–250 rpm.
3. Spin down the culture at 4 °C for 15 min at 4,000 × *g* and purify BAC DNA using Qiagen Plasmid Maxi Kit.
4. Completely resuspend bacterial pellet in 15 ml of P1 solution.
5. Add 15 ml of P2 solution and mix by gently inverting the tube and let the tube stand for 3–5 min.
6. Add 15 ml of ice-cold P3 solution, mix by gently inverting the tube and leave the tube on ice for 15 min.
7. Spin down at 4 °C for 25 min at 24,000 × *g*.
8. Transfer the supernatant to a new tube and spin down at 4 °C for 15 min at 24,000 × *g*.
9. Apply the supernatant to a Qiagen Maxi column pre-equilibrated with QBT buffer.
10. Wash the column twice with QC buffer.
11. Elute the BAC DNA from the column by apply 15 ml of QF buffer preheated to 65 °C.
12. Add 10.5 ml of isopropanol to the elute and spin at 4 °C for 30 min at 24,000 × *g*.
13. Gently discard the supernatant, add 15 ml of ice-cold 70 % ethanol and spin at 4 °C for 20 min at 24,000 × *g*.
14. Discard the supernatant and dry the pellet for 5–10 min.
15. Dissolve the pellet with 0.25 ml of TE by rotating the tube.
16. Digest 0.1 ml of the BAC DNA with PI-SceI at 37 °C for 1 h.
17. Purify the digested DNA with Qiaquick PCR purification kit. Elute the DNA with 50 μl of buffer (5 mM Tris–HCl, pH 8.0, 0.1 mM EDTA).
18. Quantify the BAC DNA with a spectrophotometer and run 1 μl on 0.6 % agarose gel for a quality check. Keep the DNA at 4 °C up to 1 week.
19. On the day of pronuclear injection, prepare BAC injection buffer (10 mM Tris–HCl, pH 7.5, 0.1 mM EDTA, 100 mM NaCl, 1× polyamine) by adding 1,000× polyamine solution. Dilute the BAC DNA in BAC injection buffer to a final concentration between 1 and 3 ng/μl (Figs. 10, 11, and 12).

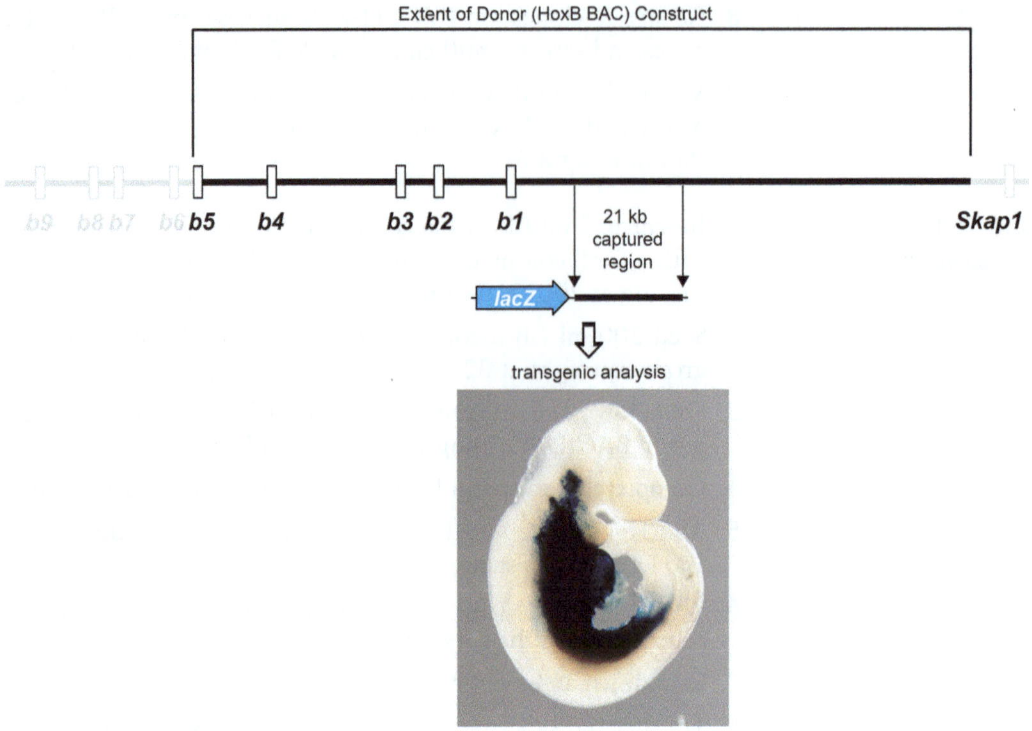

Fig. 10 Example of recombination capture followed by enhancer analysis in transgenic mice. The diagram at the top depicts the mouse *Hoxb* locus and the region encompassed by the BAC used as a donor material for this capture. The *downward arrows* below the diagram indicate the captured region and the *LacZ* reporter construct that the region was captured into. At the *bottom* is a stereo microscope image of a *lacZ* stained mouse embryo at 9.5 dpc that is transgenic for the reporter construct carrying the captured region. Further examples of this method and its use to identify novel retinoic acid response elements 3′ of *Hoxb1* that may be responsible for regions of *Hoxb1* expression in the developing heart and gut have been previously reported [13]

4 Notes

1. Homology primer design: Homology regions should be at least 35 bp and the efficiency of generating targeting events increases with homology region length. The success of producing targeting events is also reduced when the distance between the two homology regions = 0 [12]. Primer quality is important in this step. We use IDT primers as they claim that they have 99.2 % or greater base incorporation efficiency when 98.5 % is commonly reported. With longer primers this is crucial. The percent of full length oligo can be determined by the following formula (full length oligo) = $(\text{efficiency})^{n-1}$ where n = the length of the oligo. For a 75-mer this translates into 55.2 % full-length oligos at 99.2 % efficiency and 32.7 % full length at 98.5 % incorporation efficiency. Low proportions of full-length primer also cannot be overcome by just increasing the amount of PCR product used either. Imperfect PCR products

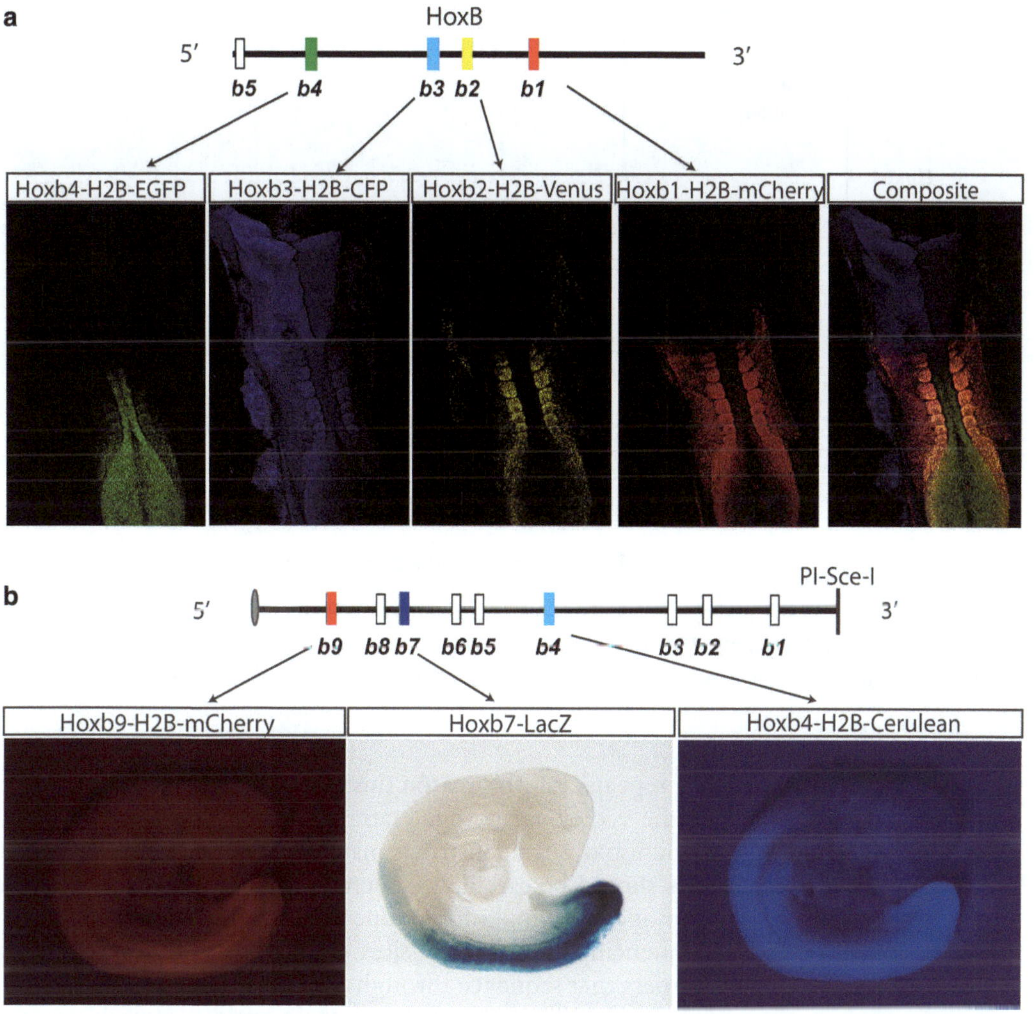

Fig. 11 Examples of transgenic mouse embryos generated with multiply targeted BACs. (**a**) *mHoxb1*, *2*, *3*, and *4* were targeted with in-frame fusions of *H2B-mCherry*, *H2B-Venus*, *H2B-Cerulean*, and *H2B-EGFP* respectively. *Top diagram* indicates fluorochrome identity and location in the mouse *Hoxb* BAC used in these experiments. Below the diagram are the spectrally separated fluorescence images for the four fluorochromes and the composite image on the far right generated from an 8.5 dpc mouse embryo transgenic for this construct. Fluorescence was spectrally separated on a Zeiss 710 microscope and a projection image generated with Imaris. (**b**) A second mouse *Hoxb* BAC was targeted with *H2B-CFP*, *LacZ*, and *mCherry*. The diagram indicates the region encompassed by the BAC and the insertion locations of the three markers. The *gray oval* at the *left* indicates the location of an inserted insulator and an inserted homing endonuclease site (PI-Sce-I) is labeled at the far right of the diagram. Below the diagram are light and fluorescence micrograph images of the expression patterns of the three markers as observed of a 9.5 dpc transgenic mouse embryo bearing this construct visualized on a Leica M205 stereoscope [9]

will compete with the perfect ones and substantially reduce the rate of successful recombination.

2. Generation and sequencing of Lox sites. Sequencing/PCR through inverted repeats as found in lox sites can be troublesome presumably as a result of secondary structures formed by

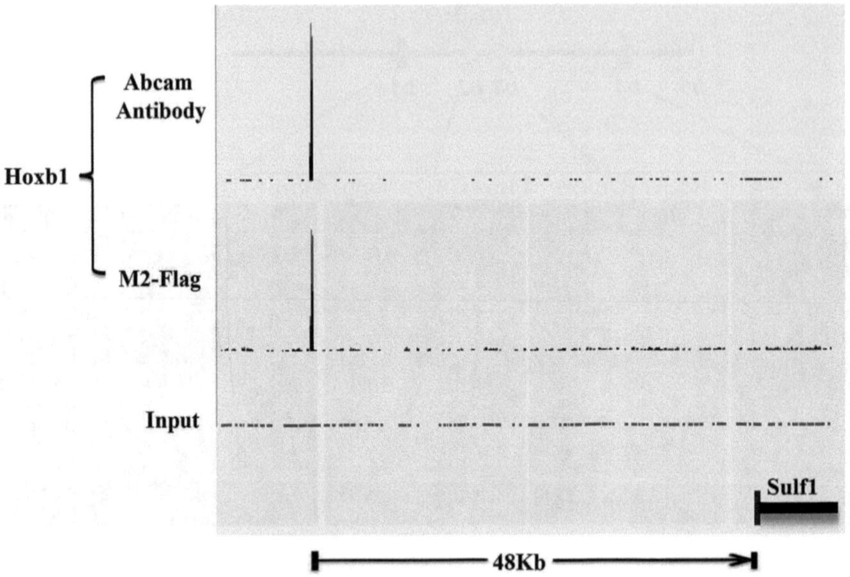

Fig. 12 Occupancy of Hoxb1 upstream of Sulf1 in 9.5 dpc embryos. ChIP results for 9.5 dpc embryos using abcam anti-Hoxb1 anti-body (ab24708) and Sigma M2 anti-flag antibody recognizing Hoxb1 with a c-terminal flag epitope in transgenic mouse embryos expressing this protein from a recombineered BAC

the repeats. For this reason this protocol suggests construction of the vector using oligos rather than PCR. While oligo construction of such constructs may be superior, the resultant vectors should be carefully sequenced as oligo production is not immune to complications due to secondary structure either. Sequencing through lox sites can be improved by choosing primers that sequence through 100–300 bp before encountering the Lox site. When the sequencing primer anneals near the Lox site (<100 bp) these sequencing runs often fail.

3. Regarding the rationale for choosing Kanamycin selection. Ampicillin is the least successful cassette we have experimented with. The antibiotic is not very stable and occasionally nonresistant colonies will grow if the media or plates are not fresh. Also, every preparation of Taq polymerase that we have used appears to contain DNA encoding ampicillin resistance that can recircularize upon recombination and produce resistance in the absence of an actual targeting event.
4. DpnI targets the sequence GA^{meth}TC in which the adenines are methylated. This occurs at all such sequences grown in E coli. The use of this enzyme allows the digestion of the remaining template in a PCR reaction which could be propagated in conjunction with the BAC being targeted and produce false positive, antibiotic-resistant clones (observable as high intensity, low mw bands in BAC digest gels). DNA produced by PCR is

not methylated at these sites and is not a target for DpnI digestion.

5. Salt must be minimized to maximize the efficiency of electroporation and prevent sparking. Resuspending purified PCR fragments in dH_2O reduces the chance of a spark, increases the resultant electroporation constant and allows for the use of larger volumes of PCR fragment in the reaction.
6. The most important factor to the success of a recombination reaction is the quality of the PCR product used for targeting. Once a high-quality PCR product has been generated, DpnI treated, and purified it can be stored at −20 °C for extended periods (We have generated targeted clones with PCR products >1 year old stored in this manner). The following are steps that we have used to increase the PCR product quality.
   - Occasionally linearization and/or purification of the selection cassette away from the backbone is necessary to produce a clean PCR product (single, nonsmeary band).
   - An optimal annealing temperature may be chosen by doing an initial PCR on a thermocycler capable of producing an annealing temperature gradient. The PCR would then be repeated at the optimal temperature.
   - If a clean band cannot be achieved by the previous methods, a smeary band may be purified and topo cloned. Sequence analysis of the topo clones can then identify a high-quality clone. This clone can then be used as a template with the original amplification primers. Additionally, a second optimization of annealing temperature can be done using the new template.
   - Once a topo clone has been generated with homology arms, these homology arms can be extended by designing new primers that extend the homology to be used on the new template. This can improve targeting efficiency by having longer homology domains and the PCR may be improved as well if there was some problem in the efficiency of the original amplification primers.
7. Expression of recombination machinery will be activated at 37 °C which will be toxic to the cells and may compromise the stability of the BACs. Pay close attention to incubation temperatures.
8. Growth of 1 ml of overnight culture to 0.4–0.6 OD takes 1.5–2.5 h for a 50 ml culture in a 250 ml baffled Erlenmeyer flask.
9. Turn on the water bath during this step so that it will be at temp (42 °C) when the culture reaches log phase.
10. At this point ice buckets should be readied, centrifuge tubes cooled on ice, dH_2O for washes put in ice water bath, cuvettes

put on ice and 2 ml tubes labeled. A cold room, if available, is a good place to stage these materials.

11. The "stickiness" of the cells at this point will be directly proportional to the amount of ions in the liquid. After the third spin there will be very few ions left and the pellet will immediately start to fall apart after the spin. To combat this, the centrifuge tube may be inverted as soon as the centrifuge run ends so that the liquid is no longer in contact with the pellet. The liquid should be removed as soon as possible. Even after inversion the pellet will start to disassociate and slide down the side of the tube.

12. At this point cuvettes should be thawed to 4 °C (store these at −20 °C to be always chilled for recombination), turn on the electroporator, set the program, setup tubes of SOC for incubation after electroporation, and collect targeting PCR products for electroporation.

13. Constants for electroporation should be between 3 and 5.2. This constant will give some indication of how effective the electroporation was. Low constants or sparks can be due to a number of factors such as high salt concentration in your purified PCR product elution, incompletely deionized wash dH_2O, bubbles in your electroporation reaction, condensation on the side of cuvettes, contamination of electro-competent cells with ice from your ice bucket and dying cells from a culture that is past log phase growth. There are probably numerous other possibilities, however, these are a few to be conscientious of.

14. Cuvettes can be reused for identical recombination reactions if the reaction is removed with a p200 pipet to be placed in SOC instead of adding SOC to the cuvette itself.

15. If primers are designed with a Tm near 60 °C they can often be used for both PCR and submitted for use as sequencing primers.

16. Products <200 bp are often hard to differentiate from primer-dimer miss-amplification. The longer the product the less efficient the reaction will be, PCR cycling will require longer extension times and the products will require a longer run to be visualized accurately on the gel confounding and delaying the analysis.

17. A 12 channel pipet and the use of reagent trays are greatly desirable, if not necessary, to process samples in large numbers.

18. We often design primers with an annealing temperature near 60 °C so that the primers can be used both for PCR and standard sequencing. The annealing temperature range we use for PCR is then determined by this primer characteristic. The PCR annealing conditions may or may not be similar depending upon how the primers were designed.

19. Extension time should be based on the expected length of the product. Calculations should be based on a polymerization rate of 1,000 bp/min.
20. Gel percent should correlate inversely with the size of the PCR product; i.e., 2 % agarose should be used for small fragments 200 bp and 1.5 or 1 % agarose for larger fragments closer to 1,000 bp. This will allow maximum separation of the bands.
21. The passage of the PCR product through the gel will be determined by both its length and the % agarose used and the length of the gel. This will have to be empirically determined for your specific setup and conditions.
22. If you are using an Owl/Thermoscientific D2 electrophoresis apparatus, a 12 channel pipetor can be successfully used to load wells made with their medium width comb by loading alternate wells. This makes numbering the photograph more difficult (1, 13, 2, 14 …) but 48 samples can be loaded per gel (96 if additional slots are added to allow the insertion of 2 additional combs in their gel plate). The conditions listed (100 V for 20 min) are designed for this gel configuration.
23. Do not vortex again after this point (p1) as it is destructive to the BAC DNA.
24. If pellet does not dry completely and alcohol remains in the tube this can alter enzyme activity and produce unpredicted banding patterns. Over-drying, which is often a problem with BAC DNA, is mostly resolved by the heat/incubation step before enzyme addition.
25. An enzyme or set of enzymes should be chosen that will produce fragments that fall mostly between 1 and 12 kb. These can be spread effectively in a large gel running overnight and differences of even a couple hundred bp are easily visible. Enzymes that have star activity are not good choices.
26. When recombineering with plasmid-based targeting vectors and BACs it is possible to create double resistant colonies in the absence of true recombination events. If any circular plasmid is left in the PCR reaction, a plasmid and BAC can co-occupy a bacteria after electroporation giving resistance to multiple antibiotics as their origins of replication are not incompatible (this is why PCR products are treated with Dpn1 to destroy remaining circular plasmid). If this has occurred, it will be visible in the results of the digest screen. A few highly concentrated bands will be present (100–1,000× intensity of BAC derived bands) in addition to the wt BAC banding pattern.
27. Phosphor imager gel imaging. Conditions for imaging ethidium bromide stained gels on an Amersham Typhoon 8600 phosphor imager. Acquisition mode = Fluorescence/ Settings = Gen. Pur. 560 long pass filter, PMT = 500–700 V,

Green laser (532 nm), normal sensitivity/Do not press sample, Pixel size = 100 μm, Focus = +3 mm. This produces a file of about 12 megabytes in 15 min for a gel that is 10 in. × 9 in. The resolution and quality of these images are far superior to those generated by gel documentation systems and allow visualization of much larger gels.

28. On more than one occasion we have found BACs that failed to release the selection cassette upon L-Arabinose induction. The only solution in these cases was to prep the BAC and retransform it into EL350/SW106, verify the transfer and repeat the L-Arabinose induction.
29. Length of the homology region: We have generated capture cassettes containing flanking homology arms between 36 and 250 bp in length. While not rigorously tested, the length of the homology arm is not a limiting factor in the successful capturing of the region of interest.
30. Capture size limitation: Using the methods describe here, we have captured fragments as large as 21 kb from BACs into plasmids.
31. Choice of unique, internal restriction site: Selection of a unique restriction endonuclease site is important. The choice should be an enzyme that readily cuts DNA under optimal conditions. Use of a restriction endonuclease with poor activity or under suboptimal conditions will result in a significant amount of background as a portion of the circularized capture plasmid will contaminate the gel extracted preparation.
32. Troubleshooting: In general this technique works very well; however, there have been a few occasions where the originally designed cassette fails to capture the region of interest. In most cases this has been resolved by redesigning the homology arms using regions adjacent to the original choice. On rare occasions this alternative capture cassette has not worked. In these latter cases, the failure has been traced back to rearrangements of the donor BAC or erroneous sequencing of the area used to design the homology arm resulting in numerous mismatches between the homology arm and its target sequence.

Acknowledgements

We thank members of the Krumlauf lab for valuable discussions, Heidi Monnin, Jenny Renyolds, and Andrea Moran and other members of the Stowers Laboratory Animal Services Facility for their care of our mice and technical assistance. All experiments involving mice were approved by the Institutional Animal Care and Use Committee of the Stowers Institute for Medical Research

(Protocol 2010-0062/2013-0114). This research was supported by funds from the Stowers Institute and in addition Mark Parrish was funded in part by a Ruth L. Kirschstein NRSA Postdoctoral Fellowship (1F32GM075611-01/5F32GM075611-02).

References

1. Sharpe J, Nonchev S, Gould A et al (1998) Selectivity, sharing and competitive interactions in the regulation of *Hoxb* genes. EMBO J 17(6):1788–1798
2. Barrow JR, Capecchi MR (1996) Targeted disruption of the Hoxb-2 locus in mice interferes with expression of Hoxb-1 and Hoxb-4. Development 122(12):3817–3828
3. Parrish M, Unruh J, Krumlauf R (2011) BAC modification through serial or simultaneous use of CRE/Lox technology. J Biomed Biotechnol 2011:924068
4. Ellis HM, Yu D, DiTizio T et al (2001) High efficiency mutagenesis, repair, and engineering of chromosomal DNA using single-stranded oligonucleotides. Proc Natl Acad Sci U S A 98(12):6742–6746
5. Copeland NG, Jenkins NA, Court DL (2001) Recombineering: a powerful new tool for mouse functional genomics. Nat Rev Genet 2(10):769–779
6. Lee EC, Yu D, Martinez de Velasco J et al (2001) A highly efficient Escherichia coli-based chromosome engineering system adapted for recombinogenic targeting and subcloning of BAC DNA. Genomics 73(1):56–65
7. Warming S, Costantino N, Court DL et al (2005) Simple and highly efficient BAC recombineering using galK selection. Nucleic Acids Res 33(4):e36
8. Pikaart MJ, Recillas-Targa F, Felsenfeld G (1998) Loss of transcriptional activity of a transgene is accompanied by DNA methylation and histone deacetylation and is prevented by insulators. Genes Dev 12(18): 2852–2862
9. Ahn Y, Mullan HE, Krumlauf R (2014) Long-range regulation by shared retinoic acid response elements modulate dynamic expression of posterior Hoxb genes in CNS development. Developmental Biology 388(1): 134–144.
10. Muyrers JP, Zhang Y, Benes V et al (2000) Point mutation of bacterial artificial chromosomes by ET recombination. EMBO Rep 1(3):239–243
11. Blomfield IC, Vaughn V, Rest RF et al (1991) Allelic exchange in Escherichia coli using the Bacillus subtilis sacB gene and a temperature-sensitive pSC101 replicon. Mol Microbiol 5(6):1447–1457
12. Zhang Y, Buchholz F, Muyrers JP et al (1998) A new logic for DNA engineering using recombination in Escherichia coli. Nat Genet 20(2):123–128
13. Nolte C, Jinks T, Wang X et al (2013) Shadow enhancers flanking the HoxB cluster direct dynamic Hox expression in early heart and endoderm development. Dev Biol 383(1): 158–173

Chapter 6

The Genetics of Murine *Hox* Loci: TAMERE, STRING, and PANTHERE to Engineer Chromosome Variants

Patrick Tschopp and Denis Duboule

Abstract

Following their duplications at the base of the vertebrate clade, *Hox* gene clusters underwent remarkable sub- and neo-functionalization events. Many of these evolutionary innovations can be associated with changes in the transcriptional regulation of their genes, where an intricate relationship between the structure of the gene cluster and the architecture of the surrounding genomic landscape is at play. Here, we report on a portfolio of in vivo genome engineering strategies in mice, which have been used to probe and decipher the genetic and molecular underpinnings of the complex regulatory mechanisms implemented at these loci.

Key words Mouse genetics, Gene regulation, In vivo genome engineering, TAMERE, STRING, PANTHERE

1 Introduction

Hox genes play an essential role in patterning the developing vertebrate embryo [1]. Apart from their ancestral roles in specifying structures along the anterior–posterior body axis, they have been co-opted to pattern a variety of structures and organs during embryogenesis, often by rewiring and diversifying their regulatory circuitries [2]. The system's flexibility necessary to fulfill such a broad range of tasks has been facilitated in part by its extensive genetic redundancy. Two successive rounds of genome duplications gave rise to four paralogous *Hox* gene clusters, containing a total of 39 genes in higher vertebrates [3, 4]. "Regulatory evolution" of paralogous vertebrate *Hox* gene clusters, i.e., neo-functionalization events produced by inducing or expanding expression domains [5], in turn seems intricately linked to gene cluster size and structure, as well as with the surrounding genomic landscape (reviewed in [6]). Consequently, probing the relationship between the genomic organization of *Hox* cluster loci and the associated transcriptional output is an important aspect to study

Yacine Graba and René Rezsohazy (eds.), *Hox Genes: Methods and Protocols*, Methods in Molecular Biology, vol. 1196, DOI 10.1007/978-1-4939-1242-1_6, © Springer Science+Business Media New York 2014

when trying to decipher the functional diversification of this gene family during the course of vertebrate evolution.

Historically, structural variants in the mouse genome have been obtained either by labor-intensive targeted ES cell work or by various and admittedly less controlled mutagenesis approaches (reviewed in [7]). However, the production of transgenic mice *via* homologous recombination in ES cells also generally leaves a DNA recombinase-recognition site behind. Such leftover sites can subsequently be used in combination to produce targeted genome modifications. Over the past 15 years, the development of a variety of in vivo genome engineering strategies has allowed the production of unprecedented number of alleles at various murine loci [8–11]. These approaches were initially designed to analyze large scale regulations at the *Hoxd* gene cluster, along with a scanning analysis using a *lacZ* transgene construct to monitor enhancer activity (*see* ref. in [12]). Recently, a reporter construct was associated with a Sleeping Beauty transposon with homing capacity [13], thus allowing for a semi-targeted saturation of a given locus with loxP sites, provided a transposon can be found (or recombined) inside [11]. The resulting insertions can thus be used for further recombination events to produce deletions and/or duplications, for example in large regulatory landscapes, including the *HoxD* locus [11, 14–16].

In this chapter, we discuss these various in vivo approaches, their applicability to address different questions concerning *Hox* gene regulation, and the prerequisites, considerations, and potential pitfalls related to their implementations. In particular, we will focus on three techniques, termed TAMERE, STRING, and PANTHERE (Targeted Meiotic Recombination; Sequential Targeted Recombination-Induced Genomic approach; Pan-genomic Translocation for Heterologous Enhancer Reshuffling), which exploit loxP sites in the genome to produce targeted, structural variants of mouse chromosomes.

2 Materials

All of the strategies presented here rely on the *Cre*/loxP system for site-specific DNA recombination, introduced to the field of mouse genetics more than 20 years ago [17, 18]. Broadly speaking, these genome engineering techniques induce small- to large-scale structural variants at genomic locations predetermined by and targeted to the locations of two selected loxP sites. LoxP sites can thus be considered the "substrate" for all subsequent experimental setups (*see* **Note 1**). Choosing the location of the loxP sites will not only determine the structure of the resulting allele, but also what conclusion can potentially be drawn from a particular experiment (*see* **Note 2**). Depending on the intended genome modification, the relative orientation of the loxP sites with respect to each other will have to be different (*see* below, Subheading 3). In addition, the use of a powerful source of transgenic *Cre* recombinase has also proven essential to the success of these approaches, in particular if the capacity for mouse breeding is restricted.

2.1 Choosing loxP Sites

1. Most currently used gene targeting vectors include loxP-flanked Neomycin-resistance (*Neo*) cassettes, to *Cre*-excise the selection marker after a successful targeting event in ES cells and thus leaving behind a single loxP site in the genome. Depending on where the cassette originally had been inserted in the gene locus, these loxP sites might however only be partly suitable as substrates, e.g. when the remaining loxP site resides inside the gene body or even in an exon. In this case, protein truncations or fusions might occur as a consequence of genome engineering, complicating the interpretation of any potential phenotypes (*see* **Note 2**). For this reason, several lines generated in the laboratory to analyze the structure–function relationships at the *HoxD* cluster were produced carrying loxP sites outside of gene bodies, either left behind by deletions of local regulatory elements [9, 19, 20], or produced de novo for this purpose [19, 20].
2. Similarly, floxed alleles for conditional gene inactivation leave behind single loxP sites that reside mostly within the gene body. However, notable exceptions can provide starting alleles for interesting experimental setups. For example, in the case of the floxed *HoxA* cluster, its deletion results in single loxP sites surrounded by the *HoxA*'s cluster endogenous regulatory landscape. Targeted translocations can then recombine a regulatory sensor into a split landscape and probe its enhancer activities (PANTHERE, *see* below, and Patrick Tschopp and Denis Duboule, unpublished results).
3. Targeted transgenes harboring *Hoxd/LacZ* fusion genes, each containing a single loxP site, have proven particularly useful to probe the regulatory potential of the *HoxD* locus (*see* **Note 3**). Besides uncovering the various position effects associated with gene order inside *Hox* clusters, subsequent relocations to different positions uncovered the potential of one and the same *Hox* promoter to respond to the different regulatory inputs, both inside and outside the gene complex [21, 22].
4. Randomly integrated *Hox* reporter transgenes containing loxP sites, for instance single genes or BACs containing larger part of any given cluster, may equally serve as substrates [23]. Here, the relocation of the transgene back into its endogenous context by means of deleting intervening sequences or translocating into the different *Hox* loci, may prove informative.

5. Different loxP sites obtained from randomly integrating regulatory sensors might be used in conjunction with *Hox* locus-specific loxP sites for subsequent genome modifications. There are a number of gene trap vectors being currently used, most of them using splice acceptor sites to trap endogenous promoter activity by coupling to a reporter gene expression. Amongst the more systematic approaches in this direction are the efforts of the "International Gene Trap Consortium" (IGTC), whose vectors contain both native and modified loxP sites in addition to FRT sites [24]. Moreover, various transposon-based systems offer similar flexibility and usability for downstream changes in genome architecture of the targeted loci [13, 25]. The tendency of certain "regulatory sensor" transposon systems to insert in *cis* in proximity to their mobilization site by local hopping ("homing") [13], can be exploited for a systematic probing of a locus' enhancer activity, global three-dimensional architecture and chromatin environment, and their impact on gene regulation (*see* **Notes 4** and **5**).

2.2 Choosing Transgenic Cre-expressor Lines

It has become clear that the *Cre* expression level in germ cells is a major determinant for successful recombination events between homologous or nonhomologous chromosomes in *trans*. While this was initially achieved by a sperm-specific (*Sycp1*) *Cre* transgenic line, the implementation of ubiquitously expressing *Cre* transgenes has greatly improved the efficiency of the various genome engineering techniques in recent years (*see* **Note 6**).

1. *Sycp1-Cre*: The first generation of TAMERE alleles (*see* below) relied upon the *Sycp1-Cre* transgenic line [26]. The efficiency of *trans*-allelic recombination was greatly increased by taking advantage of homologous chromosome pairing during prophase of meiosis I. SYCP1 is expressed during murine spermatogenesis and constitutes a part of the chromosomal synapsis-forming complex, thus providing a source of transgenic *Cre* at the right place and time. However, moving into larger and more complex genomic rearrangements required the use of a more potent *Cre* driver. Additionally, the fact that *Sycp1-Cre*-mediated *trans*-allelic recombination only occurs in the male germ line precluded screening procedures for mosaic recombination events in the soma.
2. *Hprt-Cre*: The generation and characterization of the *Hprt-Cre* transgenic line for the production of structural variant in the mouse genome greatly improved the overall efficiency and met several of the criteria outlined above [25, 27]. With the *Hprt-Cre* transgene, not only did it become possible to generate larger and more complex genomic rearrangements with high efficiency, but also the screening procedures were greatly streamlined. In "transloxer" animals, i.e., parental animals carrying

the two loxP sites and the *Cre* transgenes that give rise to F3 founder animals, somatic clones with the desired genetic modification appear, thanks to the ubiquitous and high-level expression of the *Hprt-Cre* transgene, as opposed to the previously used germ-line *Sycp-Cre*. These somatic clones in turn allow for the establishment of working genotyping conditions, using simple ear biopsies, now mostly PCR-based as opposed to Southern blots. The *Hprt-Cre* transgene has thus become the line of choice for all recent alleles produced in the laboratory (*see* **Note 7**).

3. Given the high success rate experienced with the use of the *Hprt-Cre* transgene and the apparent production of somatic clones, one might even contemplate the production of various tissue-specific mosaic rearrangements. Especially in cancer research, the need to engineer and model large-scale genome rearrangement resulting in protein fusion events or transcriptional de-regulation is apparent, as e.g. is the case for various *Hox* genes in the hematopoietic system [28]. If tissue-specific expression of a transgenic *Cre* line is deemed suitable, such somatic mosaics might occur with high enough frequencies to be testable, especially if cellular growth advantages are to be expected, as in the case of cancer-relevant rearrangements.

3 Methods

All of the three approaches outlined below rely on purely in vivo-based strategies, using successive rounds of mouse breeding and genotyping and take advantage of previously generated loxP sites, targeted or randomly integrated. Because these procedures involve long breeding schemes and are thus both time- and space-demanding, special care should be given to the prior verification of loxP position and orientation. Moreover, to rule out breakpoint artifacts and position effects in the interpretation of the results, reliable conclusions can only be drawn from multiple, similar yet slightly offset genomic rearrangements in a systematic manner. It is therefore advisable to generate entire allelic series with shared loxP sites whenever possible, such that breakpoint effects can be eliminated (Fig. 1).

3.1 Verifying loxP Sites

1. When a novel loxP site is used for the first time in the laboratory, one usually starts out by running a PCR over the targeted locus, in a way that the wild-type and mutant allele bands will be separable on a high-percentage agarose gel. Cut and gel-purify the mutant band and either subclone it or Sanger-sequence directly. This is done to verify both the correct location and, more importantly, the orientation of the loxP

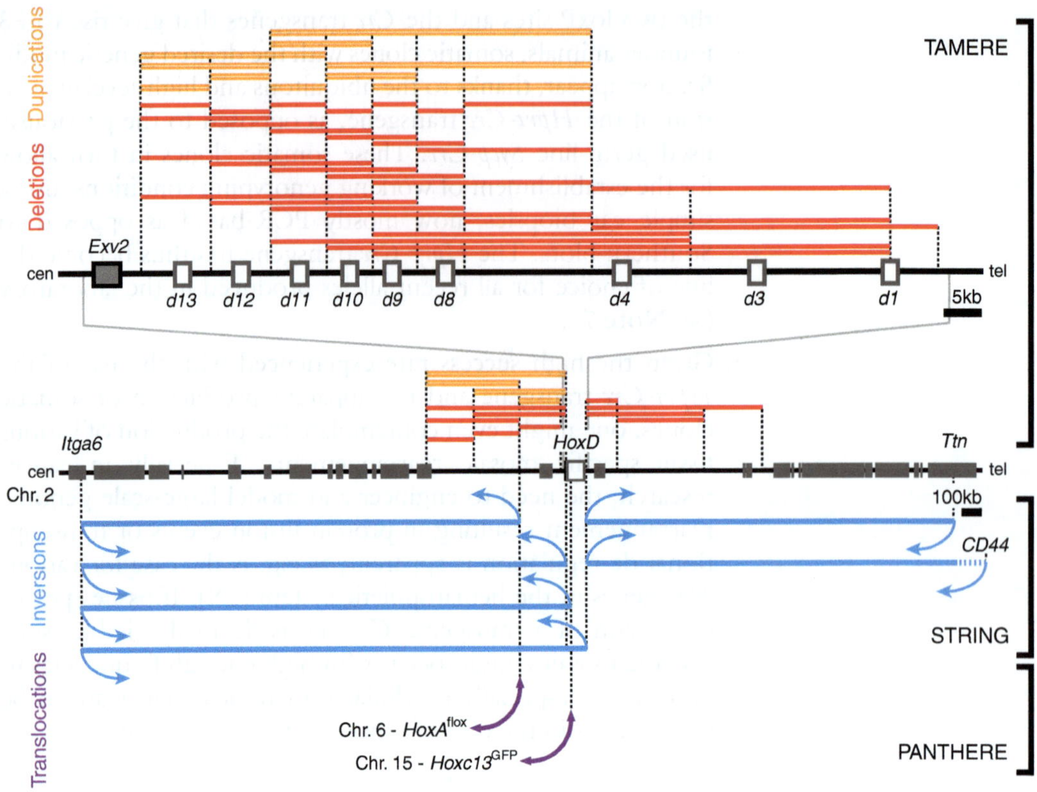

Fig. 1 An allelic series at the murine *HoxD* locus (not all alleles are shown). The upper half represents a blown-up view of the murine *HoxD* cluster (*gray lines*), depicting approximately 100 kb of DNA sequence, embedded in a 5.3 Mb stretch on chromosome 2 (below). *Vertical dotted lines* mark the position of the various loxP sites used for production of the allelic series. For TAMERE, the extent of the deleted or duplicated DNA sequences is delineated in *red single-* or *orange double-lines*, respectively. The inverted DNA fragments obtained after STRING are shown as *double-arrowed blue lines*. Breakpoints of balanced translocations produced through PANTHERE are highlighted in *purple arrows*, with their matching reciprocal breakpoints indicated next to them. For the sake of clarity, only alleles produced by the three described in vivo genome engineering techniques are shown (additional alleles include randomly integrated and targeted reporter transgenes, spanning all loci inside the cluster, as well as various BAC transgenes covering the flanking deserts)

sites, as the classical "arrow" indicating loxP sites is often confusing in many published schemes where loxP-site containing alleles are described.

2. In addition, Southern blot analyses, probing for both the vector used to introduce the loxP site as well as the surrounding target locus, is advisable before starting any crossing scheme. This holds true especially for loxP sites generated by either randomly integrated transgenes or transposon landing sites-based loxP sites (*see* also above). Southern blot analysis probing for the loxP-harboring vector can rule out the presence of multiple background insertions, while the surrounding landscape has to be checked for the correct location of integration.

3.2 Genome Engineering Techniques

1. TAMERE (Targeted Meiotic Recombination, [8]) can be utilized to generate both deletions and duplications of genomic sequences, their intervals being defined by flanking loxP sites on separate yet homologous chromosomes. The loxP sites are used to target *trans*-allelic recombination events during meiosis, catalyzed by the *Cre* recombinase. If the loxP sites are located at two distinct locations on homologous chromosomes, but in the same orientation, the targeted meiotic recombination will generate two novel alleles, one in which the loxP-intervening sequence will be deleted and a reciprocal allele where the same sequence will end up being duplicated. Given the correct genotyping strategy, both deletion and duplication alleles can be identified and recovered. In a first cross, the *Cre* transgene is combined with one of the loxP sites, and the resulting F1 progeny is crossed to an animal containing the second loxP site (Fig. 2a). F2 animals containing both loxP sites are referred to as "transloxers" and their germline will in principle contain the desired genomic rearrangements. To isolate these novel alleles, transloxers are crossed to wild-type

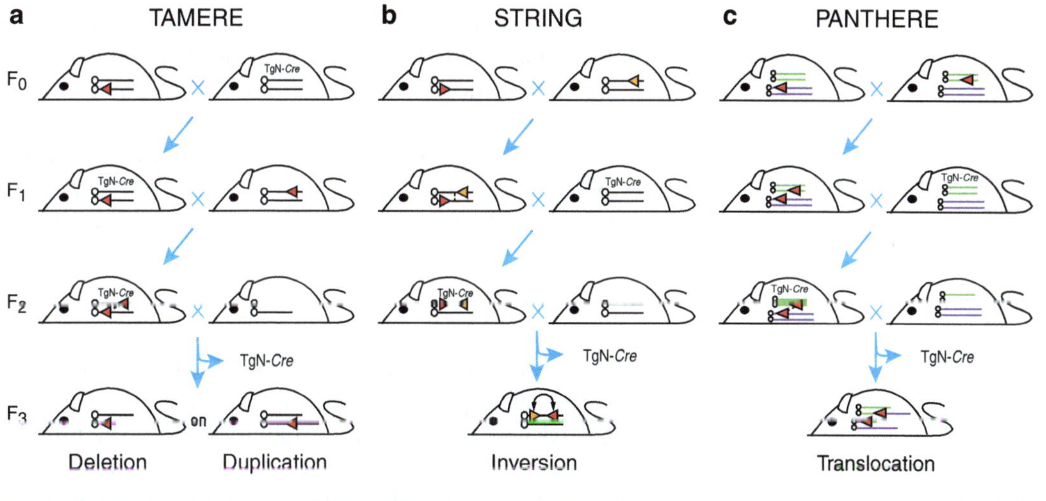

Fig. 2 Breeding schemes for TAMERE, STRING, and PANTHERE. (**a**) TAMERE. The first loxP site (*red triangle*) is crossed into the *Cre* transgene background. A second loxP site, on the homologous chromosome, is introduced. Both maternal and paternal chromosomes now carry loxP sites, albeit at different positions. F3 animals are genotyped for the appearance of deletion and/or duplication alleles, as well as the loss of the TgN-*Cre*. (**b**) STRING. Two loxP sites (*red* and *orange triangle*), either in the same, or in opposite orientations, on homologous chromosomes, are crossed together. Genotyping for both loxP sites in F2 animals identifies successful meiotic recombination between them in the F1 germline (*dotted line*). At the same time, a *Cre* transgene is introduced, which induces either a deletion (not shown), or a targeted inversion between the loxP sites. (**c**) PANTHERE. Two loxP sites on nonhomologues chromosomes (*green* and *purple*) are combined. Crossing in a strong *Cre* transgene can induce a balanced translocations between the loxP sites, which can be stably recovered in F3 animals after having lost the *Cre*. Both loxP sites must have the same orientation with respect to the centromere

animals and the F3 progeny is genotyped for both the appearance of either deletions or duplications, as well as for the loss of the *Cre* transgene (Fig. 2a). The latter is important to ensure that the animal is a de facto founder, and that not only somatic clones of the desired genomic modification are being detected. It will also prevent the maintenance of *Cre* transgenes into further breeding schemes, which may be a source of unexpected recombination.

2. STRING (Sequential Targeted Recombination-Induced Genomic approach, [9]) relies on a slightly modified rationale to combine the two loxP sites used for the subsequent generation of both inversions and deletions. Rather than using the loxP sites for targeted recombination events in *trans*, the occurrence of a randomly occurring meiotic crossover initially brings them in a *cis* configuration into the same chromosome (Fig. 2b, *see* **Note 8**). The frequency of these spontaneous crossovers naturally depends on the genetic distance between the loxP sites on the homologous chromosomes, making STRING suitable for larger rearrangements, within a megabase scale (in the mouse genome, a crossover will naturally occur with an average frequency of ca. one every megabase). Once loxP sites are in *cis*, a *Cre*-expressing transgene is crossed in, leading to *Cre*-mediated recombination in F2 animals. F3 animals are then genotyped for the loss of the *Cre*-transgene and the successful recombination event. In this way, targeted inversions between the two loxP sites can be produced, should the sites point to opposite orientation (Fig. 2b). Alternatively, if the original loxP sites were in the same orientation, large-scale deficiencies can be generated. Duplications cannot be produced using this scheme.
3. PANTHERE (Pan-genomic Translocation for Heterologous Enhancer Reshuffling, [10]), like STRING, results in large-scale genome rearrangements, more appropriate for reshuffling entire regulatory landscapes rather than modifying the internal cluster architecture as is often the case for TAMERE. The underlying principle, though, is closely related to the TAMERE approach: two loxP sites in same orientation and located on different chromosomes in *trans* are employed to target *Cre*-mediated recombination events. PANTHERE takes advantage of the strength of the *Hprt-Cre* transgene to induce recombination events between nonhomologous chromosomes, thereby generating perfectly balanced translocations. In a first step, the two loxP sites on different chromosomes are brought together, generating double-heterozygous F1 animals which in turn are crossed to the *Cre*-expressing transgenic line (Fig. 2c). F2 animals carrying all three alleles are mated with

wild-type animals and the F3 progeny is screened for the occurrence of the balanced translocation, as well as for the loss of the *Cre* transgene (Fig. 2c).

3.3 Screening Procedures

1. PCR-based screening: Most screenings are carried out using PCR methods. It is faster and cheaper than systematic Southern blot analysis and, when using the *Hprt-Cre* transgene, PCR conditions for a novel allele can already be established on somatic clones found in either ear or tail clip DNA of F2 animals. The generation of multiple alleles in concert (*see* comments above) is greatly facilitated by a streamlined genotyping procedure, i.e., reactions involving comparable PCR conditions in terms of amplicon size and primer annealing temperature for all alleles to be screened for (*see* **Note 9**).
2. Southern blot-based screening: All alleles were initially screened by Southern blot analysis, as the *Sycp1-Cre* transgenic line did not allow for the verification of PCR conditions in somatic clones of the desired rearrangement in transloxer animals. While this is still more stringent than PCR-based methods, it is also more time-consuming. In our hands, Southern blot analyses are run only for the verification of novel alleles before they are finally validated and released for experimental work (*see* below).

3.4 Verification of Novel Alleles

1. PCR: The first step is usually to run the screening PCR as well as alternative PCRs that only detect the rearranged allele(s) (Fig. 3a). Subsequently, these bands are gel-extracted, subcloned, or directly Sanger-sequenced to verify the correct rearrangement at the level of the DNA sequence over the breakpoints (Fig. 3b).
2. Southern Blot: Founder animals are always tested for the correct genomic rearrangement using Southern Blot analyses, before propagating the allele and establishing a line. Usually both sides of the breakpoints are being probed for, using standard Southern Blot protocols (Fig. 3c).
3. DNA FISH: An additional verification step can include DNA fluorescent in situ hybridization (FISH). This provides a nice tool for the visualization of large-scale genomic rearrangements, at the whole-chromosome level [9, 10]. An array of BAC probes flanking the loxP-site defined breakpoints and labeled in different colors, can be used to outline the extent of novel rearrangements in chromosome preparations of allele-derived fibroblast cell lines. Alternatively, commercially available whole chromosome paints BAC probes excluding repetitive sequences will highlight entire chromosomes, thereby making balanced translocations visible (Fig. 3d).

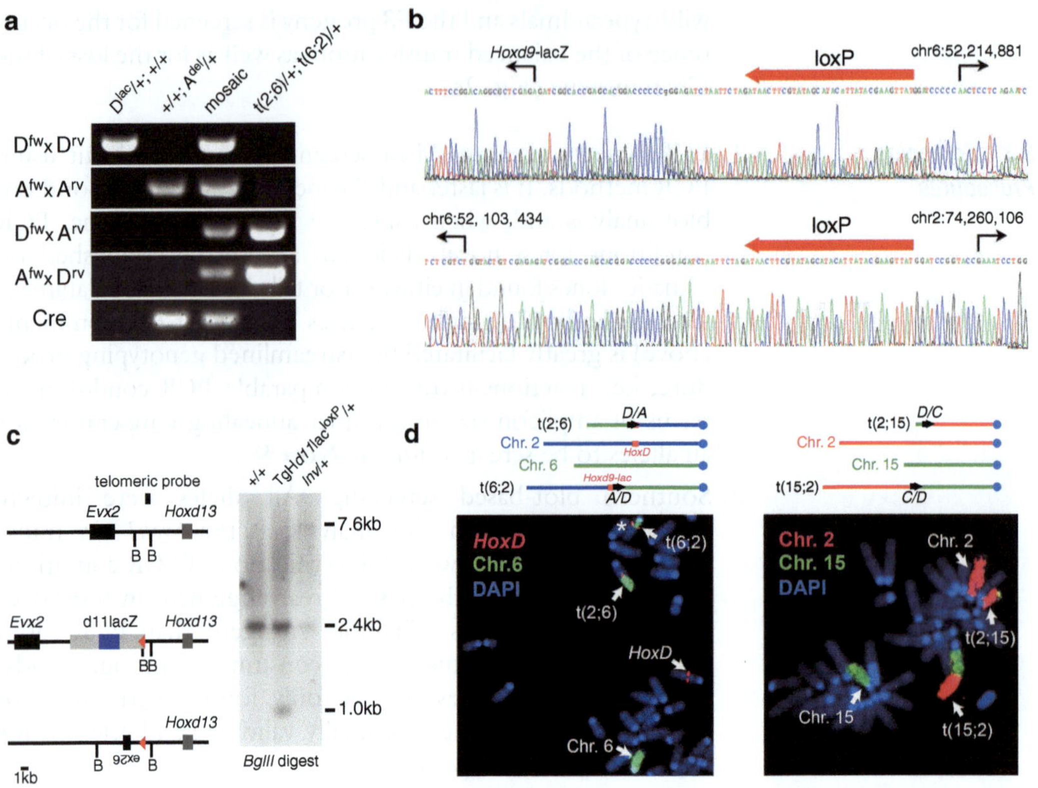

Fig. 3 Genotype validation of novel alleles (**a**) PCR screening, exemplified here with the t(2:6) translocation (PANTHERE). Pairs of primers are selected such as to amplify fragments around the recombined breakpoints. At the same time, the presence of *Cre* is also assessed to discard this transgene from future crosses (right sample). Genotyping can already be done in F1 animals (mosaic). (**b**) Sanger sequencing covering the breakpoints of the same translocation as for (**a**). The translocation is perfectly balanced. (**c**) Southern blot analysis of the inversion allele $HoxD^{Inv(TgHd11lacNsi\text{-}Itga6)}$ (STRING), documenting the genomic rearrangement at the telomeric *HoxD* breakpoint. (**d**) DNA FISH on fibroblast metaphase spread chromosomes of two translocation alleles t(2:6), left, and t(2:15), right. Visualization can rely on a combination of labeled BAC probes (e.g. spanning the *HoxD* complex) and whole chromosome paints (*left*), or two differently labeled whole chromosome paints (*right*). Note that the BAC approach is sensitive enough to pick up the 7 kb *Hoxd9*lac transgene in t(6:2) (*asterisk, left*). Data taken from [10, 37]

4 Notes

1. The introduction of the *Flp*/FRT system to the field of mouse genetics [29] offers a second, potentially alternative source for site-specific recombinase systems. This can be especially useful when considering conditional knock-out alleles: while the loxP sites will usually span coding exons, the *Neo*-resistance cassette of the targeting vector is often FRT-site flanked and resides outside the gene body. Using pairs of these FRT sites, instead of loxP sites, for the here-described genome engineering techniques

could thus avoid the complication of targeting the desired genome modification inside of gene bodies (*see* also **Note 2**). Similar considerations can be applied to the modified loxP sites often associated with gene trap vectors [24].

2. The majority of conclusions drawn from these experiments will concern the transcriptional regulation of a modified locus and hence they often include molecular and biochemical analyses of transcription. Nevertheless, as a first evaluation, mutant phenotypes arising from the genome engineering techniques will provide useful information. Their interpretation, however, can be complicated if protein products of the genes neighboring the breakpoint become modified as well during the process (e.g. truncations, fusions). Correctly matched mutant-control animals will then have to be considered. Ideally, however, one designs the modified allele such that breakpoints reside outside of gene bodies.
3. The strength of using reporter transgenes as starting alleles for genome engineering, either targeted or randomly integrated, lies in the simplicity of their transcriptional read-out. Such regulatory sensors will reveal all information associated with them already in heterozygous F1 crosses. Detection often includes standard X-Gal staining procedures, in case of *Hoxd/LacZ* fusion genes or minimal promoter-*LacZ* reporter constructs, which are quick and extremely sensitive. This sensitivity together with the stability of the beta-galactosidase product [30], however, can also lead to the over-estimation of transcriptional output, as systematic comparisons between X-Gal stainings and *LacZ* transcript levels have revealed [23]. Alternatively, other reporter constructs such as fluorescent proteins or *Cre*-fusion alleles can be employed [31, 32], the latter offering the added benefit of a lineage-tracing tool.
4. In recent years, the organization of the genome in three dimensions, inside the nucleus, has increasingly been considered for its role in gene regulation. At a global level, the entire genome seems to be folded and structured, under different modalities and at different scales, forming for example so-called topological domains at the one to two megabases scale (e.g. [33]), i.e., domains where long range enhancer-promoter interactions preferentially occur. These domains separate various regulations acting at *Hox* loci and thus may help to implement the various collinear activations of *Hox* clusters [14, 16, 34]. Therefore, breaking or rearranging such topological domains by genome engineering can be expected to significantly impact upon *Hox* transcription.
5. Likewise, *Hox* gene regulation is often accompanied by highly dynamic changes in chromatin modification patterns at target

loci [35]. As for the 3D architecture, specific histone modifications can define extended domains along and flanking vertebrate *Hox* clusters. Choosing sequentially located loxP sites offers the possibility to partition such domains, and the consequent changes at the chromatin level can in turn be probed.

6. The efficiency for the original TAMERE approach using the *Sycp-Cre* line were in the 1–5 % range, whereas the switch to the *Hprt-Cre* line resulted in an almost one order of magnitude increase in efficiency. Up to the 10 % of the screened F3 progeny now carry the desired deletion or duplication alleles. For inversions (STRING), successful rearrangements can be recovered in the 5 % range, with translocation (PANTHERE) efficiencies ranging from 0.1 to 0.5 %.
7. As outlined above, most alleles produced in the laboratory are now generated using the *Hprt-Cre* line. However, one should caution the use of strong *Cre* sources in case of inverted loxP sites, located at small distances in *cis*, as repeated inversion of a locus has been shown to give rise to cytotoxic chromosome breaks in proliferating cells [36].
8. To test for successful meiotic crossover between the two loxP sites, one simply genotypes for the presence of both sites in F2 animals. As the F1 animal carrying both loxP sites is crossed to a wild-type animal, the only way for this to occur is through a meiotic crossover in the F1 animal's germline.
9. An amplicon size of between 250 and 750 bp, and an annealing temperature of 58 °C with 35 cycles of PCR amplification have proven suitable to us for most screening procedures, given some tinkering to find the correct primer combinations. This is greatly helped by the fact that *Hprt-Cre* animals will already show mosaic recombination events in F2 animals. Otherwise, standard PCR protocols are used and as such multiple alleles can be screened for in a 96-well plate format.

Acknowledgements

Patrick Tschopp is supported by postdoctoral fellowships from the Swiss National Science Foundation (SNF), EMBO, and the *Human Frontiers Science Program* (HFSP). Work in the Duboule laboratories is supported by the Ecole Polytechnique Fédérale de Lausanne, the University of Geneva, the National Center for Competence in Research *Frontiers in Genetics*, the Swiss National Research Fund, and the ERC grant *SystemsHox.ch*. We thank Bénédicte Mascrez for the organization of the genotyping pipeline and the supervision of mouse strain production.

References

1. Krumlauf R (1994) Hox genes in vertebrate development review. Cell 78:191–201
2. Deschamps J (2007) Ancestral and recently recruited global control of the Hox genes in development. Curr Opin Genet Dev 17:422–427. doi:10.1016/j.gde.2007.07.008
3. Ohno S (1970) Evolution by gene duplication. George Alien & Unwin Ltd., London
4. Garcia-Fernàndez J (2005) The genesis and evolution of homeobox gene clusters. Nat Publ Group 6:881–892. doi:10.1038/nrg1723
5. King MC, Wilson AC (1975) Evolution at two levels in humans and chimpanzees. Science 188:107–116
6. Duboule D (2007) The rise and fall of Hox gene clusters. Development 134:2549–2560. doi:10.1242/dev.001065
7. Brault V, Pereira P, Duchon A, Hérault Y (2006) Modeling chromosomes in mouse to explore the function of genes, genomic disorders, and chromosomal organization. PLoS Genet 2:e86
8. Hérault Y, Rassoulzadegan M, Cuzin F, Duboule D (1998) Engineering chromosomes in mice through targeted meiotic recombination (TAMERE). Nat Genet 20:381–384. doi:10.1038/3861
9. Spitz F, Herkenne C, Morris MA, Duboule D (2005) Inversion-induced disruption of the Hoxd cluster leads to the partition of regulatory landscapes. Nat Genet 37:889–893. doi:10.1038/ng1597
10. Tschopp P, Fraudeau N, Béna F, Duboule D (2011) Reshuffling genomic landscapes to study the regulatory evolution of Hox gene clusters. Proc Natl Acad Sci U S A 108:10632–10637. doi:10.1073/pnas.1102985108
11. Marinić M, Aktas T, Ruf S, Spitz F (2013) An integrated Holo-Enhancer unit defines tissue and gene specificity of the Fgf8 regulatory landscape. Dev Cell 24:530–542. doi:10.1016/j.devcel.2013.01.025
12. Tschopp P, Duboule D (2011) A genetic approach to the transcriptional regulation of Hox gene clusters. Annu Rev Genet 45:145–166. doi:10.1146/annurev-genet-102209-163429
13. Ruf S, Symmons O, Uslu VV et al (2011) Large-scale analysis of the regulatory architecture of the mouse genome with a transposon-associated sensor. Nat Genet 43:379–386. doi:10.1038/ng.790
14. Montavon T, Soshnikova N, Mascrez B et al (2011) A regulatory archipelago controls Hox genes transcription in digits. Cell 147:1132–1145. doi:10.1016/j.cell.2011.10.023
15. Montavon TT, Thevenet LL, Duboule DD (2012) Impact of copy number variations (CNVs) on long-range gene regulation at the HoxD locus. Proc Natl Acad Sci U S A 109: 20204–20211.doi:10.1073/pnas.1217659109
16. Andrey G, Montavon T, Mascrez B et al (2013) A switch between topological domains underlies HoxD genes collinearity in mouse limbs. Science 340:1234167. doi:10.1126/science.1234167
17. Sauer B, Henderson N (1988) Site-specific DNA recombination in mammalian cells by the Cre recombinase of bacteriophage P1. Proc Natl Acad Sci U S A 85(14):5166–5170
18. Orban PC, Chui D, Marth JD (1992) Tissue- and site-specific DNA recombination in transgenic mice. Proc Natl Acad Sci U S A 89(15):6861–6865
19. Tarchini B, Nguyen Huynh TH, Duboule D (2005) HoxD cluster scanning deletions identify multiple defects leading to paralysis in the mouse mutant Ironside. Genes Dev 19:2862–2876. doi:10.1101/gad.351105
20. Tarchini B, Duboule D (2006) Control of Hoxd genes' collinearity during early limb development. Dev Cell 10:93–103. doi:10.1016/j.devcel.2005.11.014
21. van der Hoeven F, Zakany J, Duboule D (1996) Gene transpositions in the HoxD complex reveal a hierarchy of regulatory controls. Cell 85:1025–1035
22. Spitz F, Gonzalez F, Peichel C et al (2001) Large scale transgenic and cluster deletion analysis of the HoxD complex separate an ancestral regulatory module from evolutionary innovations. Genes Dev 15:2209–2214. doi:10.1101/gad.205701
23. Tschopp P, Christen AJ, Duboule D (2012) Bimodal control of Hoxd gene transcription in the spinal cord defines two regulatory subclusters. Development 139:929–939. doi:10.1242/dev.076794
24. Skarnes WC, von Melchner H, Wurst W et al (2004) A public gene trap resource for mouse functional genomics. Nat Genet 36:543–544. doi:10.1038/ng0604-543
25. Wu S, Ying G, Wu Q, Capecchi MR (2007) Toward simpler and faster genome-wide mutagenesis in mice. Nat Genet 39:922–930. doi:10.1038/ng2060
26. Vidal F, Sage J, Cuzin F, Rassoulzadegan M (1998) Cre expression in primary spermatocytes: a tool for genetic engineering of the germ line. Mol Reprod Dev 51:274–280.

doi:10.1002/(SICI)1098-2795(199811) 51:3<274::AID-MRD6>3.0.CO;2-M

27. Tang S-HE, Silva FJ, Tsark WMK, Mann JR (2002) A Cre/loxP-deleter transgenic line in mouse strain 129S1/SvImJ. Genesis 32:199–202. doi:10.1002/gene.10030
28. Gough SM, Slape CI, Aplan PD (2011) NUP98 gene fusions and hematopoietic malignancies: common themes and new biologic insights. Blood 118:6247–6257. doi:10.1182/blood-2011-07-328880
29. Dymecki SM (1996) Flp recombinase promotes site-specific DNA recombination in embryonic stem cells and transgenic mice. Proc Natl Acad Sci U S A 93:6191–6196
30. Zakany JJ, Kmita MM, Alarcon PP et al (2001) Localized and transient transcription of Hox genes suggests a link between patterning and the segmentation clock. Cell 106:207–217. doi:10.1016/S0092-8674(01)00436-6
31. Godwin AR, Stadler HS, Nakamura K, Capecchi MR (1998) Detection of targeted GFP-Hox gene fusions during mouse embryogenesis. Proc Natl Acad Sci U S A 95:13042–13047
32. Hérault Y, Kmita M, Sawaya CC, Duboule D (2002) A nested deletion approach to generate Cre deleter mice with progressive Hox profiles. Int J Dev Biol 46:185–191
33. Dixon JR, Selvaraj S, Yue F et al (2013) Topological domains in mammalian genomes identified by analysis of chromatin interactions. Nature 485:376–380. doi:10.1038/nature11082
34. Noordermeer DD, Leleu MM, Splinter EE et al (2011) The dynamic architecture of Hox gene clusters. Science 334:222–225. doi:10.1126/science.1207194
35. Soshnikova N, Duboule D (2009) Epigenetic temporal control of mouse Hox genes in vivo. Science 324:1320–1323. doi:10.1126/science.1171468
36. Grégoire D, Kmita M (2008) Recombination between inverted loxP sites is cytotoxic for proliferating cells and provides a simple tool for conditional cell ablation. Proc Natl Acad Sci U S A 105:14492–14496
37. Tschopp P, Duboule D (2011) A regulatory "landscape effect" over the HoxD cluster. Dev Biol 351:288–296. doi:10.1016/j.ydbio.2010.12.034

Chapter 7

Topological Organization of Drosophila Hox Genes Using DNA Fluorescent In Situ Hybridization

Frédéric Bantignies and Giacomo Cavalli

Abstract

DNA fluorescent in situ hybridization (FISH) is the method of choice to study genomic organization at the single-cell level. It has been recently used to study the topological organization of the homeotic bithorax complex (BX-C) in Drosophila as well as to describe long-range genomic interactions between the BX-C and the Antennapedia complex (ANT-C), in addition to other genomic loci. Coupled with immunofluorescence, FISH can be used to study the relative positioning of homeotic genes with nuclear subcompartments, such as Polycomb-group (PcG) bodies, transcription factories, or the nuclear lamina. Here, we describe two multicolor 3D-FISH protocols; one for whole mount Drosophila embryos or larval discs and one for Drosophila-cultured cells. Both methods can be applied to any single copy locus of interest and are compatible with immunostaining (FISH-I).

Key words Fluorescent in situ hybridization, Immunostaining, Nuclear organization, Gene positioning, Chromosome topology, Nuclear bodies, Bithorax complex, Antennapedia complex, Polycomb-group proteins, Drosophila

1 Introduction

In *Drosophila melanogaster*, Hox genes are organized in two complexes: the Antennapedia complex (ANT-C) spans approximately 400 kb and comprises five Hox genes (*labial*, *proboscipedia*, *Deformed*, *Sex combs reduced*, and *Antennapedia*) that specify anterior parts, while the bithorax complex (BX-C) spans approximately 350 kb and contains three Hox genes (*Ultrabithorax*, *abdominal A*, and *Abdominal B*) involved in the development of posterior parts. ANT-C and BX-C are located on the same chromosome arm but are separated by 10 Megabases of genomic DNA. Both complexes form large chromatin domains covered by the H3K27me3 histone modification and bound by the Polycomb-group (PcG) repressors at specific regulatory sequences. Using Chromosome Conformation Capture (3C), the topological organization of the BX-C was assayed [1]. PcG-bound regulatory elements located at

Yacine Graba and René Rezsohazy (eds.), *Hox Genes: Methods and Protocols*, Methods in Molecular Biology, vol. 1196, DOI 10.1007/978-1-4939-1242-1_7, © Springer Science+Business Media New York 2014

distances of over 100 kb within the BX-C domain are able to interact, giving rise to a topologically complex structure called "repressive chromatin hub". This hub was well illustrated by three-dimensional (3D)-FISH combined with Polycomb immunostaining (FISH-I). Two FISH probes corresponding to distant regulatory elements of the *Ubx* and *Abd-B* genes interact frequently when the genes are repressed, and this interaction occurs in nuclear subcompartments enriched in PcG proteins termed "PcG bodies". When the *Abd-B* gene is active, it is significantly displaced from the PcG compartment underscoring the functional role of this organization. Interestingly, a similar topological organization can be found for the mammalian Hox clusters [2, 3]. Long-range associations, as assayed by 3D-FISH, FISH-I, and 4C (a 3C derivative technique), also exist between the two repressed Hox complexes during development [4, 5]. Long-range interactions are however not limited to Hox genes and repressed Hox genes also associate with other PcG domains along the chromatin fiber. This indicates that they are part of a large gene interaction network and some of these long-range interactions may stabilize gene silencing. Recently, high-throughput 3C (Hi-C) data confirmed this notion of functional network of interactions among PcG target genes, and extended it further to other chromatin domains [6].

Today, 3C-based and FISH techniques appear highly complementary approaches to study higher order chromatin structures and gene interactions. 3Cs represent molecular methods to address chromosome topologies at a cell-population level, while 3D-FISH is a complementary cytological method to visualize chromosomal structures at the single cell level and will be the focus of this chapter.

We first describe the labeling of DNA probe (*see* Subheading 3.1). This protocol uses directly labeled probes by nick translation, and is well adapted to target small genomic regions (10–12 kb), encompassing promoters, gene units, or regulatory elements. We then present FISH in Drosophila embryos or larval discs (*see* Subheading 3.2) originally described in [7] and explain how to combine FISH with immunostaining (FISH-I) [4, 8]. FISH only and FISH-I are 2 and 3 day protocols respectively. We also describe FISH in Drosophila-cultured cells (*see* Subheading 3.3). Except for the production of the probes, this protocol is substantially different than the one used for Drosophila tissues (Table 1). This protocol is a more complete version of a previously published protocol [1] and is readily adaptable to mammalian cells. Finally, we briefly describe how images are acquired and processed and how data can be analyzed (*see* Subheading 3.4).

Table 1
Main steps of the 3D-FISH protocols

3D-FISH in embryos or larval discs	3D-FISH in Drosophila cells
In Eppendorf tubes	Using Poly-L-lysine-coated slides and Coplin jars
Fixation with 4 % PFA in buffer A/Heptane for 25 min (embryos) or in PBT for 20 min (larval discs)	Fixation with 4 % PFA in PBS for 10 min
Rehydratation (only for embryos)	
RNAse A in PBT for 2 h at RT	RNAse A in PBT for 1 h at RT
Permeabilization in PBS-Tr (0.3 % Triton) for 1 h at RT	Three steps of Permeabilization: – PBS-Tr (0.5 % Triton) for 10 min at RT – Three Freezing/Thaw cycles in liquid N_2 – Mild deproteinization in 0.1 M HCl for 5 min at RT
Slow transfer to 100 % pHM (containing 50 % Formamide)	Transfer to 2× SSC, 50 % Formamide
Denaturation of cellular DNA in pHM 15 min at 80 °C (denatured probes are added after)	Denaturation of cellular DNA in FHB 3 min at 78 °C (co-denaturation with probes)
Hybridization in FHB O/N at 37 °C (cover with mineral oil, performed in a thermomixer)	Hybridization in FHB O/N at 37 °C (seal with rubber cement, performed in a humid chamber)
Post-hybridization washes	Post-hybridization washes
Immunostaining (if applicable)	Immunostaining (if applicable)
DAPI staining and mounting	DAPI staining and mounting

2 Materials

2.1 Kits for Probe Direct Labeling

1. PCR purification kit (QIAGEN).
2. FISH Tag DNA kits (Invitrogen): Alexa Fluor 488 dye (catalog number F32947), Alexa Fluor 555 dye (F32948), Alexa Fluor 647 dye (F32950), Multicolor Kit (F32951) containing Alexa Fluor 488, 555, 594, 647 dyes (*see* **Note 1**). For two-color FISH, we recommend using Alexa Fluor 488 and Alexa Fluor 555 dyes, for three-color FISH, we recommend using the Alexa Fluor 647 dye in addition.

2.2 Components for Egg Laying

1. Fruit juice or vinegar agar plates.
2. Instant yeast.

2.3 Components for the Fixation of Whole Mount Embryos

1. Assemblage with nylon sieve (pore size inferior to 100 μm), alternatively cell strainers (BD Biosciences, #352340 or 352360).
2. Small glass flasks (diameter: 2.5 cm, height: 5 cm) with hermetic stopper.

3. Bleach (Sodium hypochlorite with 2.6 % of active chlorine) (*see* **Note 2**).
4. Buffer A: 60 mM KCl, 15 mM NaCl, 0.5 mM spermidine, 0.15 mM spermine, 2 mM EDTA, 0.5 mM EGTA, 15 mM PIPES, pH 7.4 (make it fresh).
5. Paraformaldehyde in powder (Prolabo, #28 794-295) or in 16 % solution ampoules (Electron Microscopy Sciences, #15710). To be dissolved to 4 % in buffer A.
6. Heptane.
7. 100 % Methanol (MeOH).

2.4 Components for FISH Hybridization

1. PBT: PBS 1×, 0.1 % Tween20 (make it fresh).
2. PBS-Tr: PBS 1×, 0.3 % Triton X-100 (make it fresh).
3. RNAse A (Sigma, R-6513): Stock solution at 10 mg/ml (100×). Store at −20 °C.
4. Deionized Formamide (Sigma, F-9037 or Calbiochem, #344 206). Store at 4 °C.
5. Dextransulfat Mw 500,000 (Pharmacia, #17-0340-01), used in the FISH Hybridization Buffer (FHB, *see* **step 8** here below): To make 10 ml of a 50 % solution, weigh 5 g of Dextransulfat in a 50 ml Falcon tube, add 7 ml of prewarmed ddH_2O at 60 °C, and agitate for at least 30 min using a magnetic stirrer. Store at 4 °C.
6. Salmon Sperm DNA, used in the FHB: Stock solution at 25 mg/ml (2.5 %, 50×). Sonicate and autoclave. Store at −20 °C. Alternatively, sheared Salmon Sperm DNA at 10 mg/ml (Ambion, AM9680).
7. Pre-Hybridization Mixture (pHM): 50 % Formamide, 4× SSC, 100 mM NaH_2PO_4, pH 7.0, 0.1 % Tween 20. Make it fresh.
8. FISH Hybridization Buffer (FHB): 50 % Formamide, 10 % Dextransulfat, 2× SSC, Salmon Sperm DNA 0.5 mg/ml (0.05 %). Store aliquots at −20 °C and prewarmed at 37 °C before use.
9. Mineral oil (Sigma, M5904).
10. CHAPS (3-[(3-cholamidopropyl) dimethylammonio]-1-propanesulfonate) (Sigma, #226947).

2.5 Components for DAPI and Mounting

1. DAPI (4,6-diamidino-2-phenylindole): Stock solution at 0.1 mg/ml in 180 mM Tris–HCl, pH 7.5. Store at −20 °C.
2. ProLong antifade or SlowFade Gold antifade (Invitrogen) (*see* **Note 3**).

2.6 Components Specific to Immunostaining (If Applicable)

1. Normal Goat Serum (NGS) (Jackson laboratories) or Normal Donkey Serum (NDS) if a primary antibody made in goat is used.

2. Blocking solution: PBS 1×, 0.3 % Triton X-100, 10 % NGS or 10 % NDS (*see* **Note 4**).
3. Secondary antibody: In the case of a two-color FISH experiment with Alexa Fluor 488 and Alexa Fluor 555 dyes, we recommend using a specific Cy5-labeled secondary antibody (Jackson laboratories) at a 1:200 dilution (i.e., 1:100 dilution of a 50 % glycerol stock solution).

2.7 Components Specific to FISH in Cells

1. PBS 1×, 0.5 % Triton X-100 (make it fresh).
2. PBS 1×, 20 % Glycerol.
3. 0.1 M HCl: 410 μl of 37 % HCl in 50 ml dH_2O (make it fresh).
4. Poly-L-lysine-coated slides (*see* **Note 5**).
5. Coverslips 22×22#1.5 (*see* **Note 6**).
6. Coplin jars (Wheaton, #900470).
7. Rubber Cement (Marabu or Elmer's).

2.8 Lab Equipments

2.8.1 For FISH in Tissues

1. Thermomixer (Eppendorf).
2. Rotating wheel.

2.8.2 For FISH in Cells

1. Thermoblock or waterbath (such as VWR digital water bath 5 L).
2. Incubators at 37 °C and 42 °C, alternatively waterbath.

2.9 Software

1. Primer3 (free software): http://bioinfo.ut.ee/primer3-0.4.0/primer3/input.htm
2. Image J (free software): http://imagej.nih.gov/ij/index.html
3. Metamorph (optional): http://www.moleculardevices.com/products/software/meta-imaging-series/metamorph.html
4. Volocity (optional): http://www.perkinelmer.com/pages/020/cellularimaging/products/volocity.xhtml
5. Huygens MLE (optional): http://www.svi.nl

3 Methods

3.1 Probe Labeling

1. Probe design: Select 12 kb of a genomic region of interest. If this is a coding region, it is convenient to use 6 kb upstream and 6 kb downstream of the transcription start site (TSS). Subdivide the sequence into contiguous 2 kb fragments, i.e., 6×2 kb and design PCR primers inside each 2 kb sequences using Primer3 software. Get 6 PCR fragments of 1.2–1.7 kb covering 12 kb from Drosophila genomic DNA (usually prepared from the source where the FISH will be performed) (Fig. 1) (*see* **Note 7**).

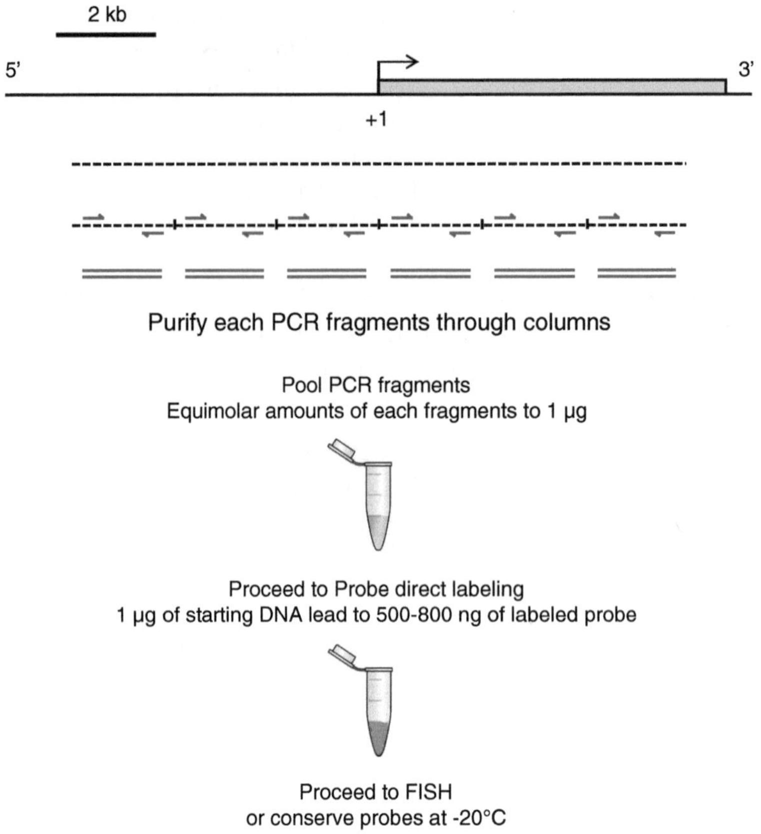

Fig. 1 Probe design and labeling. Schematic representation of a genomic region containing a gene unit. The *dashed line* represents the 12 kb selected genomic sequence centered at the TSS. This sequence is subdivided into 2 kb contiguous sequences. Forward and Reverse primers are designed for each 2 kb sequences leading to the production of 6 PCR fragments (1.2–1.7 kb in length) from genomic DNA. PCR fragments are purified, pooled, and directly labeled using the FISH Tag DNA kits (Invitrogen)

2. Probe labeling: Purify the PCR fragments using a PCR purification kit. If not pure, the PCR product can also be gel extracted. Pool equimolar amounts of PCR fragments up to 1 μg and proceed to labeling using the FISH Tag DNA kit (Fig. 1) (*see* **Note 8**). 1 μg of starting DNA will give you approximately 500–800 ng of labeled probes, i.e., enough for 5–8 FISH experiments in tissues.
3. Probe concentration: For tissues, use 80–100 ng of each probe in 30 μl of FHB/per hybridization; for cells, use 20–30 ng of each probe in 10 μl of FHB/per hybridization.
4. Probe conservation: Labeled probes can be conserved in nuclease-free water up to 3–6 months at −20 °C.

3.2 FISH in Drosophila Embryos or Larval Discs

Examples of Two-color FISH in embryos using probes in the BX-C and three-color FISH in embryos using probes in the ANT-C, BX-C, and NK-C complexes are illustrated in Fig. 2.

3.2.1 Preparation of Whole Embryos

1. Collect eggs on an agar plate with instant yeast and rinse them with dH_2O through a nylon sieve.
2. Dechorionate embryos with bleach for 5 min and rinse extensively with dH_2O. Do not let the embryo to dehydrate.
3. Fix embryos in 5 ml buffer A, 4 % Paraformaldehyde (*see* **Note 9**): Use this solution to transfer the embryos to a small glass flask. Alternatively, transfer the embryos with a paintbrush into a glass flask containing the fixing solution.
4. Add 5 ml of heptane (1:1 volume) and mix by vigorous rotation for 25 min at room temperature (RT) on a mini-shaker.
5. Remove the aqueous phase (lower phase) and keep the heptane phase (top phase). Fixed embryos will lay at the interface.
6. Devitellinize embryos by adding 5 ml of 100 % MeOH. Vortex 15 s at medium speed. Devitellinized embryos will fall down to the bottom MeOH phase.
7. Collect embryos at the bottom of the flask by pipetting with a 1 ml tip (cut the last ~0.2 cm at the end of the tip to prevent embryo damage) and transfer them into a 2 ml Eppendorf tube.
8. Rinse twice embryos in 100 % MeOH and store embryos in 100 % MeOH at −20 °C (*see* **Note 10**).

3.2.2 Preparation of Larval Imaginal Discs

1. Dissect discs in PBS 1× and transfer them into a 1.5 ml Eppendorf tube on ice. Here, it is important to do a partial dissection leaving the discs attached to the skin, the brain and/or the mouth parts of larvae (*see* **Note 11**).
2. Fix discs as soon as possible in 1 ml PBT, 4 % Paraformaldehyde (*see* **Note 9**) for 20 min at RT on a rotating wheel.
3. Wash three times discs in 1 ml PBT for 5–10 min at RT on a rotating wheel (*see* **Note 12**).

3.2.3 Hybridization to Fixed Embryos or Discs

All the following steps are performed in 1.5 ml Eppendorf tubes. Tissues (embryos or discs) should sediment at the bottom of the tube between each buffer change. If necessary, one can help sedimentation of tissues by very gentle centrifugation such as 1 min at $30 \times g$ (approximately 500 rpm).

1. Rehydrate fixed embryos (for larval imaginal discs, skip this step and proceed directly with **step 2**) by passing them through 1 ml of the following solutions (make them fresh): (1) 90 % MeOH, 10 % PBT; (2) 70 % MeOH, 30 % PBT; (3) 50 % MeOH, 50 % PBT; (4) 30 % MeOH, 70 % PBT; (5) 100 % PBT. Incubate 3–5 min at RT on a rotating wheel for each step.

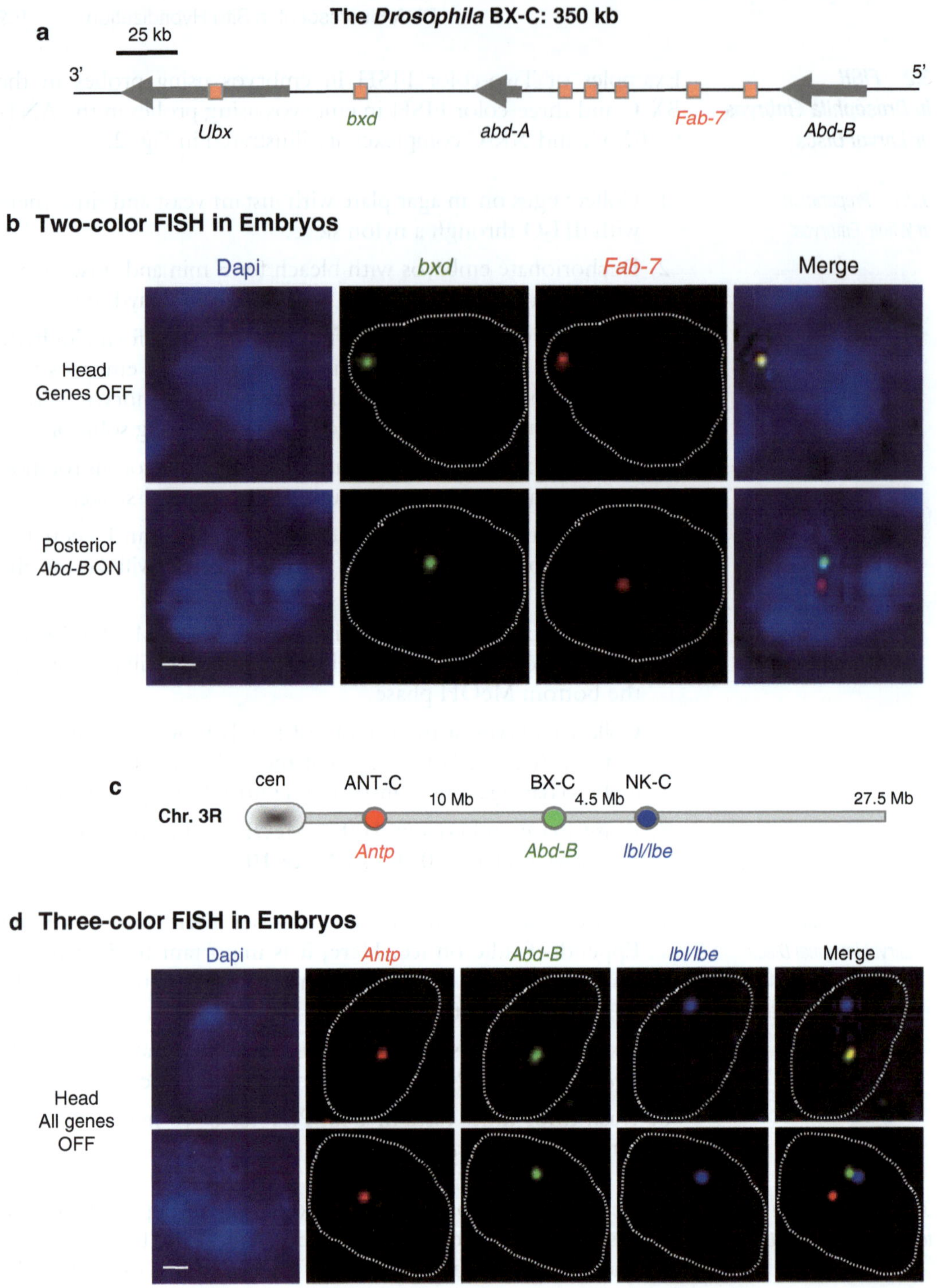

Fig. 2 Topological changes observed by two-color and three-color FISH. (**a**) Schematic linear representation of the BX-C, with the three Hox genes *Ubx*, *abd-A*, *Abd-B* represented by *gray arrows* and Polycomb-response element (PRE) represented by *orange boxes*. *bxd* and *Fab-7* are two well known PREs, which regulate the *Ubx* and *Abd-B* genes respectively and which are separated by approximately 130 kb. FISH probes targeting *bxd* and *Fab-7* PREs were used in the images below. (**b**) Two-color FISH in Embryos reveals conformation changes in the BX-C according to the transcriptional status of its genes. In the head of embryos where all genes of the BX-C are silenced, the two distant PRE are close in space and the two FISH spots are often on top to each other.

2. Incubate tissues in 1 ml PBT, 100–200 μg/ml RNAse A for at least 2 h at RT on a rotating wheel (or overnight (O/N) at 4 °C).
3. Incubate tissues or discs in 1 ml PBS-Tr 1 h at RT on a rotating wheel.
4. Transfer tissues into pHM by passing through 1 ml of the following freshly made solutions: (1) 80 % PBS-Tr, 20 % pHM; (2) 50 % PBS-Tr, 50 % pHM; (3) 20 % PBS-Tr, 80 % pHM; (4) 100 % pHM. Incubate 20 min at RT on a rotating wheel for each step.
5. Denature embryonic or larval DNA by incubation in 100 % pHM for 15 min at 80 °C using a thermomixer.
6. Denature probes (in 30 μl FHB) for 10 min at 90 °C, then transfer probes to the 80 °C thermomixer block until usage.
7. While tissues are still at 80 °C, remove as much buffer as possible with a 200 μl tip and rapidly add the denatured probes to tissues without prior cooling.
8. Transfer tubes to a thermomixer set at 37 °C and cover tissue suspensions with a drop of mineral oil.
9. Hybridization for 14–20 h, i.e., O/N, at 37 °C in a thermomixer with gentle agitation (450–500 rpm). Cover with an aluminum foil to keep in the dark.
10. Add 500 μl of 50 % Formamide, 2× SSC, 0.3 % CHAPS and remove supernatant. This helps to immediately remove the mineral oil after hybridization.
11. Post-hybridization washes by passing tissues through 1 ml of the following solutions (make them fresh): (1) 50 % Formamide, 2× SSC, 0.3 % CHAPS; repeat this wash once; (2) 10 % Formamide, 2× SSC, 0.3 % CHAPS; (3) 30 % Formamide, 70 % PBT; (4) 20 % Formamide, 80 % PBT; (5) 10 %

Fig. 2 (continued) At this level of resolution, these interactions can be interpreted by compaction or looping of the chromatin fiber. In the posterior region of embryos, the *Abd-B* gene is active and larger distances are observed between the two regulatory elements. Figure adapted from [1]. (**c**) Schematic linear representation of the Drosophila chromosome arm 3R, which contains ANT-C, BX-C and the NK homeobox gene complex (NK-C), three major PcG-bound regions. Genomic distances in Megabases between ANT-C and BX-C, and between BX-C and NK-C are indicated. The centromer region (cen) and the total length of the chromosome arm are also indicated. (**d**) Three-color FISH with probes targeting genes of these complexes reveals chromosome conformation features. In the upper panel, *Abd-B* interacts with *Antp* located proximal to the centromere, whereas in the *lower panel*, *Abd-B* interacts with *lbl/lbe* (NK-C) located distal. This reflects variable chromosome conformation inside the nucleus. Simultaneous associations between the three genes can also be observed but this is a rare event, indicating that the tropism of long-range interactions may constrain chromosome topology. The *dotted lines* in the FISH channels indicate the edge of the nucleus based on DAPI staining. Figure images correspond to deconvolved single slices from 3D stacks. Scale bars represent 1 μm. Figure adapted from [4]

Formamide, 90 % PBT; (6) 100 % PBT; (7) 100 % PBS-Tr. Perform washes (1)–(4) 20 min at 37 °C in a thermomixer with agitation (800 to 900 rpm), perform washes (5)–(7) 20 min at RT on a rotating wheel. Proceed with immunostaining if applicable (Subheading 3.2.4).

12. Counterstain DNA with PBT, DAPI (1:1,000 of the stock solution) for 10 min at RT on a rotating wheel (*see* **Note 13**).
13. Wash once in PBT for 10 min at RT.
14. Replace PBT by PBS 1× (*see* **Note 14**) and leave about 50 μl of PBS to put embryos on a slide with a cut 200 μl pipet tip. For the larval discs, dissect them from the skin, the brain and/or the mouth parts and lay them on the slide (always leave some PBS to avoid dehydration of the discs) (*see* **Note 15**).
15. On the slide, remove as much PBS as possible around tissues using a 200 μl pipet tip. Kimwipes can also be used.
16. Mount tissues with 40 μl of ProLong antifade or other mounting media. If using ProLong antifade, put slides for at least one night at 4 °C to allow curing of the mountant before microscopy analysis.

3.2.4 FISH Combined with Immunostaining (See Note 16)

Examples of FISH-I using probes against the *Antp* and the *Abd-B* genes and the Polycomb antibody are illustrated in Fig. 3.

At the end of **step 11** of Subheading 3.2.3, continue with the following steps. All steps are performed in 1.5 ml Eppendorf tubes on a rotating wheel.

1. Block tissues in 0.5 ml Blocking solution for at least 2 h at RT (or O/N at 4 °C).
2. Dilute the primary antibody at your favorite concentration in 0.5 ml of Blocking solution and incubate O/N at 4 °C (*see* **Note 17**).
3. Wash tissues as followed: Three short washes for 5–10 min in PBS-Tr 0.3 %; 3 long washes for 20–30 min in PBS-Tr 0.3 %, at RT.
4. Dilute the secondary antibody in Blocking solution and incubate for 1 h at RT.
5. Wash tissues as followed: Three short washes for 5–10 min in PBS-Tr 0.3 %; 3 long washes for 20–30 min in PBS-Tr 0.3 %, at RT.
6. Proceed to DAPI and mounting as described in Subheading 3.2.3, **step 12**.

3.3 FISH in Drosophila Cells

Examples of two-color FISH in S2 and S3 cells using probes in the BX-C are illustrated in Fig. 4.

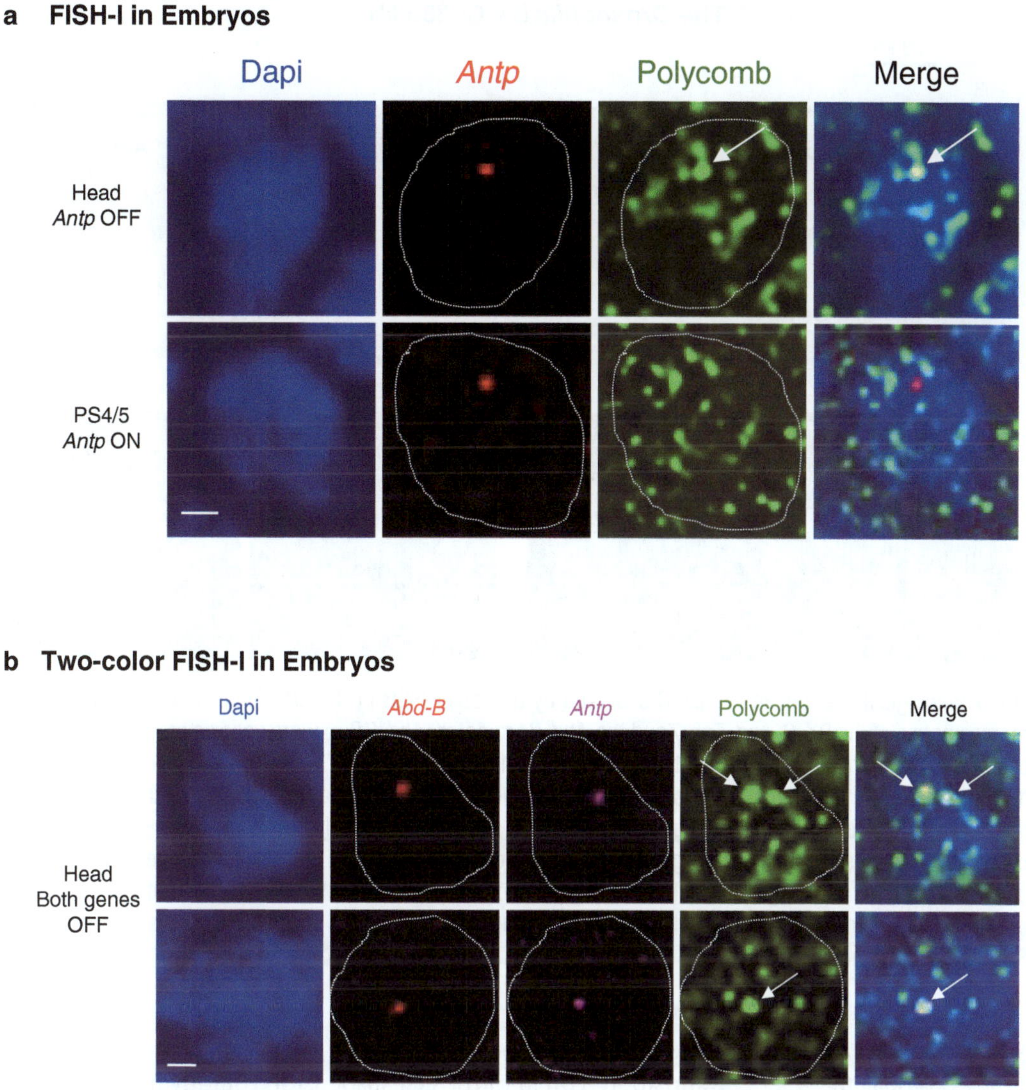

Fig. 3 Repressed Hox complexes localize in large PcG bodies and can cluster in the nucleus, as observed by FISH-I. (**a**) In the head of the embryo, the *Antp* gene of the ANT-C is in the OFF transcriptional state. The gene localizes in the majority of the case in a large PcG body (visualized here using an antibody against the Polycomb protein), indicated by a *white arrow*. In parasegment 4 and 5 (PS4/5), the gene is ON and is no longer found in a PcG body. (**b**) In the *upper panel*, in the head of the embryo, the *Antp* and the *Abd-B* genes are both silenced, and they localize in the majority of the cases in distinct large PcG bodies. In the *lower panel*, the two genes separated by 10 Mb colocalize in the same Polycomb body in approximately 15 % of the nuclei. These interactions are significant, increase with development and may stabilize PcG-dependent silencing. The *dotted lines* in the FISH channels indicate the edge of the nucleus based on DAPI staining. Figure images correspond to deconvolved single slices from 3D stacks. Scale Bars represent 1 μm. Figure adapted from [4]

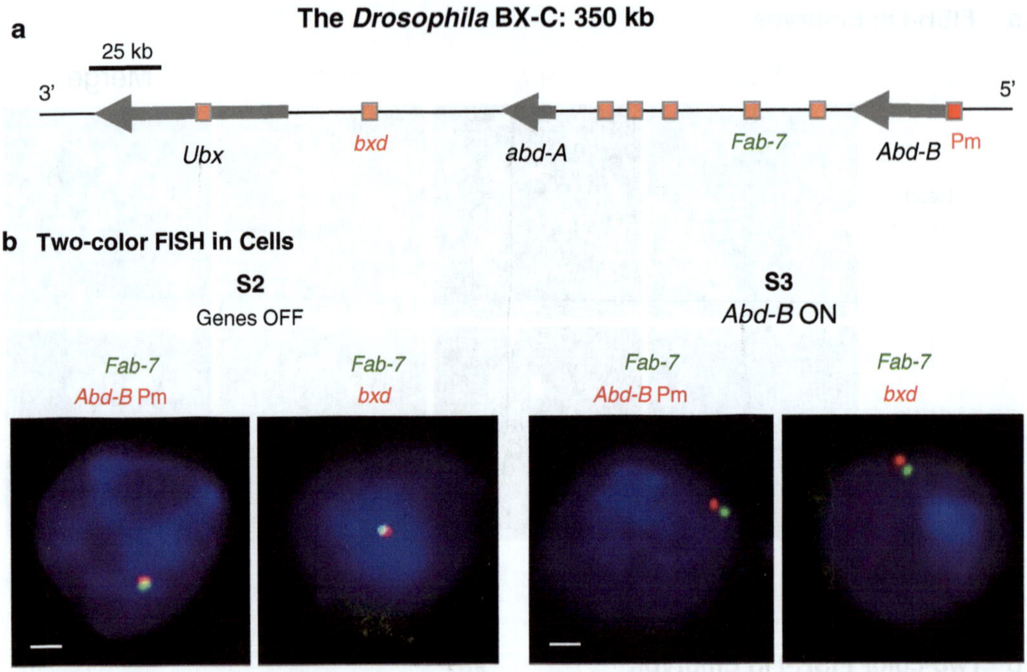

Fig. 4 Topological changes in the BX-C observed by two-color FISH in Drosophila cells. (**a**) Schematic linear representation of the BX-C. *bxd*, *Fab-7* and the *Abd-B* promoter (*Abd-B* Pm) were used as probes in the images below. *Fab-7* and *bxd* are separated by approximately 130 kb, *Fab-7* and *Abd-B* Pm by approximately 70 kb. (**b**) Two-color FISH in Drosophila cells revealing topological changes of the BX-C similar to those observed in the embryo (Fig. 2b). In S2 cells, all genes of the BX-C are transcriptionaly silent. The PRE-Pm and PRE-PRE distances are very short, often leading to well colocalized FISH spots. In S3 cells, similar to the posterior region of the embryo, the *Abd-B* gene is active. In this situation, larger distances are observed between regulatory elements. Figure images correspond to deconvolved single slices from 3D stacks. Scale Bars represent 1 μm. Figure adapted from [1]

3.3.1 Cell Immobilization on Slides (See ***Note 18****)*

1. Prepare a cell suspension of 1–2 × 10^6 cells/ml in growth media and apply 100 μl of the cell suspension roughly in the center of a slide.
2. Allow cells to adhere to the slides for approximately 1 h at 25 °C atmospheric CO_2, avoiding evaporation of the medium (*see* **Note 19**).

3.3.2 FISH Procedure on Cells

All steps until the hybridization are in a Coplin jar at RT (*see* **Note 18**).

1. Rinse attached cells in PBS 1×.
2. Fix slides in PBS 1×, 4 % Paraformaldehyde (*see* **Note 9**) for 10 min.
3. Wash three times in PBS 1× for 5 min.
4. Incubate in PBT, 100 μg/ml RNAse A for 1 h.
5. Wash once in PBS 1× for 3–5 min.

6. Incubate in PBS 1×, 0.5 % Triton X-100 (make it fresh) for 10 min.
7. Wash once in PBS 1× for 3–5 min.
8. Incubate in PBS 1×, 20 % Glycerol for 30 min.
9. Freeze in liquid nitrogen (for approximately 30 s) and thaw. As soon as the frozen layer disappears, put the slide back to PBS 1×, 20 % Glycerol. Repeat this step twice, i.e., a total of three freezing/thaw cycles (*see* **Note 20**).
10. Wash three times in PBS 1× for 5 min.
11. Incubate in 0.1 M HCl for 5 min (*see* **Note 21**).
12. Rinse three times in 2× SSC for 1 min.
13. Incubate in 50 % Formamide, 2× SSC for at least 30 min at RT before hybridization.
14. Pre-denature probes (in 10 μl FHB) for 10 min at 80 °C and put them on ice until usage.
15. Take slides out and drain off the excess of liquid.
16. In parallel, add the denatured probes to a coverslip (denatured probes in FHB can be quickly prewarmed at 37 °C to help pipetting).
17. Invert drained slides onto the coverslips such as the area containing the cells is covered and seal with rubber cement (allow rubber cement to air-dry for at least 5 min at RT).
18. Co-denature cellular DNA and probes by putting the slide (coverslip facing up) on a heating block for 3 min at 78 °C (*see* **Note 22**).
19. Hybridization for 14–20 h, i.e., O/N, at 37 °C in a dark and humid chamber.
20. Peel off Rubber Cement and let the coverslips falling into the Coplin jar containing 2× SSC (do not try to take coverslips off).
21. Wash three times in 2× SSC, 3×5 min at 37 °C with gentle shaking.
22. Wash three times in 0.1× SSC, 3×5 min at 45 °C with gentle shaking.
23. Rinse twice in PBS at RT or proceed with immunostaining if applicable (Subheading 3.3.3).
24. Counterstain DNA with PBT, DAPI (1:1,000 of the stock solution) for 10 min at RT.
25. Rinse twice with PBS.
26. Take slides out and drain off the excess of liquid. Mount with 40 μl of ProLong antifade or other mounting media.

3.3.3 FISH Combined with Immunostaining on Cells

At the end of **step 23** of Subheading 3.3.2, continue with the following steps. For **steps 2**, **3**, and **5**, drain off the excess of liquid from the slide and perform incubation between slide and coverslip to minimize the volume (20 μl) using a dark and humid chamber.

1. Wash three times in PBT, 3 × 1 min at RT.
2. Block cells with Blocking solution for 30 min to 1 h at RT.
3. Incubate primary antibody diluted in Blocking solution for 2 h at RT or O/N at 4 °C.
4. Wash four times in PBT for 1 min, 3 min and twice 5 min.
5. Incubate with secondary antibody diluted in Blocking solution for 1 h at RT.
6. Wash four times in PBT for 1 min, 3 min and twice 5 min.
7. Proceed to DAPI and mounting as described in Subheading 3.3.2, **step 24**.

3.4 Microscopy and Analysis

1. Acquisition: Confocal microscopes or any wide-field microscope equipped with a moving plate-form or objective in the *z*-axis and with 63× or 100× oil objectives are suitable for FISH acquisition. In general, we recommend collecting 3D stacks from three to five different embryos or larval discs, taking optical sections at 0.3–0.5 μm intervals along the *z*-axis. The number of optical sections will depend on the nucleus diameter. For tissues, approximately 100–200 nuclei per 3D stack can be assayed. For cells and depending on the density on the slide, up to 40–50 nuclei can be assayed.
2. Image analysis: Image J, a free software, can be used to assess 3D distance measurements between genomic loci of interest or to analyze the association of a particular locus with nuclear substructure. Alternately, Metamorph (Molecular Devices) or Volocity (PerkinElmer) software can be used.
3. Statistical analysis: The distance between two loci is measured in a large number of nuclei for a particular condition. In general, the inter-distance in 100–500 individual nuclei from different 3D stacks can be pooled and assayed. Inter-distance between two loci can then be represented as a distribution of distance categories, such as categories every 0.5 μm. In this case, a Wilcoxon test can be used to assess the significance of differences of the 3D distance distributions in between different pairs of loci or in between the same pair of loci but in different conditions. In parallel, it can be practical and meaningful to determine the percentage of colocalization or close association or pairing between two loci. For this, a cut-off is determined, which is often defined by the resolution of your microscope setting and/or the biology of your system and which can

vary from 0.3 to 1 μm. In this case, a two-tailed Fisher's exact test can be used to assess the significance of differences of colocalization in between two different conditions.

4. Figure display: Images from 3D stacks can be deconvolved using the Huygens MLE single tif procedure (Scientific Volume Imaging) or alternative deconvolution software. This help to remove background signals, especially in tissue nuclei.

4 Notes

1. The FISH Tag DNA Kits employ a two-step labeling technology—nick translation to enzymatically incorporate amine-modified nucleotides (aminoallyl dUTP) followed by chemical coupling of Alexa Fluor dyes. Over the past years, we found that this technology represents the most sensitive method.
2. After opening, a bottle of bleach is conserved for no longer than 1 month.
3. The ProLong antifade is a curing mounting medium, which is often more suitable for embryos or larval discs that do not attach to the slide. ProLong antifade does not necessitate nail polish sealing and allows longer-term storage than noncuring mounting media (slides can be stored in the dark at 4 °C for up to 3 months).
4. Alternatively to 10 % NGS or NDS, 5 % WBR (Western Blocking Reagent from Roche, #1 921 673) or 2–3 % BSA (Sigma, A3059) can be used.
5. Poly-L-lysine-coated slides: Rinse microscope slides in 100 % Ethanol for 3–5 min and allow slides to completely dry via air drying. Incubate slides in a 0.01 % poly-L-lysine solution (from a 0.1 % solution, Sigma, P8920) for 5 min in a plastic Coplin jar. Allow slides to completely air dry. Alternatively, use commercially available poly-L-lysine or Polysine-coated slides (VWR or Thermo Scientific).
6. 22 × 22#1.5 coverslips are best to use for the hybridization step. For mounting, it can be better to use 22 × 30#1.5 coverslips, especially if using nail polish for the sealing.
7. In Drosophila, we recommend using 6 PCR fragments, but 4–5 can still be enough for dependable FISH signals. In mammals, the genome is much larger, therefore we recommend to increase the size of the region selected, i.e., 20–24 kb equivalent to 10–12 PCR fragments, though less fragments might still be enough. This technique of probe labeling is however not suitable to cover large genomic region in the order of the Megabase. For larger region, one can use the recently described

"Oligopaint" technology [9] or related methods [10]. Future development of this technology should allow the use of Oligopaint for smaller regions, which might be more suitable for large scale FISH analysis.

8. All the procedures for the direct labeling of the probes are well described in the FISH Tag DNA Kit manual and we usually follow their recommendations, except for one critical step. When using 1.2–1.7 kb PCR fragments as a substrate for the nick translation reaction, we generally use less DNase I than recommended, i.e. 1.3–1.5 μl of the DNAse I working solution for a 50 μl reaction, in order to give probes ranging from 100 to 300 bp. The reaction is performed at 15 °C for at least 1.5 h but no more than 2 h. The trick for a good probe is to get at least 2 dyes incorporated per 100 bp, best if 4–6 dyes per 100 bp.
9. If using paraformaldehyde in powder, dissolve it in buffer A (for embryos), PBT (for discs), or PBS (for cells) by putting the solution in a boiling water bath for few minutes. Vortex several times to help dissolution and cool to RT before use.
10. At this stage, embryos can be stored at −20 °C for several months.
11. For larval discs, you can cut the larva in half and turn inside out the skin with two thin forceps to expose the discs to the media. For anterior discs and brain, it can be more convenient to pull tissues from the mouth hooks.
12. Unlike for embryos, there is no way to store discs for a long period of time at this stage. However, for a short period of time (1–3 days), the discs can be stored in PBT at 4 °C.
13. DAPI is five times less concentrated than for an immunostaining protocol because here the DNA has been denatured by the FISH procedure.
14. Removing traces of detergent is preferable for microscopy.
15. At this stage, tissues can be stored in PBS at 4 °C for up to 1 week if not immediately mounted.
16. This FISH-I procedure where the FISH precedes the immunostaining works well with chromatin-bound proteins such as PcG proteins (Pc, Ph) [4, 8, 1], Trithorax-group proteins (Trx-C, Ash1) (Roure, Bantignies and Cavalli, unpublished), and RNA Polymerase II using an antibody directed against the Ser2P form (H5 monoclonal antibody from Covance) (Roure, Bantignies, and Cavalli, unpublished). The protocol also works well with markers of the nuclear membrane such as lamin [11] or Wheat Germ Agglutinin conjugates (WGA, Molecular probes) (Bantignies, unpublished). However, in our hand, this procedure does not work with all nuclear

factors, including Histone marks. For some nuclear factors that do not resist to this procedure, it might be worth trying to invert the procedure, i.e., fixation 4 % Paraformaldehyde, then the immunostaining, post-fixation with 2–4 % Paraformaldehyde and then the FISH procedure. Concerning histone modifications or variants, a combined immunofluorescence and DNA FISH procedure, which partially preserved γ-H2AX and H3K27me3 labelling, has recently been described in mammalian cells [12].

17. In general, we recommend using the primary antibody two times more concentrated than for a simple immunostaining. Otherwise, if your primary antibody is a serum, we recommend doing a pre-clearing. Dilute your antibody in the blocking solution and incubate this solution with preblocked embryos or tissues for 1–2 h at RT. Then use the pre-cleared supernatant for the immunostaining procedure.
18. Alternately, cells can be immobilized on poly-L-lysine-coated coverslips. In this case, use small Coplin jars (Electron Microscopy Sciences, #72242-21) or 6-well plates for the FISH procedure. The inconvenience of using coverslips here is that they are more breakable than slides, especially during freezing/thaw cycles in liquid nitrogen. However, coverslips are required for high-resolution microscopy.
19. For mammalian cells, allow cells to adhere to the slides for 1–3 h at 37 °C, 5 % CO_2.
20. Treatment with Triton X-100 and the freezing/thaw cycles in liquid nitrogen helps to permeabilize the nuclear membrane without strongly affecting 3D chromatin architecture.
21. HCl is a mild deproteinization method, which preserves the 3D morphology of nuclei and which is preferred to pepsin treatment.
22. Use a thermoblock and return the modular heating block for maximum contact with the slides or use a waterbath with a heating block as a platform, which is often a more accurate setting.

Acknowledgement

F.B. is supported by the CNRS. Research at the G.C. lab was supported by grants from the European Research Council (ERC-2008-AdG No 232947), the CNRS, the European Network of Excellence EpiGeneSys, the Agence Nationale de la Recherche and the Association pour la Recherche sur le Cancer.

References

1. Lanzuolo C, Roure V, Dekker J, Bantignies F, Orlando V (2007) Polycomb response elements mediate the formation of chromosome higher-order structures in the bithorax complex. Nat Cell Biol 9(10):1167–1174. doi:10.1038/ncb1637, ncb1637 [pii]
2. Ferraiuolo MA, Rousseau M, Miyamoto C, Shenker S, Wang XQ, Nadler M, Blanchette M, Dostie J (2010) The three-dimensional architecture of Hox cluster silencing. Nucleic Acids Res 38(21):7472–7484. doi:10.1093/nar/gkq644, gkq644 [pii]
3. Noordermeer D, Leleu M, Splinter E, Rougemont J, De Laat W, Duboule D (2011) The dynamic architecture of Hox gene clusters. Science 334(6053):222–225. doi:10.1126/science.1207194
4. Bantignies F, Roure V, Comet I, Leblanc B, Schuettengruber B, Bonnet J, Tixier V, Mas A, Cavalli G (2011) Polycomb-dependent regulatory contacts between distant Hox loci in Drosophila. Cell 144(2):214–226. doi:10.1016/j.cell.2010.12.026, S0092-8674(10)01485-6 [pii]
5. Tolhuis B, Blom M, Kerkhoven RM, Pagie L, Teunissen H, Nieuwland M, Simonis M, de Laat W, van Lohuizen M, van Steensel B (2011) Interactions among Polycomb domains are guided by chromosome architecture. PLoS Genet 7(3):e1001343. doi:10.1371/journal.pgen.1001343
6. Sexton T, Yaffe E, Kenigsberg E, Bantignies F, Leblanc B, Hoichman M, Parrinello H, Tanay A, Cavalli G (2012) Three-dimensional folding and functional organization principles of the Drosophila genome. Cell 148(3):458–472. doi:10.1016/j.cell.2012.01.010
7. Bantignies F, Grimaud C, Lavrov S, Gabut M, Cavalli G (2003) Inheritance of Polycomb-dependent chromosomal interactions in Drosophila. Genes Dev 17(19):2406–2420. doi:10.1101/gad.269503 17/19/2406 [pii]
8. Grimaud C, Bantignies F, Pal-Bhadra M, Ghana P, Bhadra U, Cavalli G (2006) RNAi components are required for nuclear clustering of Polycomb group response elements. Cell 124(5):957–971. doi:10.1016/j.cell.2006.01.036, S0092-8674(06)00171-1 [pii]
9. Beliveau BJ, Joyce EF, Apostolopoulos N, Yilmaz F, Fonseka CY, McCole RB, Chang Y, Li JB, Senaratne TN, Williams BR, Rouillard JM, Wu CT (2012) Versatile design and synthesis platform for visualizing genomes with Oligopaint FISH probes. Proc Natl Acad Sci U S A 109(52):21301–21306. doi:10.1073/pnas.1213818110
10. Bienko M, Crosetto N, Teytelman L, Klemm S, Itzkovitz S, van Oudenaarden A (2013) A versatile genome-scale PCR-based pipeline for high-definition DNA FISH. Nat Methods 10(2):122–124. doi:10.1038/nmeth.2306
11. Brasset E, Bantignies F, Court F, Cheresiz S, Conte C, Vaury C (2007) Idefix insulator activity can be modulated by nearby regulatory elements. Nucleic Acids Res 35(8):2661–2670. doi:10.1093/nar/gkm140
12. Chaumeil J, Micsinai M, Skok JA (2013) Combined immunofluorescence and DNA FISH on 3D-preserved interphase nuclei to study changes in 3D nuclear organization. J Vis Exp 72:e50087. doi:10.3791/50087

Chapter 8

Mining the *Cis*-Regulatory Elements of Hox Clusters

Navneet Kaur Matharu and Rakesh K. Mishra

Abstract

Hox clusters have served as a favorite system to study the role of *cis*-regulatory elements at multiple layers of gene regulation. Organization and regulation of Hox genes show remarkable conservation and determine the anterior–posterior body axis across the bilaterians. Identification of a variety of regulatory regions within the complex and around it, embedded primarily in the noncoding part of the corresponding genomic region that can spread 100–150 kb, is a challenging problem. Multiple experimental and computational tools need to be employed to investigate functional features of such elements. Here we discuss parallel approaches to mine the most plausible regulatory information from the noncoding sequences of Hox clusters, among diverse species.

Key words Hox cluster, *Cis*-regulatory element (CRE), Polycomb response element (PRE), Insulator, Chromatin domain boundary, CNCS

1 Introduction

Developmental gene expression at Hox clusters is precisely regulated in a spatiotemporal manner. Such a tight regulation is brought about by the interplay of DNA regulatory elements and factors binding to them. It is of utmost interest to developmental biologists to understand the mechanisms of Hox gene regulation, and to this end, the identification of *cis*-regulatory elements (CRE) involved in Hox gene expression is a priority task. Functionally and structurally, the coding regions of homeotic genes are highly conserved. However, the essential regulatory information for correct gene expression is hidden in the almost anonymous noncoding sequences [1]. The understanding of how to make sense out of noncoding sequences of the genomes has considerably been improved by the advent of whole-genome sequence analysis approaches [1]. Genome comparisons of various species have been quite successful in pinpointing the *cis*-regulatory regions of high interest [2, 3]. A well-informed decision about the functionality of a putative *cis*-regulatory element can be taken based on its sequence

Yacine Graba and René Rezsohazy (eds.), *Hox Genes: Methods and Protocols*, Methods in Molecular Biology, vol. 1196, DOI 10.1007/978-1-4939-1242-1_8, © Springer Science+Business Media New York 2014

conservation index and by further analyzing its various "biochemical" properties, such as transcription factor-binding ability or localization of certain histone modifications [4]. The transcribing ability of noncoding regions is also a very crucial indicator of it being essential for regulation. Recently, many enhancers and chromatin-mediated cis-regulatory elements have been shown to be transcribed which is again essential for the recruitment of regulatory complex onto CRE [5–7]. Here we discuss, using Hox clusters as a paradigm, approaches to be taken to mine CRE based on sequence conservation and on conserved epigenetic entities, including histone modifications, DNA methylation, and transcription factor binding. We also briefly discuss the needs to use existing or design novel assays to assess the function of in silico-predicted CREs.

2 Methods

2.1 Mining Highly Conserved Cis-Regulatory Elements from Hox Clusters

Noncoding part of the genome has been of special interest since most of the regulatory information for gene expression resides there. Conservation is an informative tool to mine functional elements from different species. However comparing the genomes of evolutionarily close organisms may render sequence alignment tools to identify functional elements useless owing to high degree of similarity in the orthologous genomic sequences. Previous studies have shown that multispecies comparison proves promising to mine regulatory elements from noncoding regions [8]. In silico identification of noncoding cis-regulatory elements has been quite successful till date; however, in most instances, biological relevance for gene regulation remains to be tested. Recently, with the availability of epigenetic information across the genomes in multiple species, new opportunities have emerged, allowing for narrowing down noncoding regions to potential functional CREs. Below, we summarize a multistep strategy that proved successful in mining *cis*-regulatory elements from Hox clusters.

1. Comparison among the genomes is quite useful to start with (Fig. 1). Conserved noncoding sequences have very high probability to be functional. Comparison of noncoding part of Hox clusters across species shows conserved blocks which are putative *cis*-regulatory modules of comparable functions. Many of the enhancers within Hox clusters are very well conserved and underlie similar expression pattern when assayed in different species [3, 9, 10]. Enhancers are conserved over 100 bp window size for conservation with >70 % sequence identity [1, 10]. However, other regulatory elements like repressors and insulators are elusive, as they often do not have significant sequence conservation. Lowering the sequence identity stringency is not the best option: distinct conserved blocks are seen

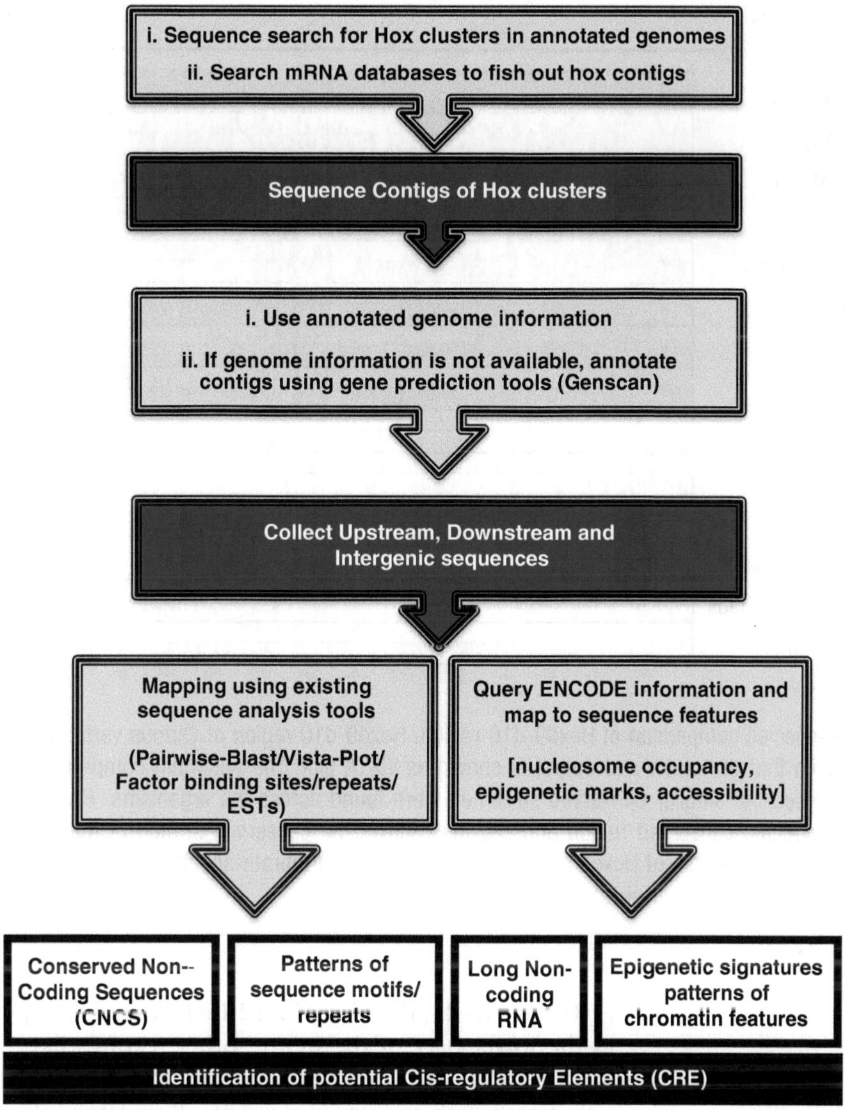

Fig. 1 In silico strategy for finding *Cis*-regulatory elements (CRE) around Hox clusters. Step-by-step work flow for in silico identification of CRE, taking different approaches to mine for conservation. Annotated conservation file can be put in any publicly available tool to generate graphical output

when distant organisms are compared with low stringency, but in case of closely related species, it results in false positives. Reducing the size of the "enhancer window" proved to be useful: a 100 bp and above window size may not yield conserved blocks even in distant species but when the window size to look for conservation is reduced, say up to 20 bp, many small stretches of conserved blocks can be found (Fig. 2). Some known regulatory elements have conserved blocks of fixed size along the phylogeny, for example polycomb response element

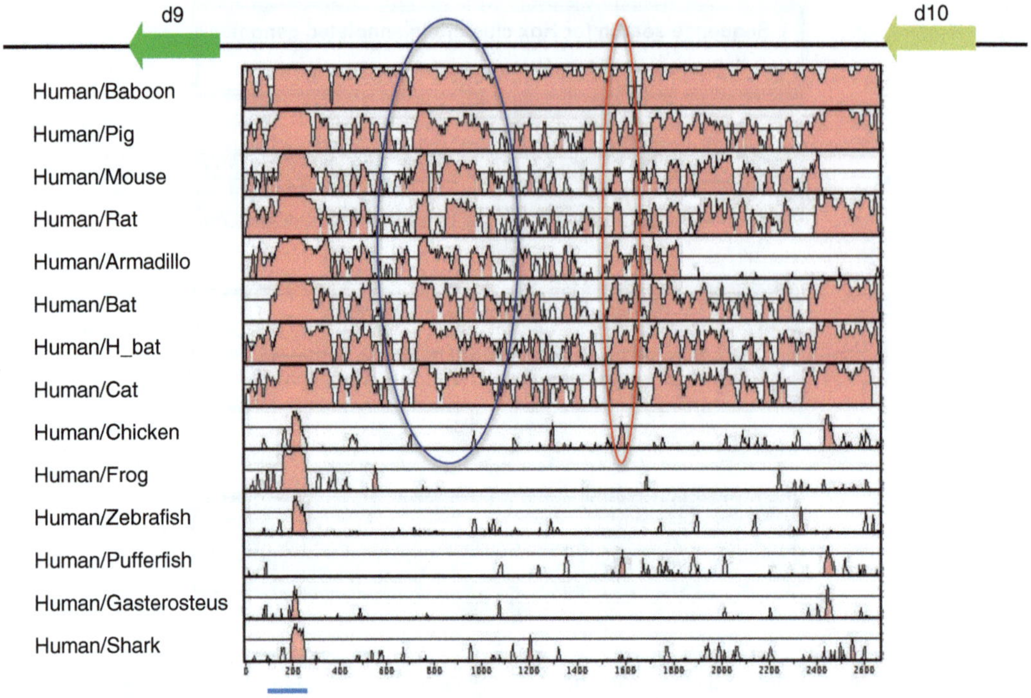

Fig. 2 Multi-species comparison of Hoxd9-d10 region. Hoxd9-d10 region of various vertebrate species was compared with that of human and output is shown as VISTA plot. The minimum comparison window is of 20 bp. Dispersed but smaller conserved stretches were found across the organisms. *Blue oval* shows the mammalian specific conserved region and *red oval* shows the conserved stretch of mammals shared by chicken also. Just upstream of Hoxd9 is the highly conserved vertebrate-specific region, shown as *blue bar*, though expanded in higher vertebrates

(PRE) at HoxD, the Hoxd11.12 PRE (Fig. 3) [11]. As shown in the lower part of the figure, other features like epigenetic modifications and chromatin accessibility states are often associated with such conserved elements. Regulatory elements may also acquire more sequence features in higher organisms and look like an inverted pyramid in sequence comparison plots (Fig. 2). These are likely to have added up more functional complexity to the core CRE.

2. Some conserved sequence blocks can be found by comparison of genomes of distant species with even higher stringency (above 70 %). These are classified as ultra-conserved elements referred to as conserved noncoding sequences (CNCS). The CNCS are very interesting to study evolutionary changes across the phylogeny. Subtle changes in these sequences account for the species-specific variations in developmental gene expression. Many ultra-conserved enhancers have single base sequence variations, which amounts for considerable difference in gene expression [12]. Several such highly conserved regulatory elements display shared domain of expression across

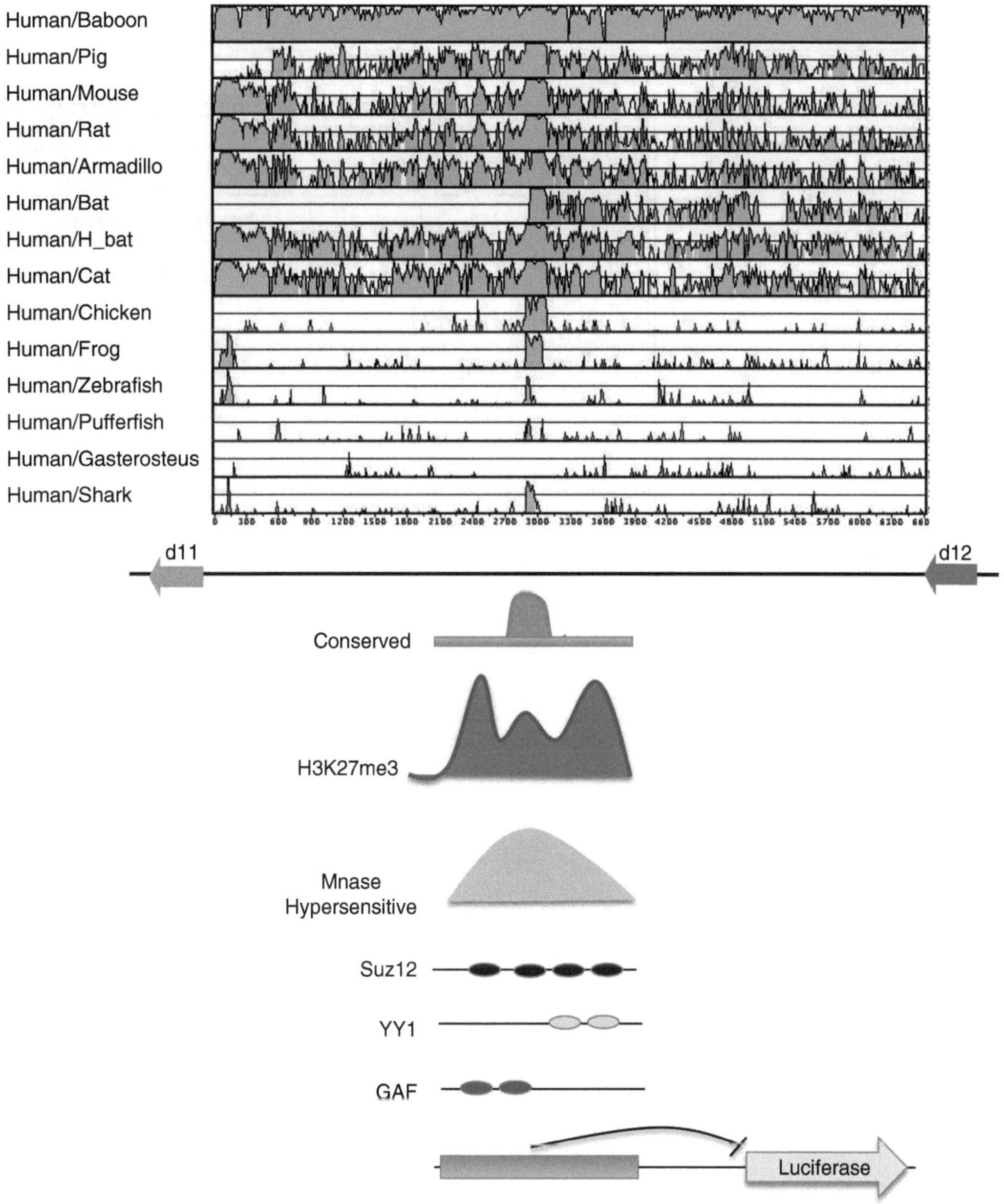

Fig. 3 Conservation of Hoxd11.d12 PRE and its associated epigenetic features. Hoxd11.d12 PRE has conserved block in all the vertebrates as seen in the upper portion, the VISTA plot. The bottom part shows various epigenetic features associated with this conserved region, for example, enrichment of H3K27me3 mark and MNase hypersensitivity (schematically represented). This region also binds to Suz12 and YY1 and has GAF occupancy in human and is shown to function as repressor in cell line-based luciferase assay system

distant organisms. For example, Dachshund (*DACH*) enhancers in vertebrates lie in the gene desert regions sharing the synteny, right from fishes to human. Despite few nucleotide changes, these enhancers are relatively conserved in their domain of expression across vertebrates [1]. On the other hand, ultra-conserved

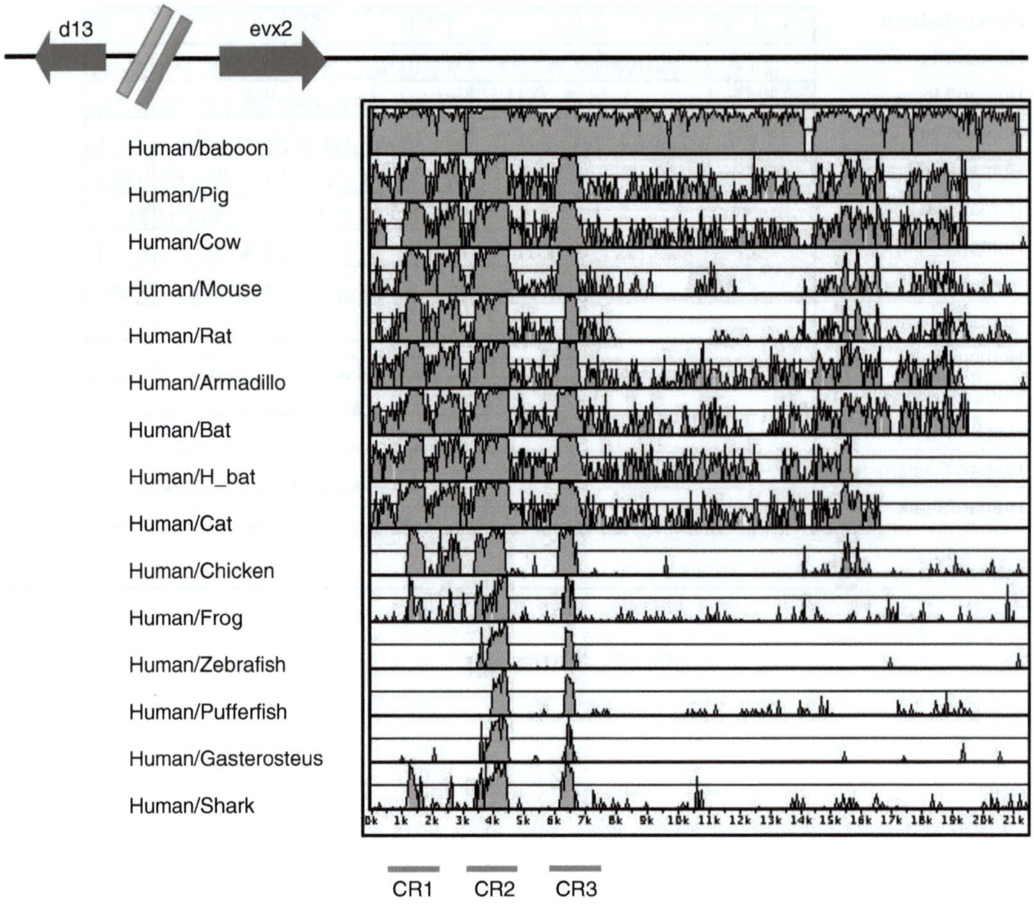

Fig. 4 Ultra-conserved region found upstream of HoxD cluster. The region upstream of Hoxd13 has three distinct conserved blocks, namely CR1, 2, and 3. As the name suggests, these regions are extremely conserved (with patches of 100 % identity) in all vertebrates (ref. 14)

enhancers with few base changes between evolutionarily closer organisms are known to impart species-specific expression patterns. For example, *HACNS1* enhancer in human and primates is very well conserved only differing in 13 base variations specific to human that confers its human-specific expression in limbs. Primate orthologue of conserved *HACNS1* is not expressed in limbs [13]. An interesting example is the upstream region of the HoxD clusters, which has 3 CNCS (CR1, CR2, and CR3) and, among vertebrates when one moves from fish to mammal, the extent of conserved blocks expands [14]. This region, showing not only extreme conservation but also steady increase in the size of conserved blocks in complex organisms, offers good candidates for studying cis-regulation of the HoxD cluster (Fig. 4).

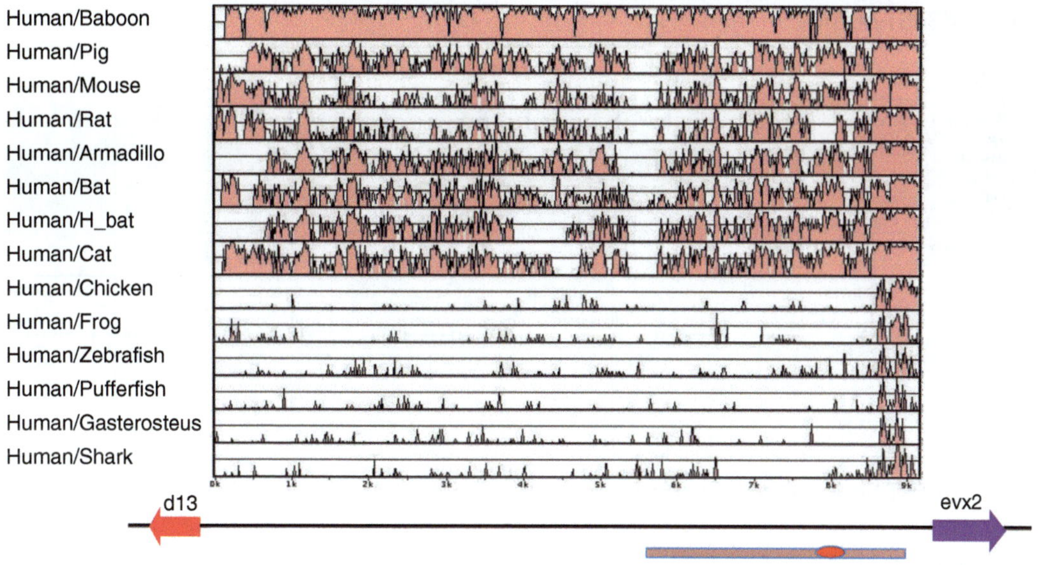

Fig. 5 Intergenic region of Evx2-d13 gene pair has evolutionarily conserved boundary function. Sequence conservation of Evx2-d13 boundary region, shown as *pink bar*, is not significant, but the presence of GAF sites (shown as *red oval*) is observed in all vertebrates. Functional assay shows conservation in terms of gene regulation and has high dependency on GAF for it to act as boundary element (ref. 16)

2.2 Mining Elusive Cis-Regulatory Elements

1. Many CREs involved in higher order chromatin-associated regulation do not show clear-cut conservation, making the task to mine them on the basis of sequence conservation difficult. To circumvent the problem, an alternative approach is to search for binding motifs of regulatory protein. A large variety of transcription factors with diverse DNA-binding specificity have well-described functions during gene expression. Public databases are available to search for transcription factor-binding sites (TRANSFAC for example), and recently this set of information has been integrated in the ENCODE database. Enrichment of particular binding motifs over a given length of DNA is a good handle to uncover elusive cis-regulatory elements. Insulators (chromatin domain boundaries), repressors, and PRE do not have significant sequence conservation along the phylogeny, but functionally they are very well conserved. Often their function relies on the conservation of binding motifs for regulatory factors, rather than apparent sequence conservation. For example, the Evx2-d13 boundary element is very elusive when sequence conservation is checked, but close to the Evx2 promoter a GAGA repeat stretch that is known to function as GAF-binding motif is conserved from zebrafish to human, although the length of the repeat may vary in different species and sometimes there may be more than one such motif (Fig. 5) [15]. Nevertheless, GAF binding plays quintessential role in conferring the boundary activity of the Evx2-d13 intergenic region [16].

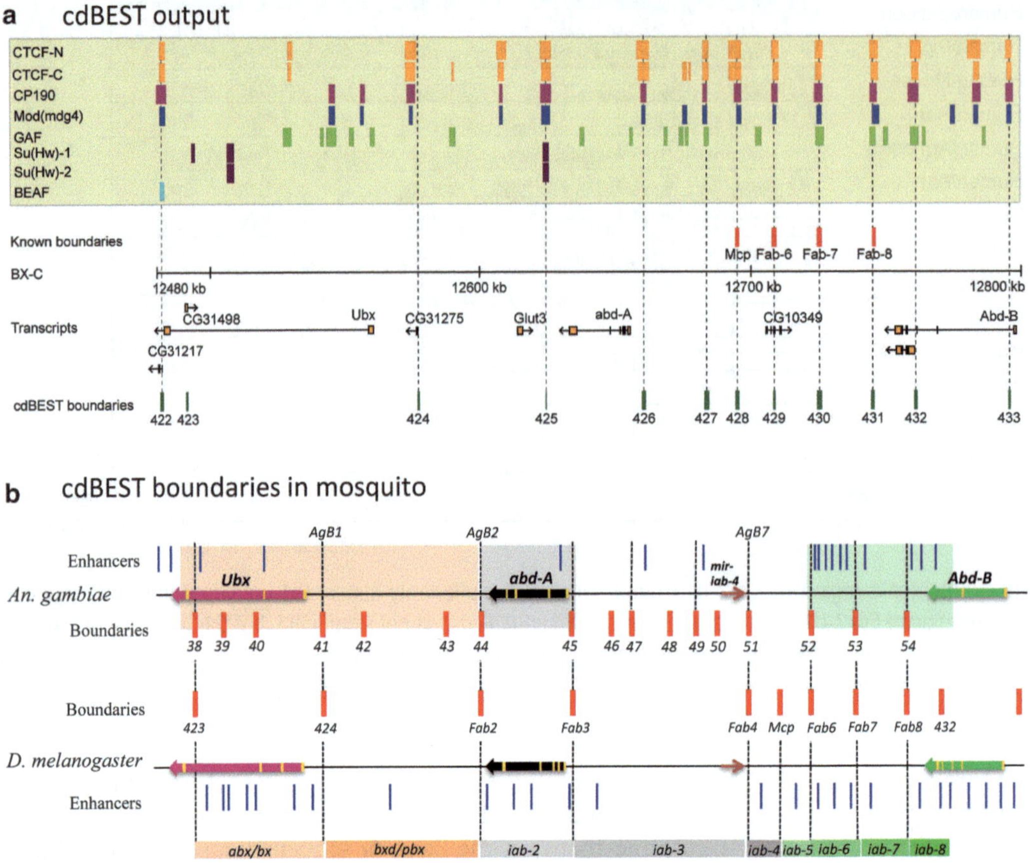

Fig. 6 Chromatin domain boundary element search tool (cdBEST) to identify insect boundary elements. (**a**) Boundary element prediction on the basis of enrichment of binding motifs for boundary factors in *Drosophila melanogaster* BX-C. All previously well-characterized boundaries were detected along with few new boundary elements in bithorax complex. (**b**) Boundary elements were detected in the mosquito Hox complex with the help of cdBEST. These boundary elements seem to punctuate the *iab*-like domains in the mosquito Hox complex (adapted from refs. 17, 18)

2. Searching for insulators and PREs in *Drosophila* can be done by recently developed cdBEST tool, which looks for the relative enrichment of boundary factor-associated protein motifs (Fig. 6a) [17]. This tool is not only useful in searching for putative boundary elements in *Drosophila* but also has been successfully used to mine boundary elements from the mosquito Hox cluster. These elements when tested in *Drosophila* show boundary activity. Interestingly, careful analysis shows that mosquito Hox cluster also has *Drosophila*-like iab domains demarcated by boundary elements (Fig. 6b) [18]. However these CREs do not show apparent sequence conservation but have "motif" conservation that conveys functional equivalence.

3. Transcription of noncoding regions is also a useful criterion to look for potential CREs. It is emerging from recent findings

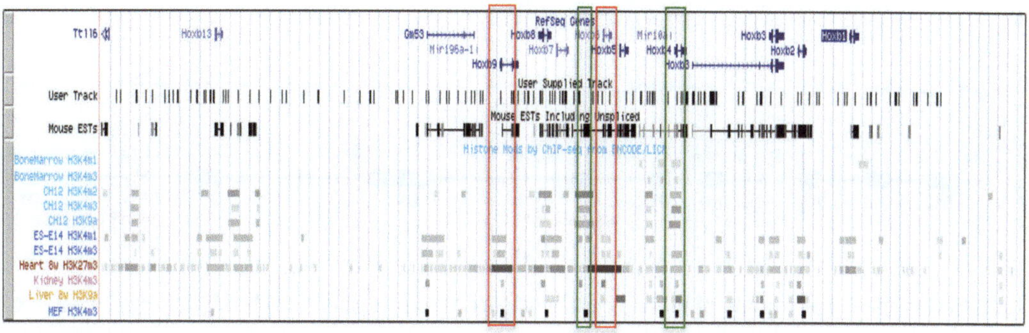

Fig. 7 Histone modification marks in mouse HoxB complex showing putative PRC2-binding and -enhancer regions. University of California, Santa Cruz (UCSC) genome browser window of mouse HoxD complex showing two H3K27me3-marked regions possibly repressed in the heart by PRC2 (*red boxes*), and two putative enhancers seen as H3K4me1-marked regions (*green boxes*) [20]

that many CREs work via noncoding RNA mediation; for example there are noncoding RNA transcription associated with enhancers and PREs [5–7]. These noncoding ESTs (expressed sequence tags) may show conserved sequence stretches across species. Transcription of these CREs may be a prerequisite for their regulatory activity and hence functional assays can be devised accordingly.

2.3 Mining CREs with Epigenetic Information

The genome of higher eukaryotes is encoded in the chromatin language in the form of histone modifications, which can be translated in expression states. The available information about the epigenetic state of genomes in different cell types is helpful when assessing the possible functionality of CREs. For example, H3K4me1 marks are significantly associated with enhancer regions and H3K27me3 is associated with the polycomb repressive complex 2 (PRC2) binding (Fig. 7) [19]. Accordingly, the CRE can be put into appropriate assay system.

1. Having checked for sequence conservation and factor-binding potential, the next yet parallel step is to annotate it with the existing epigenetic information. Overlaying the available genome-wide datasets on epigenetic modification across the genomes gives information about its putative function [4]. It may be a possibility that a noncoding region does not show any sequence conservation and at the same time factor enrichment study does not prove very helpful. It is worth looking for the epigenetic modifications present, and if these modifications are consistent in different species and in different tissues, then the confidence level of it being a CRE is higher.
2. Similarly the DNase-hypersensitivity or nucleosomal occupancy and DNA methylation are useful to look for. University of California, Santa Cruz (UCSC) genome browser is a very

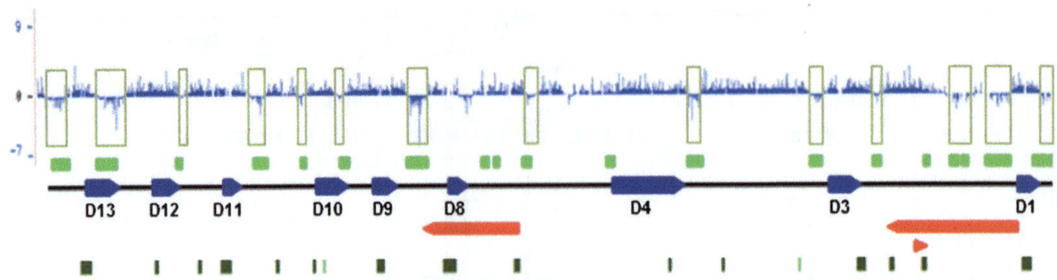

Fig. 8 Nucleosomal free regions of murine HoxD complex have boundary activity. Murine HoxD complex is evenly punctuated with relatively low H3 occupancy regions, shown as rectangles on the H3 profile and marked below as *green boxes*. These regions are highly associated with GAF (represented by *dark green boxes* at the *bottom*) and show enhancer-promoter blocking activity in cell-based assay system. These regions have conserved GAF sites (adapted from ref. 21). *Blue arrows* indicate hox genes and *red arrows* indicate noncoding RNA from this region

effective database in gathering all these information [20]. It is possible, however, that the desired genomic stretch or chromatin feature is not available in public databases. In such cases, the candidate region can be analyzed with experimental approaches to explore potential regulatory elements and to test them in functional assays. For example, a large number of cis-elements have been identified in murine Hox clusters by mapping nucleosome-free regions and testing boundary activity of selected subset of them in enhancer blocker assay (Fig. 8) [21].

2.4 Functional Assays for In Silico-Identified Potential CREs

1. Depending upon the information gathered about the epigenetic state of a CRE, it can be put into assay systems for an activator, repressor, enhancer, insulator, PRE/TRE.
2. If a CRE is transcribed but its function is not known, one way to investigate the functional relevance is to knock down the noncoding transcript in cell/time-specific manner and look for phenotype that may be related to the corresponding CRE.
3. Neighboring gene expression can be checked for differential gene expression in databases. In case of Hox gene expression it is likely that an intergenic CRE is needed for differential Hox gene expression. Developmental timing of Hox gene expression is of utmost significance. Many Hox-CREs bear stage-specific information, acting as early or late enhancers. Careful investigation of the dynamics and pattern of expression of associated gene may uncover functional features of the related CREs.

3 Concluding Remarks

In vivo assays have been developed to validate in appropriate systems regulatory elements that display diverse functions. Computational integration of diverse regulatory information opens avenues for the identification of a large number of CREs, especially for genomic loci that are well studied. One of the major challenges that lies ahead is to conceptualize novel regulatory elements and develop assays suited to study their including elements that help targeting loci to appropriate nuclear compartment.

Acknowledgements

Authors acknowledge financial support from Council of Scientific and Industrial Research (CSIR) through network programs BSC0208 and BSC0118.

References

1. Nobrega MA, Ovcharenko I, Afzal V, Rubin EM (2003) Scanning human gene deserts for long-range enhancers. Science 302(5644):413
2. Boffelli D, Nobrega MA, Rubin EM (2004) Comparative genomics at the vertebrate extremes. Nat Rev Genet 5(6):456–465
3. Lee AP, Koh EG, Tay A, Brenner S, Venkatesh B (2006) Highly conserved syntenic blocks at the vertebrate Hox loci and conserved regulatory elements within and outside Hox gene clusters. Proc Natl Acad Sci U S A 103(18): 6994–6999
4. Martin DI et al (2011) Phyloepigenomic comparison of great apes reveals a correlation between somatic and germline methylation states. Genome Res 21(12):2049–2057
5. Lai F et al (2013) Activating RNAs associate with Mediator to enhance chromatin architecture and transcription. Nature 494(7438):497–501
6. Li W et al (2013) Functional roles of enhancer RNAs for oestrogen-dependent transcriptional activation. Nature 498(7455):516–520
7. Hekimoglu B, Ringrose L (2009) Non-coding RNAs in polycomb/trithorax regulation. RNA Biol 6(2):129–137
8. Thomas JW et al (2003) Comparative analyses of multi-species sequences from targeted genomic regions. Nature 424(6950):788–793
9. Frasch M, Chen X, Lufkin T (1995) Evolutionary-conserved enhancers direct region-specific expression of the murine Hoxa-1 and Hoxa-2 loci in both mice and Drosophila. Development 121(4):957–974
10. Poulin F et al (2005) In vivo characterization of a vertebrate ultraconserved enhancer. Genomics 85(6):774–781
11. Woo CJ, Kharchenko PV, Daheron L, Park PJ, Kingston RE (2010) A region of the human HOXD cluster that confers polycomb-group responsiveness. Cell 140(1):99–110
12. Wang QF et al (2007) Detection of weakly conserved ancestral mammalian regulatory sequences by primate comparisons. Genome Biol 8(1):R1
13. Prabhakar S et al (2008) Human-specific gain of function in a developmental enhancer. Science 321(5894):1346–1350
14. Sabarinadh C, Subramanian S, Tripathi A, Mishra RK (2004) Extreme conservation of noncoding DNA near HoxD complex of vertebrates. BMC Genomics 5:75
15. Matharu NK, Hussain T, Sankaranarayanan R, Mishra RK (2010) Vertebrate homologue of Drosophila GAGA factor. J Mol Biol 400(3): 434–447
16. Vasanthi D, Anant M, Srivastava S, Mishra RK (2010) A functionally conserved boundary element from the mouse HoxD locus requires GAGA factor in Drosophila. Development 137(24):4239–4247
17. Srinivasan A, Mishra RK (2012) Chromatin domain boundary element search tool for Drosophila. Nucleic Acids Res 40(10): 4385–4395
18. Ahanger SH, Srinivasan A, Vasanthi D, Shouche YS, Mishra RK (2013) Conserved boundary elements from the Hox complex of mosquito, Anopheles gambiae. Nucleic Acids Res 41(2):804–816
19. Margueron R, Reinberg D (2011) The Polycomb complex PRC2 and its mark in life. Nature 469(7330):343–349
20. Karolchik D et al (2013) The UCSC Genome Browser database: 2014 update. Nucleic Acids Res 42(Database issue):D764–D770
21. Srivastava S, Puri D, Garapati HS, Dhawan J, Mishra RK (2013) Vertebrate GAGA factor associated insulator elements demarcate homeotic genes in the HOX clusters. Epigenetics Chromatin 6(1):8

Chapter 9

Functional Analysis of Hox Genes in Zebrafish

Franck Ladam and Charles G. Sagerström

Abstract

The zebrafish model organism is well suited to study the role of specific genes, such as *hox* genes, in embryogenesis and organ function. The ability to modulate the activity of *hox* genes in living zebrafish embryos represents a cornerstone of such functional analyses. In this chapter we outline the basic methodology for nucleic acid injections into 1–2-cell-stage zebrafish embryos. We also report variations in this method to allow injection of mRNA, DNA, and antisense oligonucleotides to either overexpress, knock down, or knock out specific genes in zebrafish embryos.

Key words *hox*, Zebrafish, Microinjection, Over-expression, Morpholino, Tol2, Transgenesis, Zinc-finger nuclease, TALE nuclease, CRISPR-Cas system

1 Introduction

hox genes act as regulators of transcription in development and disease and have been studied for decades using various model organisms or cell lines. The zebrafish (*Danio rerio*) expresses 48 Hox proteins and has been widely used to decipher their function during embryonic development. For instance, *hox* genes control anteroposterior and dorsoventral patterning of the embryo and are involved in more subtle events such as neuronal migration and specification (reviewed in [1]). The zebrafish is a useful model to study key developmental mechanisms as its embryos are easy to obtain, easy to grow, and transparent. Moreover, its genome is fully sequenced and 69 % of the genes have at least one human ortholog [2]. In this methods chapter we describe some of the techniques available to study *hox* gene function in zebrafish. The main method relies on the injection of nucleic acid into 1–2-cell-stage embryos (adapted from the zebrafish handbook [3]) to either over-express, knock down, or knock out specific *hox* genes during embryonic development. We describe (1) a general protocol for injection into 1–2-cell-stage embryos and (2) specific protocols to efficiently produce and inject nucleic acids in order to manipulate

Yacine Graba and René Rezsohazy (eds.), *Hox Genes: Methods and Protocols*, Methods in Molecular Biology, vol. 1196,
DOI 10.1007/978-1-4939-1242-1_9, © Springer Science+Business Media New York 2014

and study *hox* gene function in zebrafish. Injected embryos can be subsequently used for classical molecular biology, cellular biology, or biochemistry experiments (e.g., Western blot, in situ hybridization, immunostaining, chromatin immunoprecipitation, RT-PCR). Finally, even though the focus of this chapter is on *hox* gene function, all the described techniques are suitable for the study of any given gene.

2 Materials

2.1 General Laboratory Materials and Reagents

1. RNAse/DNAse-free DEPC-treated water.
2. 1.7 ml low-binding microcentrifuge tubes (Corning).
3. 50, 65, and 95 °C dry block heaters.

2.2 Fish Crosses, Embryo Injections, and Growth

1. Agarose.
2. 10 cm diameter plastic petri dishes.
3. Injection mold with six ramps, one 90° and one 45° beveled side (Adaptive Science Tools).
4. 0.5 % Red phenol solution (Sigma-Aldrich).
5. Borosilicate glass capillaries 1.0 OD × 0.5 mm ID/Fiber (FHC).
6. pc-10 needle puller (Narishige).
7. 1.0 l crossing tanks with dividers (Aquaneering).
8. Microloader pipet tips, for prefilling microinjection capillaries.
9. A pair of sharp stainless dissecting forceps.
10. MZ6 modular stereomicroscope with 6.3:1 zoom and eyepiece graticule (Leica) (Fig. 1).
11. Micro-manipulator M3301 (World Precision Instrument) (Fig. 1).

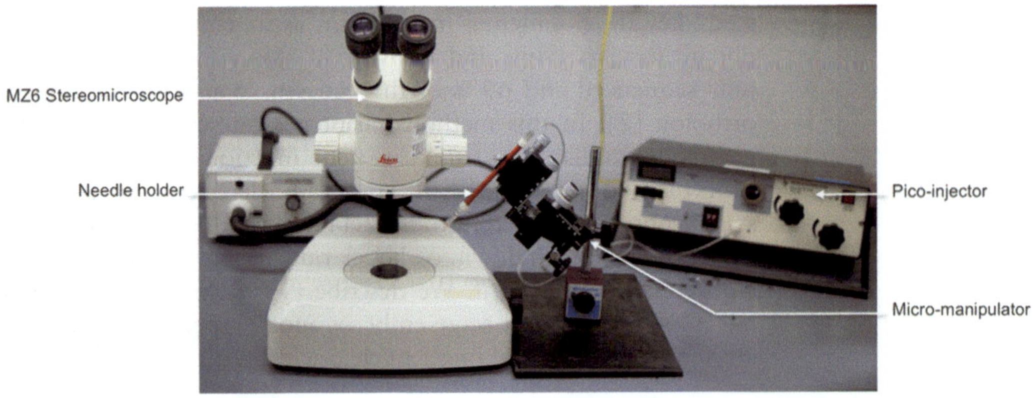

Fig. 1 Photograph of microinjection setup. Equipment required for nucleic acid injections into zebrafish embryos is labeled

12. Magnetic stand (World Precision Instrument) (Fig. 1).
13. PLI90 pneumatic pico-injector (Harvard Apparatus) (Fig. 1).
14. Pipet pump.
15. Mineral oil.
16. Plain microscope glass slides.
17. 146 mm Borosilicate glass Pasteur pipettes (Thermo Fisher Scientific).
18. Methylene blue water to grow zebrafish embryos: 0.0002 % methylene blue, 0.006 % instant ocean salts (Instant Ocean), milli-Q water.
19. Fine mesh strainer diameter 5 cm (Fante's).
20. 28.5 °C Lab-Line Imperial III incubator (Thermo Fisher Scientific).

2.3 mRNA Synthesis, MO Preparation

1. mMessage mMachine in vitro transcription kit (Ambion).
2. QIAquick PCR purification kit (Qiagen).
3. NanoDrop 3300 fluorospectrometer (Thermo Fisher Scientific).
4. RNA loading dye (Ambion).
5. 1× Tris–Borate–EDTA buffer: 220 mM Tris, 180 mM boric acid, 4 mM EDTA.

2.4 Tol2 Transgenesis

1. pTol2 kit containing Gateway-compatible vectors and the transposase-encoding vector PCS2FA-*transposase* [4, 5]. The kit is freely available from the Chien Lab [5].
2. Sterile Luria Broth bacteria medium: 25 g of powder in 1 l of milli Q water.
3. *NotI* restriction enzyme.
4. HiSpeed Plasmid Midi Kit (Qiagen).
5. 25× Tricaine solution: 15.3 mM tricaine (Sigma-Aldrich), 20 mM Tris–HCl pH 9.0.
6. Dissection spring scissor (Roboz).
7. 0.2 ml PCR tubes.
8. Lysis buffer: 10 mM Tris–HCl pH 7.5, 0.5 % sodium dodecyl sulfate, 100 μg/ml Proteinase K.
9. Phenol–chloroform–isoamyl alcohol solution.
10. 3 M Sodium acetate pH 5.2.
11. 70 and 100 % ethanol.
12. Chloroform solution.
13. One Taq Hot Start DNA polymerase (New England Biolabs).
14. 6× DNA loading dye (New England Biolabs).

3 Methods

3.1 General Protocol for Microinjection in Zebrafish

The following section describes general guidelines to achieve efficient injection into 1–2-cell-stage zebrafish embryos.

Steps 1–6 are performed on the day before injection.

1. Prepare the injection plates by pouring 15 ml of 1 % agarose melted in methylene blue water into a 10 cm plastic dish and set the injection mold on top of the hot agarose. Let the agarose solidify and remove the mold.
2. Adjust settings on the pc-10 needle puller (*see* **Note 1**).
3. Tightly set a borosilicate capillary on the needle puller. The center of the capillary should face the electric resistance filament to obtain two needles of equal size.
4. Press the start button. As the electric resistance filament warms up, one end of the capillary is pulled down to obtain two injection needles.
5. Prepare the solution to inject (*see* Subheadings 3.2–3.4) by mixing the desired amount of material (for instance 1 μg of mRNA), 1 μl of 0.5 % red phenol solution, and RNAse/DNAse-free water to a 10 μl final volume into a 1.7 ml microcentrifuge tube (*see* **Note 2**).
6. At the end of the day, place zebrafish at a ratio of two males for three females into the inner chamber of a crossing tank filled with fish water, using the divider to keep males and females separated. Leave the fish overnight at 28 °C in the dark (*see* **Note 3**).
7. The following morning as the light comes on, mix the males and females together in the inner chamber filled with fresh water. As eggs are laid and fertilized they will sink to the bottom of the outer chamber of the tank (*see* **Note 4**).
8. Using a p2 pipet and Microloader pipet tips, backload 2 μl of the injection solution (from **step 5**) into the injection needle.
9. Set the filled needle on the needle holder. Cut off the very tip of the needle using a thin pair of dissecting forceps, monitoring the process under the stereomicroscope (*see* **Note 5**).
10. Press the footswitch of the pico-injector and make sure that a drop of red liquid is forming at the tip of the needle.
11. Quantify the amount of material contained in a drop by injecting into 100 μl of mineral oil set on a microscope glass slide. Measure the radius of the drop with the calibrated graticule on the stereomicroscope. Use the formula $4/3\pi r^3$ to determine the volume of the sphere and therefore the volume injected. Using the volume of the drop and the original concentration of your solution, calculate the amount of material contained in one drop.
12. Adjust the pressure and duration parameters on the pico-injector to obtain the drop size you need for your experiment (Fig. 1).

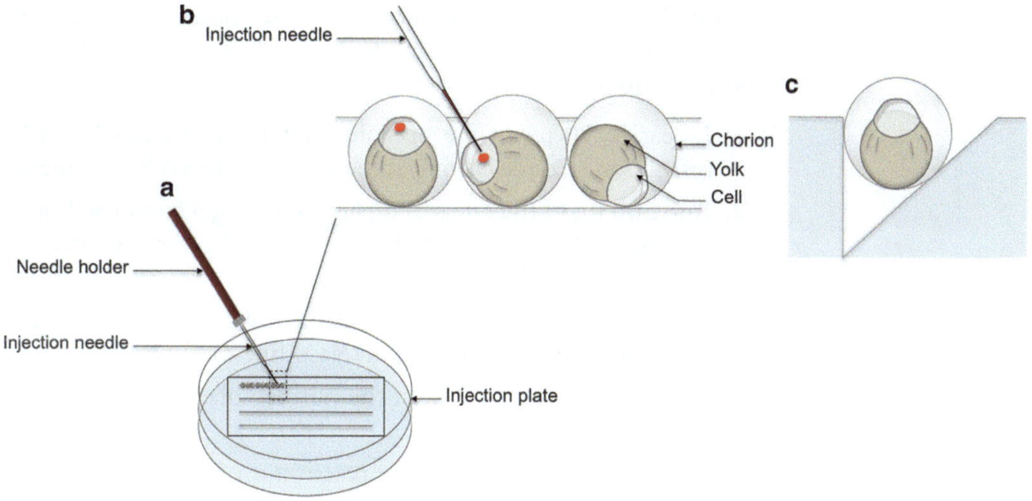

Fig. 2 Schematic illustrating injection into one-cell-stage zebrafish embryos. (**a**) Fertilized one-cell-stage embryos are placed on the agarose injection-mold ramps. (**b**) Close-up of zebrafish embryos showing the injection needle and the injected solution (*red dot*). Note that the needle needs to get through the chorion and the cell membrane before injecting the solution. (**c**) Side view of a one-cell-stage zebrafish embryo set on an agarose mold ramp

13. Collect the embryos into a 10 cm plastic dish using a mesh strainer and a squirt bottle filled with methylene blue egg water (Fig. 2, *see* **Note 6**).
14. Using a pipette, place the collected embryos on the injection plate and orient them on the ramps using forceps or a p200 pipet tip so that the cell is to the top (Fig. 2, *see* **Note 7**).
15. Carefully introduce the needle into the cell and inject the desired amount of material by pressing the footswitch. Make sure that the red stain remains in the cell when pulling the needle out of the embryo (Fig. 2).
16. Move the needle to the next embryo and repeat **step 15** to inject as many embryos as needed (*see* **Note 8**).
17. Place the injected embryos into a 10 cm plastic dish containing methylene blue water and incubate at 28 °C until reaching the desired developmental stage (*see* **Note 9**).

3.2 mRNA Injections

Exogenous gene expression in zebrafish can be accomplished by injecting 5′ 7-methyl guanosine capped messenger RNAs (mRNAs) directly into 1–2-cell-stage embryos. Capped mRNA in vitro transcription necessitates the prior cloning of the coding cDNA of interest into an expression vector allowing for in vitro transcription by RNA polymerases SP6, T3, or T7. Capped mRNAs mimic endogenous mRNAs and, unlike expression plasmids, allow the expression of the transgene throughout the embryo (*see* **Notes 10** and **11**). Injection of capped mRNA can also be used to express non-endogenous proteins such as those needed for genome engineering methods (e.g., zinc-finger nucleases and TALE nucleases, *see* Subheading 3.5).

1. Linearize 10 μg of the plasmid containing the cDNA with a restriction enzyme cutting at the 3′ end of the coding sequence.
2. Run 500 ng of the linearized plasmid on a gel composed of 1 % agarose melted in 1× Tris–borate–EDTA buffer and check for complete digestion.
3. Purify the linearized plasmid using the PCR purification kit and elute in 30 μl of RNase/DNase-free milli-Q water.
4. Quantify by loading 2 μl of the purified plasmid on the nanodrop spectrophotometer.
5. Run an in vitro transcription reaction on 2 μg of the digested DNA, following the mMessage mMachine in vitro transcription kit instructions.
6. Digest the DNA template by adding 2 μl of RNase-free DNase to the reaction (provided in the mMessage kit).
7. Purify the capped mRNAs using the RNAeasy RNA purification kit and elute in 30 μl of RNase/DNase-free water.
8. Quantify by loading 2 μl of the purified capped mRNA onto a nanodrop spectrophotometer.
9. Mix 300 ng of the capped mRNAs with water and RNA loading dye to a 1× final concentration, heat up to 65 °C for 5 min, and set on ice for another 5 min.
10. Load on a gel composed of 1 % agarose melted in 1× Tris–borate–EDTA buffer to assess the quality of the RNA sample (*see* **Note 12**).

3.3 Morpholino Antisense Oligonucleotide Injections

Morpholino antisense oligonucleotides (MOs) consist of nucleic acid bases bound to morpholine rings instead of deoxyribose rings [6]. They function by steric blockade of mRNA-dependent processes such as splicing and translation. MOs are usually synthesized to be complementary to mRNA translation start site or to exon–intron boundaries (*see* **Note 13**). MOs are very stable and easy to work with and sequences can be designed and synthesized at will by the Gene Tools company (Gene Tools, LLC). They are usually provided as a white dried powder.

1. Resuspend the MO powder in sterile milli-Q water to make a 3 mM stock solution (*see* **Note 14**).
2. Make a 1 mM MO solution ready for injections by mixing 3 μl of 3 mM MO, 1 μl 0.5 % red phenol dye, and 5 μl of milli-Q water in a 1.7 ml tube (*see* **Note 15**).
3. Heat up the MO solution to 65 °C for 5 min and set on ice for another 5 min.

3.4 DNA Injections for Transgenesis

As discussed earlier, mRNA injection into zebrafish embryos allows widespread, but transient, expression of a specific gene. However, detailed analyses of gene function may require driving expression in a time- and tissue-restricted fashion. This can be achieved through the expression of a transgene under the control of a tissue-specific promoter/enhancer region. Yet, plasmid injection generates low, transient, and mosaic expression of the transgene and is therefore often not suitable for gene studies. Thus, the following section describes a simple protocol to achieve efficient and stable transgene integration into the zebrafish genome using the Tol2 system. The Tol2 system allows the transposition into the genome of a transgene flanked by sequences (called 5′ and 3′ arms) derived from the Tol2 transposable element (originally identified in medaka fish) when co-injected with an mRNA expressing a transposase into 1–2-cell-stage zebrafish embryos [7–10]. This method efficiently integrates large DNA fragments and, in the case of *hox* genes, has been successfully used to insert a 70 kb DNA fragment containing the pufferfish (*Fugu rubripes*) *hoxAa* cluster (*hoxa5* through *hoxa13*) into the zebrafish genome [11].

3.4.1 Preparation of the pTol2-Transgene Vector

The plasmid to be injected should consist of the transgene cloned into the pTol2 vector (referred below as *pTol2-transgene* plasmid).

1. Purify the pTol2-transgene plasmid, from exponentially growing bacterial culture in Luria Broth medium, using the QiaQuick Midi prep Kit. Elute the plasmid in milli-Q water. Do not use the EB solution provided in the kit (*see* **Note 16**).
2. Quantify by loading 2 μl of the purified plasmid onto a nanodrop spectrophotometer.

3.4.2 Preparation of the Transposase mRNA

1. Digest the PCS2FA-*transposase* vector with the *NotI* restriction enzyme.
2. Purify the digested plasmid and in vitro transcribe the *transposase* mRNA as described in Subheading 3.2.

3.4.3 Injection into 1–2-Cell-Stage Embryos

1. Mix 1 μg of pTol2-transgene with 1 μg of the *transposase* mRNA, 1 μl of 0.5 % red phenol solution, and RNAse/DNAse-free water to a 10 μl final volume into a 1.7 ml microcentrifuge tube (*see* **Note 17**).
2. Inject into 1–2-cell-stage embryos.

3.4.4 Validation of Transgene Insertion

Validation of the transgene insertion into the genome can be achieved by polymerase chain reaction (PCR) amplification of purified zebrafish genomic DNA (*see* **Note 18**). This method is performed on injected embryos grown to the adult stage [12].

1. In a small plastic box, place two or three fish into tricaine diluted in fish water to a 1× concentration. Wait until the fish stop swimming, but gill movements persist.

2. Collect one fish with a plastic spoon and place it on a 10 cm plastic dish.
3. Hold the very tip of the caudal fin with a pair of forceps, clip a small piece of fin using a pair of dissection spring scissors, place in a 0.2 ml PCR tube, and proceed to the next fish (*see* **Note 19**).
4. Add 100 μl of lysis buffer and incubate for 4 h at 50 °C (*see* **Note 20**).
5. Vortex briefly.
6. Incubate the samples at 95 °C for 10 min to inactivate the Proteinase K contained in the lysis buffer.
7. Pipette 50 μl of the lysate into a new 1.7 ml tube and add 150 μl of milli-Q water.
8. Add 400 μl of phenol–chloroform–isoamyl alcohol solution and vortex vigorously.
9. Spin at 16,000 × *g* for 15 min at 4 °C.
10. Pipette the aqueous top layer into a new 1.7 ml tube, add 1 volume of chloroform, and vortex vigorously.
11. Spin at 16,000 × *g* for 5 min at room temperature.
12. Pipette the aqueous top layer into a new 1.7 ml tube.
13. Add 0.1 volume of 3 M pH 5.2 sodium acetate and 2 volumes of 100 % ethanol.
14. Place the samples at −20 °C for 30 min.
15. Centrifuge at 16,000 × *g* for 15 min at 4 °C.
16. Wash the DNA pellet with 1 ml of 70 % ethanol.
17. Centrifuge at 16,000 × *g* for 5 min at 4 °C.
18. Dry the pellet by leaving the tube open at room temperature for 5–10 min.
19. Resuspend the DNA pellet in 50 μl milli-Q water.
20. Run a PCR reaction by mixing the following in a 0.2 ml PCR tube (*see* **Note 21**):
    - 5 μl of 5× *One Taq* standard reaction buffer.
    - 0.5 μl of 10 mM dNTPs.
    - 0.5 μl of 10 μM forward primer.
    - 0.5 μl of 10 μM reverse primer.
    - 0.125 μl of *One Taq* Hot Start DNA polymerase.
    - 100 ng of purified genomic DNA.
    - Nuclease-free water to 25 μl.
21. Mix 15 μl of the PCR product with 3 μl of 6× DNA loading dye and load on a gel composed of 1 or 2 % agarose melted in 1× Tris–borate–EDTA buffer.

3.5 Genome Engineering Technologies

As mentioned previously, the use of MOs for loss-of-function experiments has various drawbacks. Within the past few years, three different methods—zinc-finger nucleases (ZFNs), TALE nucleases (TALENs), and the CRISPR-Cas system—have been employed to allow specific and stable disruption of zebrafish genes. These technologies involve ectopic expression, in the embryo, of a nuclease moiety targeted to the gene of interest by a sequence-specific recognition module (detailed descriptions of each technique are available elsewhere [13–15]). Gene disruption by ZFNs and TALENs is achieved by injecting mRNA (as in Subheading 3.2) encoding a chimeric protein composed of the FokI nuclease fused to the recognition module (a zinc finger or a TALE protein, respectively). Mutagenesis by the CRISPR-Cas system requires the injection of a guide RNA and mRNA encoding the Cas9 nuclease (as in Subheading 3.2). Identification of mutant founders is then accomplished using PCR and/or sequencing analogously to how transgenic founders are identified (Subheading 3.4.4). These techniques have been used in zebrafish to knock out genes by generating mutations (insertions or deletions) but also to introduce specific DNA sequences (knock-in) or induce chromosomal deletions or inversion [16–18].

4 Notes

1. Recommended settings are provided by the manufacturer [19]. In our experiments we routinely use the following settings: STEP1, N°1 heater set to 69.5 °C, tensile force corresponding to 4 weights.
2. The solution can be used right away or frozen at −80 °C for later use. Avoid freeze-thaw cycles.
3. Using the crossing tanks described here, a ratio of two males for three females gives good yields of embryos; however other ratios might be considered to increase the number of embryos.
4. It is important to keep the adults separated from the embryos, as adults will eventually feed on the embryos.
5. The size and shape of the needle tip is a critical parameter. It should be as thin as possible but still be able to get through the chorion.
6. Collect the embryos before the first mitotic division occurs (i.e., before 35–45 min post-fertilization).
7. You can also place the embryos so that the cell is to one side (Fig. 2). We do not recommend working with the cell under the yolk as this may result in clogging of the needle.
8. As you inject embryos keep a steady drop size by adjusting the pico-injector parameters.

9. Regularly remove dead embryos to avoid growth of bacteria and mold.

10. Plasmid DNA injection in zebrafish embryos results in mosaic expression of the transgene. Moreover, since protein synthesis requires transcription and translation inside the cells, the expression of the transgene takes longer than when mRNA is injected. For those reasons RNA injections are more suitable for transient overexpression of a specific gene in zebrafish embryos.

11. Injection of mRNA might induce nonspecific effects due to nonphysiological expression levels of the protein. Therefore, it is recommended to determine the optimal amount of mRNA to inject by doing a titration experiment and choosing the lowest concentration of mRNA that produces a specific phenotype. Moreover, we recommend injecting a nonfunctional protein-coding RNA as a control condition. Also, because the injection is made at the one-cell stage, the mRNA of interest may be expressed at places and times the endogenous mRNA is not. In the case of Hox proteins, this is of particular importance as their endogenous pattern and timing of expression are tightly regulated during development (e.g., the first *hox* gene (*hoxb1b*) is expressed 6 h post-fertilization and is restricted to the hindbrain and the most posterior part of the embryo [20–22]). Hence, misexpression of Hox proteins may lead to homeotic transformations [21–24], often resulting in defective embryonic development. Therefore other methods to study *hox* function should also be considered (*see* Subheadings 3.3 and 3.4).

12. While this does not allow for determining the actual size of the mRNA, it is a simple way to assess for RNA integrity and detect any potential DNA or RNA contaminants.

13. The use of MOs to study gene function in zebrafish has several drawbacks. (a) MOs may have nonspecific effects and several controls should be employed to confirm the biological significance of any MO-based phenotype [25]. First, one should consider the use of a nonfunctional MO. Such MOs usually have mutations introduced into the sequence of the original MO so that its target recognition is altered. Second, off-target effects may result in activation of a p53-dependent apoptotic response. Therefore, co-injection of the gene-specific MO with a p53 targeting MO should be used to determine if an apoptotic phenotype is p53 mediated. Third, a titration experiment should be done in order to determine the optimal amount of MO leading to a specific phenotype and minimal death of the embryos. Fourth, a rescue experiment should be performed by co-injecting the MO and the target mRNA (modified to be unaffected by the MO). (b) The efficiency of the knockdown is often hard to assess due to the lack of available antibodies to study protein expression in zebrafish. (c) The effect of MOs is transient. (d) MOs are comparatively expensive.

14. Keep at −20 °C.
15. MO concentration and volume can be modified according to your needs.
16. Quality of the plasmid is a critical parameter, as poor-quality DNA preparation leads to embryo death.
17. The relative amounts of pTol2-transgene vector and PCS2FA-transposase mRNA should be experimentally determined to obtain an efficient integration of the transgene into the genome.
18. Several copies of the transgene are often found on different chromosomes throughout the genome and the number of transgene copies therefore decreases as the transgenic fish are outbred to a wild-type background. Moreover, transgene insertion is random and may result in gene disruption. Thus, to avoid nonspecific effects and ensure steady transgene expression, one should consider a few recommendations: (a) outbreed the transgenic fish to a wild-type background through several generations; (b) determine the transgene copy number by southern blot [9]; (c) localize the transgene insertions in the genome [9]; and (d) raise and work with at least two different transgenic lines.
19. Do not cut too close to the body as you would hurt the fish. A 3 mm-by-3 mm piece of fin is usually enough to obtain sufficient amounts of DNA.
20. Samples can also be incubated overnight at 50 °C.
21. Optimal PCR parameters (i.e., annealing temperature, duration of elongation, number of cycles) depend upon the primers and the sequences amplified. Primers should allow specific amplification of the transgene. If the gene to be overexpressed is found endogenously in the genome one might consider using primers located within sequences specific to the Tol2 construct.

Acknowledgements

This work was supported by NIH grants NS038183 and HD065081 to CGS.

References

1. Alexander T, Nolte C, Krumlauf R (2009) Hox genes and segmentation of the hindbrain and axial skeleton. Annu Rev Cell Dev Biol 25:431–456
2. Howe K, Clark MD, Torroja CF et al (2013) The zebrafish reference genome sequence and its relationship to the human genome. Nature 496:498–503
3. Westerfield M (2007) The zebrafish book, 5th edition; a guide for the laboratory use of zebrafish (Danio rerio). University of Oregon Press, Eugene
4. Villefranc JA, Amigo J, Lawson ND (2007) Gateway compatible vectors for analysis of gene function in the zebrafish. Dev Dyn 236: 3077–3087

5. Kwan KM, Fujimoto E, Grabher C et al (2007) The Tol2kit: a multisite gateway-based construction kit for Tol2 transposon transgenesis constructs. Dev Dyn 236:3088–3099
6. Ekker SC, Larson JD (2001) Morphant technology in model developmental systems. Genesis 30:89–93
7. Kawakami K (2007) Tol2: a versatile gene transfer vector in vertebrates. Genome Biol 8(Suppl 1):S7
8. Suster ML, Abe G, Schouw A et al (2011) Transposon-mediated BAC transgenesis in zebrafish. Nat Protoc 6:1998–2021
9. Suster ML, Kikuta H, Urasaki A et al (2009) Transgenesis in zebrafish with the Tol2 transposon system. In: Cartwright EJ (ed) Methods in molecular biology, vol 561. Humana Press, Totowa, NJ, pp 41–63
10. Kikuta H, Kawakami K (2009) Transient and stable transgenesis using Tol2 transposon vectors. In: Lieschke GJ, Oates AC, Kawakami K. (eds) Methods in molecular biology, vol 546. Humana Press, Totowa, NJ, pp 69–84
11. Suster ML, Sumiyama K, Kawakami K (2009) Transposon-mediated BAC transgenesis in zebrafish and mice. BMC Genomics 10:477
12. Kimmel CB, Ballard WW, Kimmel SR et al (1995) Stages of embryonic development of the zebrafish. Dev Dyn 203:253–310
13. Urnov FD, Rebar EJ, Holmes MC et al (2010) Genome editing with engineered zinc finger nucleases. Nat Rev Genet 11:636–646
14. Gaj T, Gersbach CA, Barbas CF III (2013) ZFN, TALEN, and CRISPR/Cas-based methods for genome engineering. Trends Biotechnol 31:397–405
15. Hwang WY, Fu Y, Reyon D, Maeder ML et al (2013) Efficient genome editing in zebrafish using a CRISPR-Cas system. Nat Biotechnol 31:227–229
16. Xiao A, Wang Z, Hu Y et al (2013) Chromosomal deletions and inversions mediated by TALENs and CRISPR/Cas in zebrafish. Nucleic Acids Res. doi:10.1093/nar/gkt464
17. Gupta A, Hall VL, Kok FO et al (2013) Targeted chromosomal deletions and inversions in zebrafish. Genome Res 23:1008–1017
18. Bedell VM, Wang Y, Campbell JM et al (2012) In vivo genome editing using a high-efficiency TALEN system. Nature 491:114–118
19. Narishige Web News (2007) http://news.narishige-group.com/pdf/news001en.pdf. Accessed 4 Dec 2013
20. Alexandre D, Clarke JD, Oxtoby E et al (1996) Ectopic expression of Hoxa-1 in the zebrafish alters the fate of the mandibular arch neural crest and phenocopies a retinoic acid-induced phenotype. Development 122:735–746
21. Vlachakis N, Choe SK, Sagerström CG (2001) Meis3 synergizes with Pbx4 and Hoxb1b in promoting hindbrain fates in the zebrafish. Development 128:1299–1312
22. McClintock JM, Carlson R, Mann DM et al (2001) Consequences of Hox gene duplication in the vertebrates: an investigation of the zebrafish Hox paralogue group 1 genes. Development 128:2471–2484
23. Choe S-K, Zhang X, Hirsch N et al (2011) A screen for hoxb1-regulated genes identifies ppp1r14al as a regulator of the rhombomere 4 Fgf-signaling center. Dev Biol. doi:10.1016/j.ydbio.2011.05.676
24. Bruce AE, Oates AC, Prince VE et al (2001) Additional hox clusters in the zebrafish: divergent expression patterns belie equivalent activities of duplicate hoxB5 genes. Evol Dev 3:127–144
25. Gerety SS, Wilkinson DG (2011) Morpholino artifacts provide pitfalls and reveal a novel role for pro-apoptotic genes in hindbrain boundary development. Dev Biol 350:279–289

Chapter 10

Transgenesis in Non-model Organisms: The Case of *Parhyale*

Zacharias Kontarakis and Anastasios Pavlopoulos

Abstract

One of the most striking manifestations of Hox gene activity is the morphological and functional diversity of arthropod body plans, segments, and associated appendages. Among arthropod models, the amphipod crustacean *Parhyale hawaiensis* satisfies a number of appealing biological and technical requirements to study the Hox control of tissue and organ morphogenesis. *Parhyale* embryos undergo direct development from fertilized eggs into miniature adults within 10 days and are amenable to all sorts of embryological and functional genetic manipulations. Furthermore, each embryo develops a series of specialized appendages along the anterior–posterior body axis, offering exceptional material to probe the genetic basis of appendage patterning, growth, and differentiation. Here, we describe the methodologies and techniques required for transgenesis-based gain-of-function studies of Hox genes in *Parhyale* embryos. First, we introduce a protocol for efficient microinjection of early-stage *Parhyale* embryos. Second, we describe the application of fast and reliable assays to test the activity of the *Minos* DNA transposon in embryos. Third, we present the use of *Minos*-based transgenesis vectors to generate stable and transient transgenic *Parhyale*. Finally, we describe the development and application of a conditional heat-inducible misexpression system to study the role of the Hox gene *Ultrabithorax* in *Parhyale* appendage specialization. Beyond providing a useful resource for Parhyalists, this chapter also aims to provide a road map for researchers working on other emerging model organisms. Acknowledging the time and effort that need to be invested in developing transgenic approaches in new species, it is all worth it considering the wide scope of experimentation that opens up once transgenesis is established.

Key words Arthropods, Crustaceans, *Parhyale hawaiensis*, *Minos* transgenesis, Microinjections, Heat-shock promoter, Conditional gene misexpression, Hox genes, *Ultrabithorax*, Appendage development

1 Introduction

The enormous diversity of body plans, segments, and associated appendages is considered one of the cornerstones underlying the evolutionary success of the arthropod phylum, comprising chelicerates, myriapods, crustaceans, and insects [1]. Among these extant arthropod groups, crustaceans (barnacles, copepods, ostracods, shrimps, lobsters, crabs, and their kin) exhibit the most impressive

Yacine Graba and René Rezsohazy (eds.), *Hox Genes: Methods and Protocols*, Methods in Molecular Biology, vol. 1196, DOI 10.1007/978-1-4939-1242-1_10, © Springer Science+Business Media New York 2014

diversity in appendage morphology and function that is evident not only across crustacean species but also within the same species. Until recently, it has not been possible to explore the molecular and cellular basis of segmental specialization and appendage diversification in any crustacean species due to the lack of suitable experimental approaches. During the last decade, the amphipod *Parhyale hawaiensis* has emerged as the most powerful available crustacean model for developmental genetic and molecular cell biology studies [2, 3].

Considering the sister group relationship of crustaceans and insects, it is not surprising that *Parhyale* research has benefited enormously from the wealth of methodologies and knowledge available in insect models like *Drosophila melanogaster*, *Tribolium castaneum*, and others. In this respect, *Parhyale* studies are contributing to our understanding of how developmental mechanisms have changed or remained conserved over macroevolutionary time scales [4–8]. More importantly, *Parhyale* has proven not only suitable for comparative developmental studies but also a powerful model system in its own right. Current research demonstrates the strengths of *Parhyale* in addressing key processes in animal development, like germ layer specification, cell fate specification, head and central nervous system development, organ morphogenesis, and regeneration [9–14]. It should also be noted that amphipods exhibit diverse lifestyles and associated morphological and physiological adaptations. In addition to *Parhyale*, ongoing developmental studies in other amphipod species (e.g., *Orchestia cavimana* [15, 16], *Caprella scaura* [17], *Gammarus minus* [18]) hold great promise in establishing the Amphipoda as a group where one could study evolution on smaller time scales.

Parhyale is a marine amphipod crustacean with a worldwide tropical distribution living in shallow aquatic habitats [2, 3]. It was put forward as an attractive model organism by William Browne and Nipam Patel in the late 1990s. Since then, several labs in the USA and Europe have joined a growing *Parhyale* community, attracted by the easiness to grow and maintain this species in dense cultures, the relatively short generation time (2 months), and the accessibility of embryos at all stages of development all year round. *Parhyale* embryogenesis and early cell lineages have been described in detail [2, 19, 20], and an increasing number of experimental resources are being developed at a fast pace. *Parhyale* embryos can be subjected to various embryological manipulations, developmental genetic techniques, and molecular and cell biology approaches, including cell microinjection, isolation and ablation [10, 21, 22], cell lineage tracing [19, 20, 23], in situ hybridization and immunohistochemistry [24–26], RNA interference and morpholino-based gene knockdown [12, 27], transposon and integrase-mediated genetic transformation [11, 28], conditional gene misexpression [14], and live imaging using transmitted light or fluorescence microscopy ([10, 19, 29]; see also http://www.cell.com/pictureshow/lightsheet2). Transcriptomic and genomic resources have also been made available by high-throughput sequencing of BAC clones and cDNA libraries [30–32], as well as by ongoing efforts to sequence and assemble the relatively large *Parhyale* genome that is estimated at 3 Gb.

Parhyale is a sexually dimorphic species; sexually mature males are easily distinguishable from females by their enlarged grasping appendages on the third thoracic segment [2]. Males grasp and hold females for one or more days until mating occurs. The released females molt and oviposit 5–50 eggs (depending on the age) in a ventral brood pouch. Adults breed year-round every 2–3 weeks and can be set routinely in single crosses to generate inbred lines. The embryos in each brood develop almost synchronously and can be dissected from females at any stage and cultured in Petri dishes in artificial seawater. After 10–11 days of embryogenesis at 25–26 °C, the hatched juveniles resemble miniature adults since *Parhyale* is a direct developer. Hatchlings increase in size through successive molts and need about 7–8 weeks at 25–26 °C to reach sexual maturity.

The optical properties of *Parhyale* embryos allow detailed microscopic inspections of constituent cells with exceptional spatial and temporal resolution ([10, 19, 29]; see also http://www.cell.com/pictureshow/lightsheet2). Early cleavages of fertilized eggs are holoblastic [2, 20]. The first cleavage is slightly unequal and generates two blastomeres, each contributing to the ectodermal and mesodermal lineages of either the left or the right side of the animal. The second cleavage is also slightly unequal, while the third cleavage is highly unequal producing a stereotyped arrangement of four macromeres and four micromeres that are uniquely identifiable based on their relative position, size, and contacts. Already at the 8-cell stage, the fate of each blastomere is restricted to a single germ layer (although blastomeres exhibit some regulative capacity too [22]); three macromeres give rise to the ectoderm, the fourth macromere gives rise to the visceral and anterior mesoderm, two micromeres form the somatic mesoderm, one micromere forms the endoderm, and one micromere forms the germline [2, 20]. Later cell divisions segregate the superficial layers of the embryo from the yolk and cells aggregate to form the embryo rudiment.

Ectodermal cells in the growing embryo become organized into the head lobes anteriorly and into regular rows posteriorly, forming a grid-like pattern typical for amphipods and other malacostracan crustaceans [2, 20, 33]. Similar to *Drosophila* and other arthropods, *Parhyale* embryos exhibit initially a parasegmental organization; each row of cells in the grid corresponds to one parasegment [2, 34]. The parasegmental rows undergo stereotyped divisions that, together with the progressive addition of more parasegmental rows

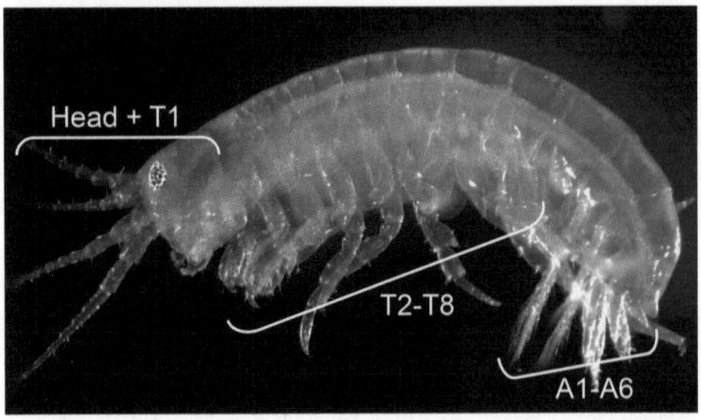

Fig. 1 The body plan of the crustacean amphipod *Parhyale hawaiensis*. The head is fused to the first thoracic segment (T1), the thorax is composed of seven segments (T2–T8), and the abdomen consists of six segments (A1–A6)

posteriorly, lead to the axial elongation of the embryo. Subsequent cell divisions disrupt the regularity of the grid and contribute to formation of segmental units and appendage bud outgrowths. *Parhyale* axial patterning and growth occur in an anterior-to-posterior progression; more anterior structures form and elaborate earlier than more posterior ones. As appendages extend and differentiate along their proximodistal axis, the yolk gets sequestered into the developing midgut, and the head region becomes clearly distinguishable from the trunk of the embryo. Towards the end of embryogenesis, the pigmented compound eyes form, the dorsal heart starts beating, and muscles start twitching shortly before hatching [2].

The segmented body of *Parhyale* consists of the head, the thorax, and the abdomen [2, 14] (Fig. 1). The head is composed of six segments bearing five pairs of sensory and feeding appendages (antennae 1, antennae 2, mandibles, maxillae 1, maxillae 2) and is fused to the first thoracic segment (T1) bearing a pair of segmented feeding appendages, known as maxillipeds. The next seven thoracic segments (T2–T8) develop larger segmented appendages; T2 and T3 bear subchelate grasping appendages, called gnathopods, and T4–T8 bear elongated walking appendages called pereopods. The abdomen is subdivided into two regions with three segments each: the pleon with three pairs of biramous paddling appendages (A1–A3 pleopods) and the urosome with three pairs of reduced thickened appendages (A4–A6 uropods). This striking specialization of *Parhyale* appendages along the anterior–posterior body axis offers exceptional material to study the molecular and cellular basis of organ patterning, growth, and morphogenesis.

This chapter introduces step by step all the methodologies and techniques that we established for transgenesis-based functional studies of Hox genes in *Parhyale* [14, 28]. First, we describe a robust and easily adaptable protocol for microinjection of early-stage *Parhyale* embryos.

Second, we describe the application of quick and sensitive extrachromosomal assays, known as excision and transposition assays, to examine whether the selected DNA transposon (in this case the *Minos* transposon) can be mobilized efficiently in a transposase-dependent manner in the cellular environment of interest (in this case in early-stage *Parhyale* embryos) [28, 35, 36]. The excision assay tests the ability of a *Minos* transposon to excise from a donor plasmid, when provided with a suitable source of transposase (in this case with the *Minos* transposase provided as mRNA or as plasmid). The transposition assay tests the ability of the *Minos* element to transpose by the cut-and-paste mechanism from its original site in the donor plasmid into a new site in a target plasmid in a transposase-dependent manner.

Third, we present the use of *Minos*-based transposon vectors for the generation of stable transgenic and transient transgenic *Parhyale* [28, 36]. Stable transgenesis (a.k.a. germline transformation) involves insertion of exogenous DNA constructs (transgenes) carried by the transposon into the germline of the injected animal (G0), so that transgenes can be transmitted down the generations. The offspring (G1) that will emerge from fertilization of a transformed gamete will be stably transformed (transgenic); all its somatic cells will carry the same insertion(s) of the transgene. The transformation efficiency is typically relatively low; only a small percentage of the injected G0s will acquire a small percentage of transformed gametes, and will give rise to a small percentage of transformed progeny. For this reason, *Minos* vectors are engineered to carry a transformation marker gene, in addition to the transgene of interest (Fig. 3b). Expression of the transformation marker allows the straightforward identification of transgenic individuals among a large number of untransformed animals. Nowadays, the most commonly used transformation markers in arthropod transgenesis, including *Parhyale*, combine fluorescent proteins with an artificial *cis*-regulatory element responsive to the Pax6 transcription factor, known as *3xP3* [28, 37 39].

In *Parhyale* transgenesis, expression of the transformation marker genes and of the transgenes is not only detected in transgenic animals (G1s, G2s, etc.) but also in a substantial fraction of the injected G0 animals [14, 28]. The ability to obtain genomic integration events of *Minos* in early-stage *Parhyale* embryos enables to produce G0 animals with very low levels of mosaicism (transient transgenesis). In such a transient transgenic animal, a large proportion (or even all) of the descendant cells from the injected blastomere carry the *Minos* insertion(s). It should be stressed here that transposition is a stochastic process; the stage and cell at which the injected transposon integrates into the genome, as well as the number and genomic loci of integration, vary dramatically among injected embryos. Therefore, transient transgenic *Parhyale* embryos can exhibit diverse patterns and levels of marker gene/transgene expression [14]. Despite this caveat, injections at the 1-cell stage can produce bilateral marker gene/transgene expression, while single blastomere injections at the 2-cell stage often result in unilateral expression of marker genes/transgenes expressed in the ectoderm and somatic mesoderm [28]. The prescreening of G0s for marker gene/transgene expression is an extremely useful feature in *Parhyale* transgenesis. It provides an early and accurate indication about the potential success or failure of the experiment, and also provides information about transgene expression and function, months before stable transgenics are available for analysis.

Fourth, we describe a gain-of-function approach based on conditional heat-inducible misexpression of transgenes, and the application of this system to study the role of the Hox gene *Ultrabithorax* in *Parhyale* appendage specialization [14]. This method allows to assess the function of the gene of interest by misexpressing it in cells and at developmental stages that normally do not experience its product. A number of other transgenesis-based functional approaches not discussed in this chapter have also been realized in *Parhyale*, including trapping of genes in unbiased genetic screens and trap conversion by targeted integration of new constructs into trapped loci [11].

2 Materials

2.1 Microinjection of Early-Stage Parhyale Embryos

1. Sea salt (Tropic Marin).
2. Flake food for tropical fish (TetraMin).
3. Crushed coral.
4. Air or water pumps.
5. Artificial seawater (ASW): Dissolve sea salt in purified water (about 34 g/l) to a specific gravity of about 1.022.
6. Filtered artificial seawater (FASW): ASW filtered through a 0.22 μm filter.
7. Filtered artificial seawater with antibiotics and antimycotics (FASWA): FASW with penicillin-streptomycin (diluted 1/100) and fungizone-amphotericin B (diluted 1/200).
8. Penicillin-Streptomycin (Gibco).
9. Fungizone-Amphotericin B (Gibco).
10. Blunt dissecting forceps.
11. Fine paintbrushes.
12. Micropipettes and tips.
13. Pasteur pipettes.
14. Collection baskets: Can be made from cut 50 ml Falcon tubes with a fine Nytex mesh sealed at one opening or bought ready-made (Corning).
15. Plain 150, 90, and 60 mm Petri dishes.

16. Tissue culture grade coated 60 and 35 mm Petri dishes (Nunc).
17. Clove oil (Sigma).
18. Bovine serum albumin.
19. Glass slides.
20. Agarose.
21. Ready-made microinjection needles (Eppendorf Femtotips).
22. Borosilicate glass capillaries 1 mm O.D.×0.58 mm I.D. with inner filament (Clark Electromedical Instruments).
23. Microloading pipette tips (Eppendorf).
24. Incubator set at 25–26 °C.
25. CO_2 station.
26. Flaming-Brown micropipette puller (Sutter Instrument Company).
27. Beveler (Narishige).
28. Injector (Narishige or Eppendorf).
29. Micromanipulator (Leica or Narishige).
30. Dissecting stereoscope with external light source.
31. Upright microscope.

2.2 Testing the Activity of DNA Transposons with Excision and Transposition Assays

1. Materials described in Subheading 2.1.
2. Nuclease-free ddH_2O.
3. Microcentrifuge tubes.
4. Filter micropipette tips.
5. Plasmid midi or maxi prep kit (Qiagen or Nucleobond).
6. NotI restriction endonuclease (NEB).
7. T7 mMESSAGE mMACHINE kit (Ambion).
8. 3 M sodium acetate solution pH 5.2.
9. Isopropanol.
10. Absolute ethanol.
11. 70 % ethanol RNase-free (mix absolute ethanol with DEPC-treated ddH_2O).
12. Phenol red solution (Sigma).
13. Holmes–Bonner solution: 100 mM Tris–HCl pH 7.5, 10 mM EDTA, 350 mM NaCl, 2 % SDS, 7 M urea (store at −20 °C, mix well before use).
14. Disposable RNase-free tubes and pestles (Kontes).
15. Phenol:chloroform:isoamyl alcohol 25:24:1.
16. Chloroform:isoamyl alcohol 24:1.
17. High-specificity/sensitivity *Taq* DNA polymerase and PCR buffer (Taq 2000 from Agilent or Platinum from Invitrogen).

18. Purified custom oligos.
19. 10 mM dNTP mix.
20. DNA molecular weight ladder.
21. pGEM-T Easy vector system (Promega).
22. High-efficiency *E. coli* electrocompetent cells (ElectroMAX DH5α-E cells from Invitrogen).
23. SOC medium.
24. LB medium.
25. LB + Cm plates: LB plates with 30 μg/ml chloramphenicol.
26. LB + Cm + Suc plates: LB plates with 30 μg/ml chloramphenicol and 10 % sucrose.
27. LB + Cm + Tet plates: LB plates with 30 μg/ml chloramphenicol and 12 μg/μl tetracycline.
28. Gel electrophoresis setup.
29. Benchtop cooling microcentrifuge.
30. Nanodrop spectrophotometer.
31. Standard microbiological equipment: Culture flasks and tubes, disposable sterile pipettes, 37 °C shaker and incubator, centrifuge with rotor and tubes.
32. PCR thermal cycler.

2.3 Transposon-Based Stable and Transient Transgenesis in *Parhyale*

1. Materials described in Subheadings 2.1 and 2.2.
2. Fluorescence stereoscope equipped with appropriate filter sets for detection of fluorescent proteins.
3. Equipment and reagents for Southern blot analysis, detailed in [28, 40, 41].
4. Equipment and reagents for inverse PCR, detailed in [40, 42, 43].

2.4 Conditional Heat-Inducible Misexpression of Hox Genes in Transient and Stable Transgenic *Parhyale*

1. Materials described in Subheadings 2.1–2.3.
2. Cloning reagents: PCR reagents, RACE kit, restriction enzymes, ligase, phosphatase.
3. Equipment and reagents for Northern blot analysis, detailed in [28, 40, 41].
4. RNA extraction reagent (TRIzol from Invitrogen) or kit (Qiagen).
5. Reverse transcription kit (SuperScript III from Invitrogen).
6. DNase I (amplification grade from Invitrogen).
7. Real-time PCR kit (SYBR Green I from Roche).
8. Equipment and reagents for in situ hybridization and antibody staining, detailed in [14, 24–26].

9. Glutaraldehyde.
10. Osmium tetroxide.
11. Formaldehyde.
12. 10× PBS pH 7.4: 18.6 mM NaH_2PO_4, 84.1 mM Na_2HPO_4, 1.75 M NaCl (store at room temperature).
13. Triton X-100.
14. Hoyer's medium/lactic acid (1:1) solution.
15. Sylgard 184 silicone elastomer kit (Dow Corning).
16. Fine dissecting forceps.
17. Incubator set at 37 °C.
18. Real-time PCR instrument.
19. Heating plate set at 60 °C.
20. Compound microscope.
21. Critical point dryer.
22. Sputter coater.
23. Scanning electron microscope.

3 Methods

3.1 Microinjection of Early-Stage Parhyale Embryos

3.1.1 Collection of Parhyale Embryos

Parhyale are maintained easily in dense cultures using standard aquarium equipment in plastic containers with a bottom layer of crushed coral covered in artificial seawater (2–3 l ASW in 10–20 l container). Large cultures with thousands of animals are maintained on standard fish flake food or other diets at 22–26 °C, are aerated with a submerged water or air pump, and are kept waste free with phosphate- and nitrate-absorbing bags and weekly to monthly water changes (for more details refer to [3] or visit http://www.extavourlab.com/protocols/Parhyale%20hawaiensis%20culture.pdf). A few of these cultures provide daily accessibility to hundreds of embryos at all stages of development all year round.

1. Adult *Parhyale* form mating pairs. The day before injecting, collect 50 or more pairs from the main cultures with a basket or by sucking them up using a Pasteur pipette with a large opening. Transfer pairs into 150 mm Petri dishes with ASW and few corals.
2. Many of the collected *Parhyale* pairs will have separated the day of injections. Gravid females with eggs in their ventral brood pouch are easily identifiable by eye. Transfer gravid females into a basket immersed in FASW using a Pasteur pipette or by picking up with blunt forceps the piece of coral they are sitting on.
3. Anaesthetize gravid females by bubbling CO_2 gas for 30–60 s into the filtered artificial sea water (FASW) or by transferring

the basket into FASW with 1:2,000 diluted clove oil. Wait for 1–2 min until completely still and transfer anaesthetized females with a paintbrush (or with a Pasteur pipette or by grabbing the females with blunt forceps from their antennae) into 60 mm tissue culture Petri dish in filtered artificial seawater with antibiotics and antimycotics (FASWA).

4. Hold each female gently on its back under a dissecting scope with one pair of blunt forceps. Place another pair of blunt forceps (or a blunted rounded end of a Pasteur pipette) between the embryos and the pouch, and nick the pouch along the ventral midline by lifting the forceps. Then place the forceps below the eggs and push them gently out of the nicked pouch. Continue with the rest of anaesthetized females, collecting all embryos into the 60 mm tissue culture Petri dish in FASWA. Transfer dissected females into another Petri dish with ASW, wait until they are fully awake and mobile, and return them into the main *Parhyale* cultures.
5. Sort the collected embryos under the dissecting scope according to their developmental stage [2].
6. Repeat **steps 2–5** every 4 h. This way you can collect a large number of 1-cell-stage embryos and stage them according to experimental requirements (*see* **Note 1**).

3.1.2 Preparation of Needles for Microinjection

The quality of microneedles is one the most important parameters for successful microinjections. Commercially available microneedles are well suited for microinjection of *Parhyale* embryos, but are relatively expensive. Alternatively, microneedles can be prepared from glass capillaries on a Flaming-Brown needle puller, and then beveled on a rotating microbeveler (*see* **Note 2**).

1. We prepare microneedles from borosilicate glass capillaries on a Sutter P-87 puller with a box filament using the following settings: heat 850, pull 10, velocity 150, time 250, pressure 680. These parameters need to be adjusted for each puller, filament, and capillary type used (*see* **Note 2**).
2. For each pull cycle, mount a glass capillary onto the puller. Start the program to turn on heating of the filament. The glass heats up and a weak pull draws the glass out until it reaches the programmed velocity. The heat turns off and after a time delay the hard pull is executed. A puff of air is delivered at a certain pressure and period of time to quickly cool the glass. Remove needles from the puller and store in a 150 mm Petri dish on a stripe of plasticine.
3. Pulled needles are then beveled to reproducibly sculpt pointy tips for smooth eggshell penetration. Alternatively, the tip can be broken off randomly by touching it against a glass surface or by cutting it with scalpel or forceps.

4. To bevel needles, place the microbeveler under a dissecting scope and focus on the rotating platform. Clean the dust of the rotating platform by wiping with ethanol and spraying pressurized air. Load a microneedle at an angle of 30°. Use the beveler's controller to move the needle down until the tip just touches the platform. Move a bit further down so that the needle is just slightly bent. Leave beveling for at least 30 min before using the needle for microinjection.

3.1.3 Preparation of Agarose Steps

Parhyale eggs are extremely sensitive to desiccation and should be kept wet during microinjection. Placing the embryos on agarose steps provides a convenient solution for immobilizing the embryos while keeping them wet (Fig. 2a).

1. Prepare the template for the agarose steps by sticking two glass slides together with tape, so that their long edges protrude 1 mm. Place slides into a 90 mm Petri dish.

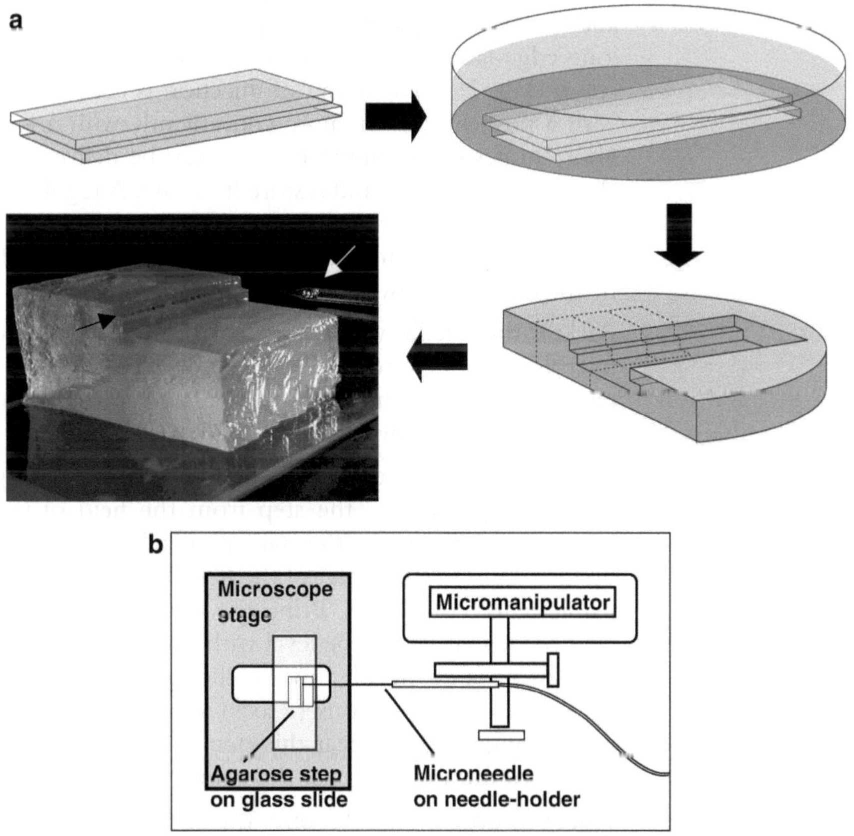

Fig. 2 Setup for *Parhyale* embryo microinjection. (**a**) Preparation of custom-made agarose steps as described in Subheading 3.1.3. *Parhyale* embryos (*black arrow*) are kept wet on the agarose step that faces the microneedle coming from the side (*white arrow*). (**b**) The *Parhyale* microinjectory is composed of a microscope focused on the agarose step with the embryos, a micromanipulator controlling the movement of the microneedle through a needle holder, and an injector (not shown) that delivers small volumes of the injection mix through the microneedle

2. Dissolve 2 % agarose in FASW by bringing to boil under continuous stirring. Let it cool down to 40–50 °C and pour 30 ml into Petri dish with slides.
3. Wait until agarose has solidified and carefully detach the slides from the agarose. Remove unwanted agar with a scalpel keeping the part that has formed a narrow step.
4. Agarose steps can be stored at 4 °C covered in FASWA.

3.1.4 Microinjections

1. The *Parhyale* microinjectory is composed of an injector, a micromanipulator, and an upright microscope equipped with a 5× or a 10× dry objective (an inverted microscope or dissecting scope could do as well) (Fig. 2b). The injector allows small liquid volumes to be delivered precisely through the microneedle by applying a regulated pressure for a certain period of time. The pressure is supplied from a compressed gas cylinder containing air or nitrogen. The microneedle is mounted on a needle holder, the movement of which is controlled by a micromanipulator with three knobs or a joystick to move the needle in the *x*-, *y*-, and *z*-axes. The agarose step is placed on a glass slide so that its long edge faces the needle coming from the side of the microscope (Fig. 2a). Triggering of injection is accomplished either with a push button or more conveniently with a foot switch. Short surges of maximum pressure can be delivered to clear a clogged microneedle and restore flow rate. A regulated holding pressure (balance pressure) is applied to the microneedle in between injections to prevent dilution of the injected material by the inflow of seawater due to capillary forces.
2. Backfill the needle with 2–3 μl of the injection mix using a microloading pipette tip. Mount the filled needle onto the needle holder and the micromanipulator at a small downward angle or horizontally (*see* **Note 3**).
3. Bring the agarose step into focus using the microscope focusing knobs. Withdraw the step from the field of view (away from the needle) using the stage controllers. Bring the needle tip into focus in the center of the field of view using the micromanipulator controllers. Bring the agarose step back into the field of view; it should be level with the tip of the needle. The movement of the needle towards the egg (*x*-axis) and all minor up and down corrections (*z*-axis) should be done using the micromanipulator. Repeat this step whenever the eggs or the needle go out of focus over the course of injections.
4. Pipet one drop of FASWA onto the agarose step. The tip of the needle should be covered with FASWA. Looking through the eyepieces, apply maximum pressure to the needle. Repeat a few times until the injection mix flows out of the needle. Keeping the needle tip underwater, adjust the holding pressure (balance) to prevent inflow of seawater into the needle. A slight outflow of the injection mix is acceptable.

5. Remove the needle tip out of FASWA (with *x*-axis micromanipulator controller) and adjust the injection pressure or time of each pulse so that it delivers the right amount to be injected (about 100 pl in the case of early-stage *Parhyale* embryos; *see* **Note 4**). Remove the needle tip from FASWA only for a short while to check the flow rate. Always return and keep it underwater to avoid clogging.
6. Cut the end of a pipette tip and fit it onto a micropipette. Coat the plastic tip by sucking up a solution of bovine serum albumin (BSA) to prevent eggs from sticking to the plastic. Use the micropipette with the coated tip to transfer the embryos to be injected from the Petri dish onto the agarose step.
7. Suck up 2–10 embryos (depending on the experience) in 3–4 μl of FASWA and pipet them onto the agarose step. If required, arrange the embryos on the step one next to the other with a fine paintbrush.
8. Use the stage controllers to move the agarose step sideways to center each egg for injection. Center the first egg and move the needle tip onto its middle with the *x*-axis micromanipulator controller. The eggshell should retract slightly and then expand again engulfing the needle tip. Apply the injection pressure while staying close to the egg cortex, and then withdraw needle from the embryo with the *x*-axis micromanipulator controller. The water surface tension will hold the embryo on the agarose step.
9. Center the next embryo with the stage controllers and proceed as described in the previous step. To correct the contact point of the needle tip on the egg, move the needle up (for a higher contact point) or down (for a lower contact point) using the *z*-axis controller of the micromanipulator. Do not forget to check the flow rate frequently as described in **step 5**. When finished, use a fine paintbrush to transfer embryos to a 35 mm tissue culture Petri dish in FASWA.
10. Repeat **steps 7–9** for all embryos. Transfer about 30 injected embryos in each 35 mm coated Petri dish and incubate them at 25–26 °C. Surviving injected embryos should be transferred every second day to a new 35 mm tissue culture dish with FASWA.
11. Under optimal microinjection conditions, at least 30 % of injected *Parhyale* embryos should hatch 10–13 days later.

3.2 Testing the Activity of DNA Transposons with Excision and Transposition Assays

Both activity assays involve co-injecting a donor plasmid (carrying the *Minos* transposon) and a target plasmid—with or without a source of transposase—into early-stage *Parhyale* embryos, incubating these for 24 h and extracting total nucleic acids. The excision assay is based on PCR reactions to determine whether *Minos* has excised from the donor plasmid using primers flanking the

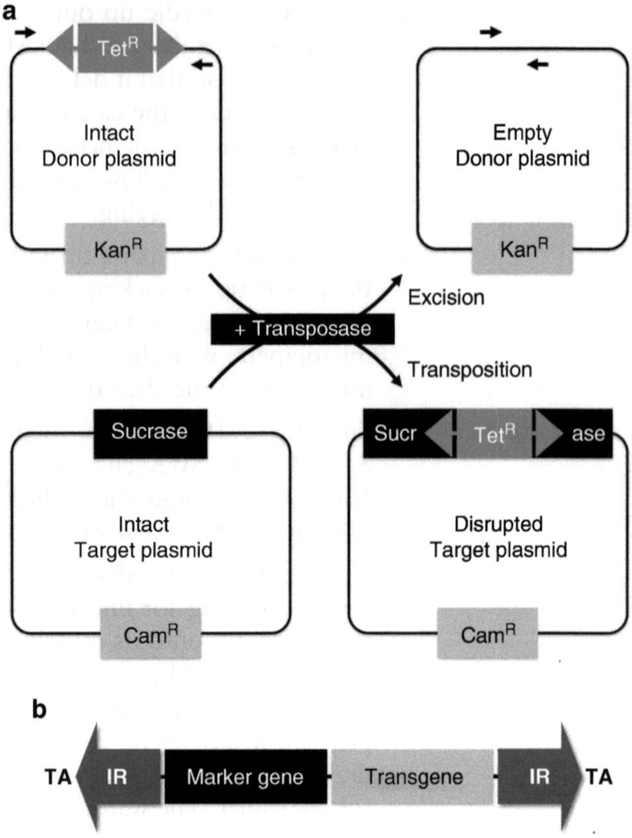

Fig. 3 Overview of activity assays and transgenesis vectors based on the *Minos* DNA transposon. (**a**) The excision and transposition assays test the ability of the *Minos* transposon to excise from a donor plasmid, when provided with a source of *Minos* transposase, and transpose by the cut-and-paste mechanism into a target plasmid. Please refer to Subheading 3.2 for more details. (**b**) Schematic representation of engineered transgenesis vectors: the *Minos* inverted repeats (IR) flank the transgene of interest and a transformation marker gene that allows identification of transgenic individuals. *Minos* transposition occurs exclusively into a TA target dinucleotide that is duplicated upon insertion on either side of the inverted repeats

transposon on the donor plasmid (Fig. 3a). The transposition assay involves transforming bacteria with the recovered plasmids and screening these for the presence of target plasmids containing the transposed *Minos* element based on bacterial marker selection (Fig. 3a).

3.2.1 Preparation of Plasmid DNA

The quality of injected plasmid DNA is important for efficient *Minos* mobility in transgenic experiments and in excision and transposition assays. The majority of purified plasmid DNA molecules should be in the supercoiled configuration. Plasmids prepared on a medium or a large scale from commercially available ion-exchange columns are used routinely in our laboratories.

1. Extract plasmid DNA from 50 to 500 ml of bacterial culture according to the manufacturer's instructions and dissolve purified plasmid DNA in nuclease-free water at a concentration of at least 1 μg/μl.
2. Centrifuge dissolved plasmid DNA at >12,000 × *g* for 30 min at 4 °C to precipitate any insoluble particles that might clog the microneedle.
3. Pipet plasmid DNA into a new tube and store at −20 °C. Plasmids used in multiple rounds of injections over several days should be stored in aliquots to avoid multiple freeze-thaw cycles.

3.2.2 Preparation of Capped mRNA

In vitro synthesized capped mRNA encoding the *Minos* transposase (helper mRNA) is our preferred transient source of transposase in *Parhyale* transgenic experiments and in excision/transposition assays (*see* **Note 5**). The plasmid template pBlueSK-MimRNA contains a T7 promoter driving the expression of the *Minos* transposase coding sequence with 5′ and 3′ UTR flanking sequences from the *Drosophila hsp70* and *inflated* (αPS2 integrin) genes, respectively [43, 44].

1. Set up a digest in a tube at 100 μl with the NotI restriction enzyme to linearize 5–10 μg of the pBlueSKMimRNA plasmid template. After 2 h of incubation, analyze 500 ng by 1 % agarose gel electrophoresis (next to uncut vector) to confirm complete linearization.
2. Extract linearized pBlueSKMimRNA once with equal volume (100 μl) phenol–chloroform–isoamyl alcohol and once with chloroform–isoamyl alcohol. Add 1/10th volume (10 μl) sodium acetate solution and two volumes (200 μl) ice-cold absolute ethanol. Incubate at −80 °C for at least 30 min and precipitate linearized plasmid by centrifuging at >12,000 × *g* for 20 min at 4 °C.
3. Discard supernatant and wash pellet with RNase-free 70 % ethanol. Air-dry pellet for 2–3 min until it becomes completely transparent and dissolve it in 10 μl nuclease-free water. Quantify concentration on a nanodrop spectrophotometer and use as template for capped mRNA synthesis using the T7 mMESSAGE mMACHINE kit (Ambion).
4. In an RNase-free tube pipet 10 μl 2× NTP/CAP mix, 2 μl 10× T7 reaction buffer, 1 μg linearized plasmid, and 2 μl T7 enzyme mix, and add nuclease-free water to a final reaction volume of 20 μl. Incubate for 2 h at 37 °C. Remove template plasmid by adding

1 μl RNase-free DNase I to the reaction and incubating for 15 more minutes at 37 °C.

5. Stop reaction by adding 115 μl nuclease-free water and 15 μl ammonium acetate stop solution. Extract once with equal volume (150 μl) phenol–chloroform–isoamyl alcohol and once with chloroform–isoamyl alcohol. Add equal volume (150 μl) isopropanol, mix well, make aliquots depending on the use (about 20–30 μl), and store at −20 °C. In our experience, each in vitro transcription reaction yields about 30 μg of capped *Minos* transposase mRNA.
6. Just before use in microinjections, precipitate mRNA by spinning at >12,000 × *g* for 20 min at 4 °C. Discard supernatant and wash pellet with RNase-free 70 % ethanol. Air-dry pellet until it becomes completely transparent and dissolve it in about 5 μl nuclease-free water (actual volume depends on the yield and final concentration needed in downstream application). Quantify on a nanodrop spectrophotometer.

3.2.3 Preparation of the Microinjection Mix

Prepare 10–20 μl of the injection mix just before microinjection. Special care should be taken in the preparation of the mix to avoid clogging of the fine needle tip.

1. Donor plasmids carrying the *Minos* transposons are injected at a concentration never exceeding 1 μg/μl, usually between 300 and 500 ng/μl.
2. The mRNA or the plasmid DNA encoding the *Minos* transposase (referred to as helper mRNA or helper plasmid, respectively) are usually injected at a ratio 1:2 to 1:5 relative to the donor plasmid, i.e., at a final concentration of 100–300 ng/μl.
3. In the case of *Parhyale*, the different injection components are mixed together in water. All injection mixes also include the inert dye phenol red (1/10 dilution of stock), which allows better visualization of the injected material.
4. Centrifuge injection mix at >12,000 × *g* for 20 min at 4 °C to precipitate any insoluble material that might clog the needle, and keep tube on ice throughout microinjections. Use 2–3 μl of the injection mix to backfill the needle using a microloading pipette tip fitted onto a micropipette.

3.2.4 Microinjections and Nucleic Acid Extraction

1. Collect 1- to 16-cell-stage *Parhyale* embryos and proceed with microinjections as described in Subheading 3.1.
2. Inject pools of at least 50 embryos with each one of the following mixes (*see* **Note 6**):

# Mix	Donor plasmid pMiLRTetR(L)	Transposase source	Target plasmid pBC/SacRB
1	–	–	–
2	150 ng/μl	–	300 ng/μl
3	150 ng/μl	mRNA (75 ng/μl)	300 ng/μl
4	150 ng/μl	mRNA (150 ng/μl)	300 ng/μl
5	150 ng/μl	mRNA (300 ng/μl)	300 ng/μl
6	150 ng/μl	Plasmid (300 ng/μl)	300 ng/μl

3. Incubate pools of injected embryos for 1 day at 25–26 °C.
4. Transfer each pool of embryos into a microcentrifuge tube with a micropipette. Remove excess seawater and proceed with the next step or flash freeze in liquid nitrogen and store at −80 °C.
5. In each tube, add 100 μl of Holmes–Bonner solution and homogenize with a pestle for 1 min.
6. Add 100 μl more Holmes–Bonner solution and extract with 200 μl of phenol–chloroform–isoamyl alcohol, mixing gently on a rotating platform for 10 min.
7. Centrifuge at >12,000 × *g* for 5 min at room temperature. Transfer upper aqueous phase to a new tube.
8. Repeat extractions (**steps 6**, 7), twice with phenol–chloroform–isoamyl alcohol and twice with chloroform–isoamyl alcohol.
9. Add 1/10th volume (20 μl) sodium acetate solution and two volumes (400 μl) ice-cold absolute ethanol. Incubate at −80 °C for at least 30 min and precipitate nucleic acids by spinning at >12,000 × *g* for 20 min at 4 °C.
10. Discard supernatant and wash pellet twice with 70 % ethanol.
11. Air-dry pellet for 2–3 min until it becomes completely transparent and dissolve it in 10 μl of nuclease-free water.
12. Quantify on a nanodrop spectrophotometer and analyze 1 μl from each sample by 1 % agarose gel electrophoresis. The genomic DNA and ribosomal RNA bands should be visible on the gel.

3.2.5 *Minos* Excision Assay

1. For the excision assay, use an equal amount of extracted nucleic acids (about 20 ng) from each sample as template in PCR reaction with a high-specificity/sensitivity Taq DNA polymerase according to the manufacturer's instructions with primers MiR-hydei (5′-TGCATTCTCTATGCT-3′) and MiL-Lorist (5′-CCAGCTGGCTTATCGAAA-3′) [35]. For example, set

50 μl reactions in 1× Taq buffer, 1.25 Units Taq, 200 μM each dNTP, 0.5 μM each primer, and 1.5–2 mM $MgCl_2$ using the following cycling program.

Initial denaturation:	98 °C for 2 min	
Main cycling program:	Denaturation	98 °C for 30 s
(35 cycles)	Annealing	55 °C for 30 s
	Extension	72 °C for 60 s
Final extension:		72 °C for 5 min

2. Analyze 10 μl of each PCR reaction by 1 % agarose gel electrophoresis next to a DNA molecular weight ladder.
3. Amplification of the non-excised *Minos* transposon from intact pMiLRTetR(L) donors produces a 2.2 kb band, whereas amplification from "empty" donors after *Minos* excision produces a 211 bp band (*see* **Note 7**). The relative abundance of the 2.2 kb and 211 bp bands between samples provides a qualitative assessment of *Minos* excision activity between conditions assayed [28, 35] (*see* **Note 8**).
4. To verify the specificity of *Minos* excision, gel-purify the 211 bp band, clone it in a T-Vector, and sequence individual clones. Transposase-mediated excision leaves behind characteristic footprints in empty donor plasmids consisting of the four terminal nucleotides of either end of the *Minos* transposon flanked by the duplicated TA target site (TAcgagTA or TActcgTA; [35, 45]).

3.2.6 *Minos* Transposition Assay

1. For the transposition assay, use an equal amount of extracted nucleic acids from each sample (about 50 ng) to transform high-efficiency *E. coli*-competent bacteria by electroporation according to the manufacturer's instructions.
2. Transfer each electroporated bacterial suspension in 1 ml SOC medium in a 15 ml snap-cap tube and shake for 1 h at 225 rpm at 37 °C.
3. Spread 5 % of cells (50 μl) on LB + Cm plates and the rest 95 % of cells (950 μl) on LB + Cm + Suc plates (*see* **Note 9**). Incubate plates at 37 °C until colonies reach a diameter of 1–2 mm (about 16 h).
4. Make replica patches of chloramphenicol and sucrose-resistant colonies grown on LB + Cm + Suc plates onto LB + Cm + Tet replica plates and grow overnight at 37 °C (*see* **Note 9**).
5. Count the number of colonies recovered on LB + Cm plates, and the number of chloramphenicol, tetracycline, and sucrose triple-resistant clones grown on the LB + Cm + Tet replica plates.

6. For each injected mix, calculate transposition efficiency as the percentage of target plasmids disrupted by *Minos* insertion. In this calculation, divide the number of chloramphenicol, tetracycline, and sucrose triple-resistant clones (disrupted targets) with the number of colonies grown on LB + Cm multiplied by 19 (total number of target plasmids assayed) (*see* **Note 10**).
7. Inoculate liquid LB + Cm + Tet cultures and extract plasmid DNA on a small scale from a subset of chloramphenicol, tetracycline, and sucrose triple-resistant clones that are considered independent interplasmid transposition events. Validate clones by digesting with the NotI restriction enzyme 0.5–1 μg of each extracted plasmid DNA (*see* **Note 11**). Check by 1 % agarose gel electrophoresis for the presence of two diagnostic bands with a total size of about 8 kb (6 kb pBC/SacRB + 2 kb *Minos* transposon).
8. Verify the specificity of *Minos* transposition by determining the insertion site of the *Minos* transposon in each disrupted target plasmid. Sequence the nucleotides flanking the *Minos* left and right inverted repeats with primers 309_reverse (5′-GATTCCGT TACATTAGTTGC-3′) and 1500_forward (5′-TAAGTATGATA GTAAATCAC-3′), respectively [35]. *Minos* transposition occurs exclusively into a TA target dinucleotide that is duplicated upon insertion and should be found on either side of the sequenced inverted repeats (Fig. 3b) [35, 45].

3.3 Transposon-Based Stable and Transient Transgenesis in Parhyale

The methodologies described in this Subheading apply to any *Minos*-based construct that needs to be inserted into the *Parhyale* genome. However, the original demonstration of germline transformation in a new species of interest is typically done with a simple vector made of the transformation marker flanked by the transposon's inverted repeats. For this reason, we describe here mainly the use of the donor plasmid pMi{3xP3-DsRed} [28]. This plasmid contains the *Minos*{*3xP3-DsRed*} transposon, in which the *Minos* inverted repeats flank the *3xP3-DsRed* transformation marker (*see* **Note 12**). Many more *Minos*-based vectors are available containing other fluorescent proteins under *3xP3* control, like pMi{3xP3-mTFP1}, pMi{3xP3-EYFP}, pMi{3xP3-EGFP}, and others [36]. As described in Subheading 3.4.2, all these *Minos* vectors have unique restrictions sites for cloning the transgenes to be delivered into the *Parhyale* genome.

1. Prepare the donor plasmid pMi{3xP3-DsRed} and *Minos* transposase helper mRNA according to Subheadings 3.2.1 and 3.2.2, respectively.
2. Prepare 10–20 μl of the injection mix containing 500 ng/μl (or 300 ng/μl) of the donor plasmid, 300 ng/μl (or 100 ng/μl) of the helper mRNA, and 0.1 volumes phenol red as described in Subheading 3.2.3.

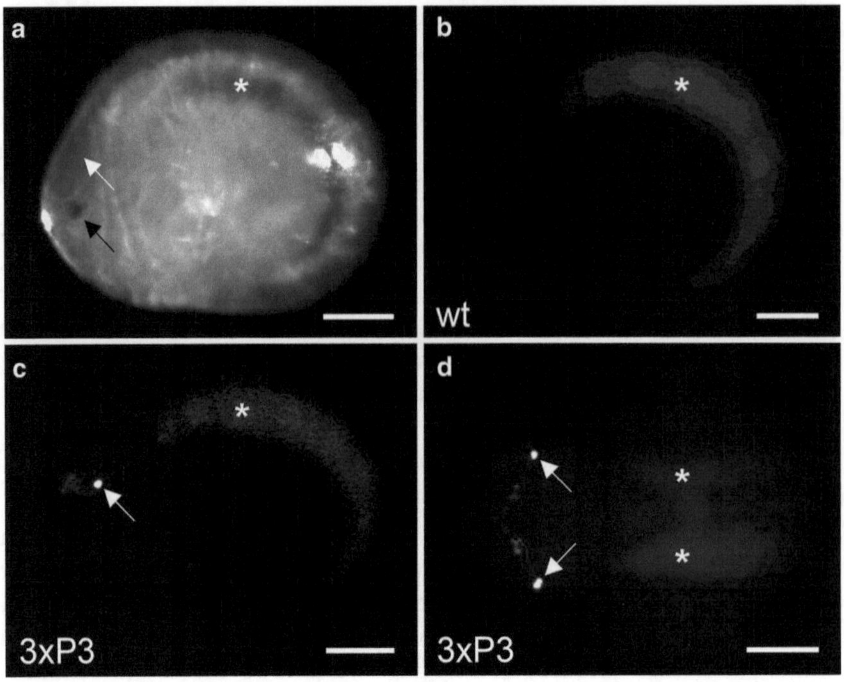

Fig. 4 *3xP3*-driven transformation marker gene expression in transgenic *Parhyale* embryos. (**a**) Bright-field image of a late-stage *Parhyale* embryo. (**b**) Fluorescent image of a wild-type late-stage *Parhyale* embryo. (**c**, **d**) Fluorescent images of late-stage transgenic *Parhyale* embryos. Yolk autofluorescence produces a dorsal crescent of fluorescence in the gut (*asterisks*) in transgenic and non-transgenic embryos. Transgenic embryos exhibit the *3xP3*-driven spot of strong fluorescence in each side of the head (*white arrows*) located posterior to the pigmented compound eye (*black arrow*). Fluorescently labeled neuronal projections are visible in panel (**d**) extending from each spot (cell body) towards the brain. Panels (**a–c**) are lateral views, panel (**d**) is dorsal view, anterior is to the *left* in all panels. Scale bars are 100 μm

3. Collect at least 200 1- and 2-cell-stage *Parhyale* embryos and proceed with microinjections as described in Subheading 3.1. In the case of 2-cell stage embryos, inject the smaller blastomere that will give rise to the germline [20].
4. Transfer injected embryos in FASWA in 35 mm tissue culture Petri dishes, aiming for about 30 embryos per dish. Label the lids with the mix injected, the stage of injection and number of embryos in the dish, and the date and time of injection.
5. Incubate dishes with embryos at 25–26 °C. Check the dishes daily and remove dead embryos to avoid microbial contamination. Surviving injected embryos should be transferred every second day (using a micropipette with a BSA-coated plastic tip) into a new 35 mm tissue culture dish in FASWA.
6. Screen late-stage *Parhyale* embryos 9–10 days after injection (their formed compound eyes should be pigmented; Fig. 4a) for *3xP3*-driven DsRed expression under a fluorescence stereoscope equipped with the DsRed filter set. Screen both sides of each embryo. In each side, *3xP3* drives a spot of fluorescence

in the posterior head region behind the compound eye (Fig. 4b, c) [28]. In strongly expressing *Parhyale* embryos, it is also possible to detect fluorescence in a neuronal projection extending from this spot (the cell body) towards the brain (Fig. 4d) (*see* **Note 13**).

7. Store embryos exhibiting the characteristic *3xP3* pattern in separate 35 mm dishes in FASWA. Detection of *3xP3* expression in a G0 is a good predictor for its germline transformation and its potential to produce transgenic G1s! In our experience, about 30–40 % of embryos injected at the 1-cell and 2-cell stage with pMi{3xP3-DsRed} were transient transgenics with bilateral or unilateral *3xP3* fluorescent patterns [28] (*see* **Note 14**).
8. From day 10 onwards, screen dishes twice daily for embryos hatched. Under optimal microinjection conditions, at least 30 % of *Parhyale* embryos should hatch 10–13 days after injection.
9. Using a micropipette with a BSA-coated plastic tip, transfer hatchlings into Petri dishes in FASW with a couple of pieces of coral and only few pieces of ground fish flakes. Keep hatchlings with bilateral or unilateral *3xP3* expression individually in 60 mm tissue culture Petri dishes. Hatchlings without *3xP3* expression can also be grown separately in 60 mm Petri dishes or in groups of five animals in 90 mm Petri dishes. Label lids accordingly.
10. Change surviving G0s regularly every 4 days into new Petri dishes with fresh FASW and ground fish flakes using a micropipette or by picking up with blunt forceps the piece of coral they are sitting on. To reduce plastic consumption, recycle used Petri dishes by cleaning them with tap and purified water (no soap water).
11. Repeat **step 10** for about 2 months until G0s grow to about 1 cm and reach sexual maturity. At this stage, G0 males are distinguishable from G0 females based on the size of grasping appendages on the third thoracic segment, which are greatly enlarged in males [2]. Females can also be distinguished by their paired ovaries (oblong and opaque) visible through the cuticle in the dorsal thorax.
12. Set single backcrosses of sexually mature G0s to 2–3 similar-sized wild-type *Parhyale* adults of the opposite sex in 90 mm Petri dishes in FASW with a couple of pieces of coral and ground fish flakes. To reduce the amount of labor, non-*3xP3*-expressing G0s (that have a lower probability of transformed germlines) can be first screened in intercrosses of G0 individuals. Change FASW and food regularly as described in **step 10**. Assign a unique ID to each G0 and label lid with the sex and ID of the G0(s) crossed (e.g., Female#1, Male#2, etc.).
13. Check crosses daily for gravid females. Pick up gravid females and grow them separately for 9 days in 60 mm Petri dishes in FASW with few pieces of coral and ground fish flakes. Label lid with the sex and ID of the associated G0.

14. Dissect 9–10-day-old embryos (G1s) from each gravid female in a 35 mm tissue culture Petri dish with FASWA as described in **steps 3** and **4** in Subheading 3.1. The same female can be crossed again one day later as described in **step 12**.

15. Screen G1s for *3xP3*-driven DsRed expression under a fluorescence stereoscope. Stable transgenic animals will display a bilateral *3xP3* fluorescent pattern (Fig. 4d). Discard non-*3xP3*-expressing G1s. Screen at least 50 G1s from each G0 and discard G0s that do not produce transgenic progeny.

16. Grow *3xP3*-expressing G1 siblings (derived from the same G0 parent) to adulthood separately or in groups as described in **steps 9–11**.

17. The ratio of *3xP3*-expressing to non-expressing G1s provides a first hint about the abundance of *Minos* insertions transmitted by their G0 parent; the higher the number of *Minos* insertions in the germline of a G0, the higher the proportion of its transgenic G1 progeny. The actual number of insertions transmitted can be identified by Southern blot analysis on genomic DNA isolated from pools of transgenic G1 siblings [28, 43]. Analyze *Minos*{*3xP3-DsRed*} insertions by digesting genomic DNA with SacI and using as probe the *DsRed* coding sequence. For detailed protocols on genomic DNA preparation, digestion, Southern blotting, and hybridization, please refer to these other sources [28, 40, 41] (*see* **Note 15**).

18. To analyze the segregation and stability of *Minos* insertions, compare the Southern blot pattern between individual transgenic G1 parents and G2 offspring [28]. To do this, backcross adult G1s individually to 2–3 similar-sized wild-type *Parhyale* as described in **step 12**. Assign a unique ID to each G1 and label lid with the sex and ID of the G1 crossed (e.g., transgenic G1 siblings from G0 Female#1 can be labeled Female#1.1, Male#1.2, etc.). Screen for transgenic G2 progeny as described in **steps 13–15** and grow them to adulthood separately or in groups as described in **steps 9–11**. Carry out Southern blot analysis on genomic DNA isolated from a single adult G1 parent and from single adult G2 offspring as described in **step 17**. Each G2 offspring should exhibit a subset or all of the bands (but not different bands) present in the G1 parent.

19. Assess the specificity of *Minos* transposition from the donor plasmid into the *Parhyale* genome by the cut-and-paste mechanism.
    - First, confirm integration of intact transposons into the genome: Check that all bands detected in the Southern blots described in **steps 17** and **18** exceed a minimum size expected for intact transposons [28, 43].
    - Second, confirm specific integration of the *Minos* transposon without flanking sequences from the donor plasmid: Check

that the Southern blot pattern of transgenic animals differs from the Southern blot pattern of the donor plasmid [28, 43]. If required, repeat hybridization of the Southern blot membranes using the plasmid backbone as probe.

- Third, verify the target site specificity of the *Minos* transposase for the TA dinucleotide: Recover the DNA sequences upstream and downstream of *Minos* insertions by inverse PCR and sequencing as detailed elsewhere [40, 42, 43]; the inverted terminal repeats should be flanked by the characteristic TA dinucleotide, followed by sequences unrelated to those of the plasmid backbone.

20. Establish *Parhyale* transgenic lines by repeated rounds of inbreeding to drive transposon insertion(s) to homozygosity (*see* **Note 16**). Select G1 siblings with the strongest *3xP3* expression that are presumably inheriting multiple common *Minos* insertions, and grow them to adulthood as described in **steps 15** and **16**. Set intercrosses between two G1 siblings of the opposite sex (e.g., Female#1.1 × Male#1.2) as described in **steps 12–14**. Select G2 siblings with the strongest *3xP3* expression that are presumably homozygous for one or more of the *Minos* insertions. Grow selected G2s to adulthood and use two animals of the opposite sex as founders to establish the transgenic line.
21. Establish three or more independent transgenic lines in parallel, derived from different G0s. In the long term, keep the lines that exhibit homogeneous transgene expression among sampled individuals. Keep small-scale cultures in small plastic containers on a bottom layer of crushed coral covered in ASW at 22–26 °C (100–200 ml ASW in 0.5–1 l container). No aeration is required if the seawater and food are changed regularly, at least once a week.

3.4 Conditional Heat-Inducible Misexpression of Hox Genes in Transient and Stable Transgenic Parhyale

The establishment of genetic transformation in *Parhyale* has opened several possibilities for functional genetic approaches in this emerging model organism. We describe here a gain-of-function approach based on conditional heat-inducible misexpression to study the role of the Hox gene *Ultrabithorax* (or any other developmental regulatory gene of interest) in *Parhyale* appendage specification.

3.4.1 PCR-Based Isolation of Parhyale Heat-Inducible cis-Regulatory Sequences

Note that these protocols are only briefly outlined here, because extensive descriptions can be found in other sources [14, 40].

1. Amplify by degenerate PCR from *Parhyale* genomic DNA part of the coding sequence of a *heat-shock protein 70* gene (*Phhsp70*) with primers Hsp70F (5′-ACIACITAYTCITGYGTIGG-3′) and Hsp70R (5′-AAIGGCCARTGYTTCAT-3′).

2. Verify by Northern blot analysis using as probe the amplified coding sequence the conditions for *Phhsp70* heat inducibility in wild-type *Parhyale* incubated for 1 h at different temperatures. This analysis shows that *Phhsp70* transcripts become strongly induced at 37 °C and are not detectable at lower temperatures tested [14].
3. Recover the heat-inducible *cis*-regulatory sequence upstream of *Phhsp70* start codon (5′UTR, promoter sequence and heat-responsive enhancer) piecemeal by repeated rounds of inverse PCR (*see* **Note 17**).
4. Amplify by conventional PCR from *Parhyale* genomic DNA a contiguous fragment containing the heat-inducible *cis*--*regulatory* sequence (called *PhHS*) with primers Phhsp70F (5′-TTACTGTAACCGCAGGGGCAAAAGA-3′) and Phhsp70R (5′-ACAGCATCCTTCACGTCTCCTCCAA-3′).

3.4.2 Analysis of cis-Regulatory Sequences with Reporter Constructs in Transgenic Parhyale

1. Clone *PhHS* upstream of the *DsRed* fluorescent reporter and the *SV40* polyadenylation sequence in the versatile subcloning vector pSLfa1180fa to generate plasmid pSL-PhHS-DsRed [14, 46]. To place any other gene of interest under *PhHS* control for heat-inducible misexpression in *Parhyale*, remove the *DsRed* coding sequence by NcoI/NotI digest of pSL-PhHS-DsRed, and replace it with the coding sequence of interest digested with NcoI (or BspHI or PciI) encompassing the start codon in its 5′ end and with NotI (or PspOMI) after the stop codon in its 3′ end (*see* **Note 18**).
2. Digest pSL-PhHS-DsRed with AscI, gel-purify the *PhHS-DsRed-SV40polyA* reporter cassette, and clone it in an AscI-digested *Minos* vector (e.g., pMi{3xP3-EGFP}) to generate donor plasmid pMi{3xP3-EGFP;PhHS-DsRed}. The resulting transposon construct contains the *3xP3-EGFP* transformation marker and the *PhHS-DsRed-SV40polyA* reporter flanked by the *Minos* inverted repeats.
3. Prepare 10–20 μl of an injection mix containing 300 ng/μl of the donor plasmid pMi{3xP3-EGFP;PhHS-DsRed}, 100 ng/μl of the *Minos* transposase helper mRNA, and 0.1 volumes phenol red as described in Subheadings 3.2.1–3.2.3.
4. Microinject at least 200 1- and 2-cell-stage *Parhyale* embryos as described in Subheading 3.1, and grow injected embryos in FASWA to late stages as described in **steps 4** and **5** in Subheading 3.3.
5. Screen 9–10-day-old G0 embryos for transformation marker gene expression (*3xP3*-driven EGFP fluorescence) and single out embryos exhibiting the characteristic *3xP3* pattern as described in **steps 6** and **7** in Subheading 3.3.

6. Prescreen both *3xP3*-expressing and non-expressing G0s for transgene activity, i.e., for heat-inducible *PhHS*-driven DsRed fluorescence. To heat-shock *Parhyale*, pipet embryos with a BSA-coated plastic tip into 35 mm tissue culture Petri dishes with prewamed FASWA at 37 °C. Incubate for 1 h at 37 °C and transfer dishes with heat-shocked embryos back to 25–26 °C. Screen embryos for DsRed fluorescence 12 h after heat-shock.
7. Establish independent transgenic lines with *Minos*{*3xP3-EGFP*;*PhHS-DsRed*} insertions as described in **steps 8–17** and **20–21** in Subheading 3.3. During inbreeding, select G1 and G2 siblings with the strongest *3xP3-EGFP* and *PhHS-DsRed* (after heat-shock) expression.

3.4.3 Characterization of the Parhyale Heat-Inducible System

The properties of the heat-inducible *PhHS* system are assessed in transgenic *Minos*{*3xP3-EGFP*;*PhHS-DsRed*} *Parhyale* lines. The reader is also referred below to other sources for more detailed protocols.

1. Determine the spatiotemporal aspects of heat-inducible embryonic expression; transgene expression with *PhHS* in *Parhyale* embryos can be induced robustly from early germband stage onwards uniformly in all cells and tissues [14].
2. Examine the on kinetics of *PhHS* at the transcriptional level with quantitative RT-PCR. Extract total RNA from pools of about ten late-stage transgenic embryos heat-shocked for varying periods of time at 37 °C as described in [14]. Perform relative quantification of *DsRed* transcript levels (relative to the highest expressing sample) on reverse-transcribed cDNA with primers that hybridize on *SV40polyA* (SV40F 5′-CCACATTTGTAGAGGTTTTACTTGC-3′ and SV40R 5′-TGAGTTTGGACAAACCACAACTA-3′). In each sample, normalize *DsRed* transcript abundance against housekeeping *Parhyale* ribosomal genes *PhRpL21* and *PhRpL32* (with primer pairs PhRpL21F 5′-CCGAGGCTTCAAGAAGAATG-3′ and PhRpL21R 5′-AAAATCCGGCCTCGTACTCT-3′; PhRpL32F 5′-CCAGCATTGGTTATGGTTCA-3′ and PhRpL32R 5′-TTGAGCTTAGCCTTGCCATT-3′).
3. Examine the off-kinetics of *PhHS* at the transcriptional level with quantitative RT-PCR on samples collected at various time points after a 1-h heat-shock at 37 °C. Analysis of the on/off kinetics of *PhHS* shows that misexpressed transcripts peak 1–2 h after the start of heat-shock at 37 °C, have a half-life of about 6 h, and fade away within 10–12 h after the end of the heat-shock [14].
4. Analyze misexpressed transcript and protein accumulation/localization by whole-mount in situ hybridization and anti-

body staining, respectively, in heat-shocked transgenic *Parhyale* embryos [14, 24–26]. This time-course study shows that nascent transcripts appearing as nuclear dots are detectable from 0 to 2 h after a 1-h heat-shock at 37 °C. Transcription ceases about 2 h post-heat-shock, and cytoplasmic transcripts and protein accumulate 3–4 h post-heat-shock (*see* **Note 19**).

3.4.4 Cloning, Expression, and Functional Analysis of *Parhyale* Hox Genes

The steps described here focus on the Hox gene *Ultrabithorax* (*PhUbx*) and its role in *Parhyale* appendage specialization.

1. Extract total RNA from mixed-stage *Parhyale* embryos and reverse transcribe into cDNA as described elsewhere [14].
2. Use this cDNA to amplify by PCR part of the *PhUbx* homeobox with degenerate primers encoding the conserved ELEKEF and WFQNRR amino acid sequences of the homeodomain. Clone the PCR product in a T-Vector and sequence individual clones.
3. Recover the full coding sequence of *PhUbx* by 5′ and 3′ RACE. Clone RACE products in a T-Vector and sequence individual clones. This analysis identified two *PhUbx* splice variants, I and II, which differed in their first few N-terminal amino acids [12]. Each of these isoforms was studied separately following the steps described below [14]. For simplicity, we will refer collectively to both isoforms here as *PhUbx* (*see* **Note 20**).
4. Analyze *PhUbx* expression pattern during normal embryogenesis by in situ hybridization using *PhUbx*-specific riboprobes, and by immunostaining using raised PhUbx-specific antibodies or the cross-species-reactive monoclonal FP6.87 [12, 24–26]. PhUbx mRNA and protein are expressed at high levels in the walking appendages of thoracic segments 4–8 (T4–T8) and at lower levels in the T2 and T3 gnathopods, but are not expressed in the T1 maxillipeds and more anterior head appendages (Fig. 6a) (*see* **Note 21**).
5. Examine appendage morphology in the thorax and posterior head of wild-type *Parhyale* hatchlings by scanning electron microscopy (SEM) and cuticle preparations (Fig. 5a) [14]. Thoracic segments T4 to T8 develop elongated segmented walking appendages (pereopods), each with a cuticular plate (coxal plate) and a gill (except T8) at its base. The more anterior T2 and T3 gnathopods (that facilitate mating and grasping) differ from pereopods by the distinct size and shape of their segments, the presence of characteristic sensory bristles, and in the case of T2 also by the absence of a gill and the different shape of the coxal plate. The developing maxillipeds on T1 are extensively modified to function in feeding. Maxillipeds develop a main segmented limb branch with the same number of segments as the more posterior thoracic appendages, but are

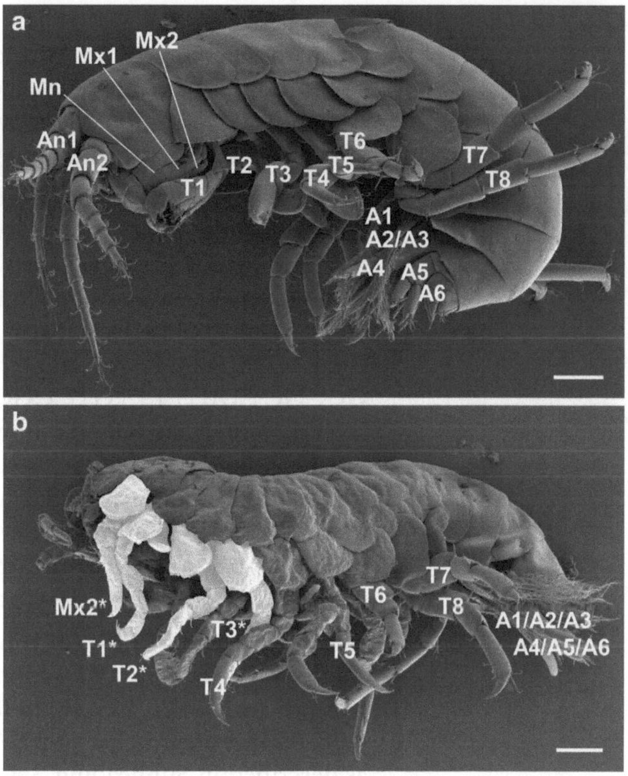

Fig. 5 Scanning electron microscopy of wild-type and homeotically transformed *Parhyale* hatchlings. (**a**) Appendage morphology of a wild-type *Parhyale*. (**b**) Homeotic transformation of feeding and grasping appendages towards walking legs (shown in white and marked with an *asterisk*). Abbreviations of appendages indicated on the *left side* of each specimen: An1 (antenna 1), An2 (antenna 2), Mn (mandible), Mx1 (maxilla 1), Mx2 (maxilla 2), T1–T8 (thoracic appendages 1–8), A1–A6 (abdominal appendages 1–6). Both panels show lateral views with anterior to the *left*. Scale bars are 100 μm

highly reduced in size, medially fused, and lack coxal plates and gills; maxillipeds develop prominent proximal outgrowths to manipulate food, similar to the more anterior gnathal appendages (maxillae 2 and 1).

- For SEM analysis, fix hatchlings in 1 % glutaraldehyde in FASW for 1 h and then in 1 % osmium tetroxide in FASW for 1 h, wash several times in FASW, and dehydrate through an ethanol series. If required, store specimens in 90 % ethanol; otherwise proceed immediately with three washes in absolute ethanol, critical point drying, coating with gold or platinum, and observation under a scanning electron microscope.
- For cuticle preparations, fix hatchlings in 3.7 % formaldehyde in FASW for 1 h and wash several times in 1× PBS with

0.1 % Triton X-100 (PTx). Dissect individual appendages with fine forceps in PTx on a Sylgard plate, dehydrate through an ethanol series, mount in Hoyer's medium/lactic acid (1:1) solution, clear overnight on a 60 °C plate, and observe under a compound microscope.

6. Clone the *PhUbx* coding sequence under the control of the heat-inducible *PhHS* in a *Minos* transposon, as detailed in Subheading 3.4.2, **steps 1** and **2**. For example, the *PhUbx-II* coding sequence was amplified from a full-length cDNA clone with primers PhUbxII_BspHI_F (5′-TTAGTCATGAACT CCTACTTTGAAC-3′) and PhUbx_NotI_R (5′-TATTGCG GCCGCTTAGTTTTGTCCGGGGTT-3′), digested with BspHI/NotI, and cloned downstream of *PhHS* in NcoI/ NotI-digested pSL-PhHS-DsRed. The resulting plasmid pSL-PhHS-PhUbxII was digested with AscI, and the gel-purified *PhHS-PhUbxII-SV40polyA* cassette was cloned into AscI-digested pMi{3xP3-EGFP} to generate donor plasmid pMi{3xP3-EGFP;PhHS-PhUbxII} [14].
7. Microinject 1,500–2,000 *Parhyale* embryos at the 1- and 2-cell stage with an injection mix containing 300 ng/μl of the donor plasmid pMi{3xP3-EGFP;PhHS-PhUbx}, 100 ng/μl of the *Minos* transposase helper mRNA, and 0.1 volumes phenol red as described in Subheading 3.4.2, **steps 3** and **4**.
8. Subject injected embryos to daily heat-shocks (or every 12 h) for 1 h at 37 °C, starting from stages 12–13 onwards (72 h of embryogenesis at 25–26 °C) and continuing until stage 28 (day 9 of embryogenesis at 25–26 °C) [2].
9. Grow embryos to hatching, anaesthetize them as described in **step 3** in Subheading 3.1.1, and examine their morphology under a stereomicroscope. Keep note of the number of embryos injected, the number of embryos hatched, and the number of hatchlings with wild-type and abnormal phenotypes [14].
10. Examine appendage morphology in affected hatchlings by scanning electron microscopy and cuticle preparations as described in **step 5**. Classify the types and frequencies of abnormal phenotypes observed [14]. In the case of *PhUbx* misexpression, the following appendage transformations were observed, sorted out in descending order of frequency: maxilla 2-to-maxilliped transformation, antennae-to-thoracic appendage transformation, maxilla 2/maxilliped-to-gnathopod transformation, and maxilla 2/maxilliped/gnathopod-to-walking appendage transformation (Fig. 5b) (*see* **Note 22**).
11. To associate the induced homeotic transformations with the pattern and intensity of *PhUbx* misexpression, repeat microinjection and heat-shock of about 500 embryos as described in **steps 8** and **9**, up to stages 23–24 (days 6–7 of embryo-

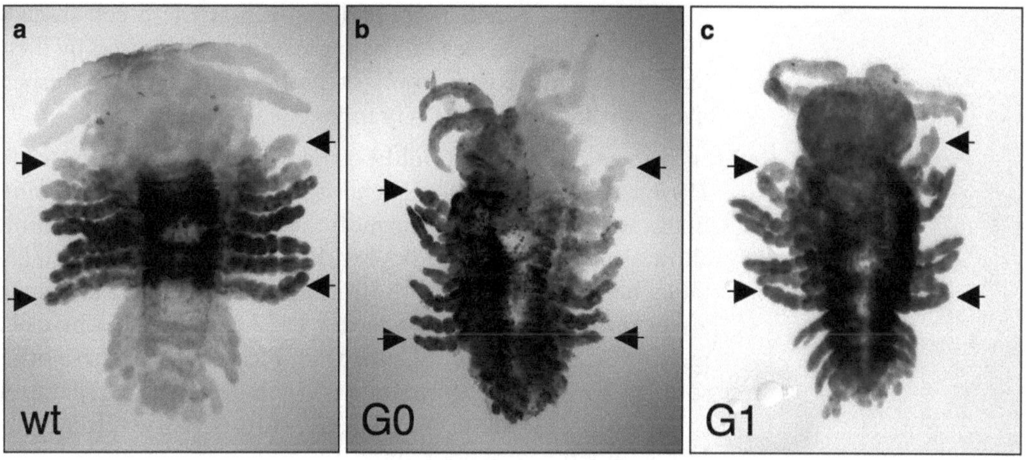

Fig. 6 PhUbx expression in wild-type and transgenic *Parhyale*. (**a**) Wild-type expression pattern of *PhUbx* detected by in situ hybridization. (**b**) Transient transgenic G0 embryo exhibiting unilateral ectopic PhUbx expression in half the head region detected by FP6.87 antibody staining. (**c**) Stable transgenic G1 embryo exhibiting uniform ectopic PhUbx expression detected by FP6.87 antibody staining. In each embryo, *arrows* indicate the normal anterior (in T2) and posterior (in T8) border of PhUbx expression in developing appendages. Note that the FP6.87 monoclonal antibody detects also the abdominal-A Hox protein in the more posterior abdominal segments. In all panels, anterior is to the *top*

genesis at 25–26 °C) when appendage morphogenesis is almost complete [2]. Fix surviving embryos 3–4 h after the last heat-shock and analyze PhUbx expression pattern and levels by immunostaining using a raised PhUbx-specific antibody or the cross-species-reactive monoclonal FP6.87 (Fig. 6b) [14, 24–26]. To also observe embryo morphology, counterstain the DNA of immunostained embryos (e.g., by DAPI staining) and mount embryos in 70 % glycerol for microscopy (*see* **Note 23**).

12. To achieve homogeneous rather than mosaic *PhUbx* misexpression, establish stable transgenic lines with the *Minos*{*3xP3-EGFP*;*PhHS-PhUbx*} transposon as described in Subheading 3.3. The effects of ectopic *PhUbx* can be then analyzed in these transgenic lines with the approaches described in **steps 8–11** (Fig. 6c). However a number of important considerations should be taken into account when misexpressing pleiotropic genes like *PhUbx* that function at many different stages and processes during development [14] (*see* **Note 24**).

4 Notes

1. Freshly oviposited 1-cell-stage embryos are extremely fragile and should be dissected from the brood pouch 1–2 h later once they have hardened. After oviposition, *Parhyale* embryos spend 0–4 h at the 1-cell stage, 4–6 h at the 2-cell stage,

6–7.5 h at the 4-cell stage, and 7.5–9 h at the 8-cell stage at 26 °C [2]. These times can be increased or decreased by incubating the embryos at lower or higher temperatures, respectively, within the 18–30 °C range.

2. In principle, needles should be rigid enough to penetrate the *Parhyale* eggshell without bending, should have a fine pointy tip to avoid yolk leakage through the induced hole, and should allow sufficient injection flow without frequent clogging. The table below shows how the different adjustable parameters affect tip morphology on a Sutter P-87 puller with a box filament (Sutter Instrument Company). Please refer to these other sources for more information on microneedle preparation ([47]; Sutter Pipette Cookbook: http://www.sutter.com/PDFs/pipette_cookbook.pdf).

Parameter	Range	Increase	Decrease
Heat	0–999	Smaller tips Longer taper	Larger tip Shorter taper
Pull	0–255	Smaller tips Longer taper	Larger tip Shorter taper
Velocity	0–255	Smaller tips	Larger tip
Time	0–255	Shorter taper	Longer taper
Pressure	0–730	Shorter taper	Longer taper

3. Never try to mount or change the needle when applying pressure, because this might shoot the needle. When removing a needle always disconnect the needle holder from the injector and keep it upright, directed away from you and your colleagues.

4. One can estimate the amount injected by injecting the aqueous mix into oil and measuring the diameter of the drop under the microscope.

5. The helper mRNA provides a ready-to-use transposase source that has been shown to increase *Minos* transposition rates in various arthropod species compared to helper plasmids [43, 44]. It also alleviates the need to characterize functional promoters to drive the expression of transposase from helper plasmids [28].

6. For the excision and transposition assays, we describe here the use of the original donor plasmid pMiLRTetR(L) that carries a 2 kb *Minos* transposon with the tetracycline resistance gene [28, 35]. Any plasmid carrying a *Minos* transposon can be tested for excision in the excision assay with appropriate flanking primers to amplify by PCR a small diagnostic band.

However, amplification efficiency of the non-excised *Minos* transposon in PCR reactions will depend on transposon size, i.e., smaller transposons will be amplified more efficiently than bigger ones. The use of pMiLRTetR(L) as donor plasmid is required for the transposition assay that is based on bacterial marker selection. The helper plasmid can carry the *Minos* transposase coding sequence under the control of any *cis*-regulatory sequence of interest, like a constitutive or a heat-inducible promoter isolated from *Parhyale* or other species.

7. Additional bands of intermediate size may appear in samples containing the 2.2 kb band [28]. Most likely, these bands represent extra conformations of the 2.2 kb band containing the long *Minos* inverted repeats.

8. The excision and transposition assays are very convenient to assess and compare various aspects of *Minos* transposition, like *Minos* activity in new species of interest, the mobility of transposons with different sizes, and the activity of alternative transposase sources (e.g., mRNA helpers with different 5′ and 3′ UTRs or helper plasmids with different promoter elements). Importantly, injection of the donor plasmid without transposase provides a first clue about the possibility of *Minos* cross-mobilization by endogenous *Minos*-related transposases encoded by the targeted species. This possibility needs to be investigated and excluded in every new species of interest, because it has important implications for the stability of *Minos* insertions and propagation of established transgenic lines.

9. The *Minos* transposition assay involves screening for interplasmid transposition events of the *MiLRTetR* transposon from the pMiLRTetR(L) donor plasmid into the sucrase gene of the pBC/SacRB target plasmid [28, 35]. The screen is done in bacteria by positive selection for the chloramphenicol and tetracycline resistance genes carried by the pBC/SacRB target plasmid and the *MiLRTetR* transposon, respectively, and by rescuing *E. coli* lethality induced by the sucrase gene (*sacB* from *Bacillus subtilis*) in the presence of sucrose. The function of the sucrase gene can be disrupted due to insertional mutagenesis by the *MiLRTetR* transposon. The chloramphenicol, tetracycline, and sucrose triple-resistant clones are selected stepwise by first screening for chloramphenicol and sucrose double-resistant clones, and then screening these for chloramphenicol and tetracycline resistance (removing sucrose-resistant clones that are not caused by *Minos* insertion). Selection in the first step for chloramphenicol and tetracycline double-resistant clones is not recommended, because it is very stringent and may lead to loss of true positives.

10. Multiplying with this factor of 19 normalizes for the fact that 5 % of transformed bacteria are grown on LB-Cm plates and 95 % of transformed bacteria (19 times more cells) are subjected to the triple selection. For example, if 5 % of cells spread on LB + Cm plates produced 500 colonies, while 95 % of cells spread on LB + Cm + Suc plates and then replica patched onto LB + Cm + Tet produced 95 triple-resistant clones, the calculated transposition efficiency is 1 %.
11. The pBC/SacRB target plasmid and the *MiLRTetR* transposon each carries a NotI site. The NotI digestion pattern will vary between digested triple-resistant clones depending on the landing TA site, but the total size should be about 8 kb (except in the case of multiple *Minos* insertions in the same target plasmid) [28, 35].
12. The choice of the fluorescent protein to be coupled to *3xP3* depends on the signal-to-noise ratio detected in the embryos and tissue of interest. Originally, *DsRed* was chosen over *EGFP* in *Parhyale*, because embryos exhibited a stronger *3xP3*-driven signal and lower embryo autofluorescence during DsRed detection compared to EGFP detection. Once the *3xP3* pattern became known, we have employed alternative, spectrally distinct fluorescent markers under *3xP3* control. This way, we have expanded the number of different transgenes that can be combined in the same transgenic *Parhyale* animal.
13. The *Parhyale 3xP3* pattern differs from that of transgenic insects where fluorescence is driven in the photoreceptors of compound eyes and in other tissues [28, 39, 48]. This discrepancy may result from the basal promoter sequences that the *3xP3* element is coupled to, which are derived from a *Drosophila hsp70* gene. When *3xP3* is cloned in the vicinity of *Parhyale hsp70* promoter sequences, expression is not only detected in the posterior head region but also in the photoreceptors and optic lobes of *Parhyale* [14]. In either case, *3xP3* represents a very convenient transformation marker gene: first, it is a reliable marker to identify transgenic *Parhyale*, and second, it exhibits a highly localized expression pattern that does not interfere with the detection of transgene expression.
14. Transient transgenic embryos injected at the 1-cell stage can exhibit either bilateral or unilateral expression of the marker gene/transgene depending on the timing of *Minos* insertion into the genome. Transient transgenic embryos injected at the 2-cell stage can exhibit only unilateral expression of the marker gene/transgene, because each blastomere at this stage is fated to give rise either to the left or the right half of the ectoderm and somatic mesoderm [20, 28]. The percentage of transient transgenic embryos recovered varies between experiments and tends to decrease with increasing transposon size.

15. In our transgenic experiments with pMi{3xP3-DsRed}, some *3xP3*-expressing G0 parents gave rise exclusively to *3xP3*-expressing G1 progeny [28]. Southern blot analysis on pools of transgenic G1 siblings revealed that their G0 parents carried dozens of *Minos* insertions in their germlines. Each G1 had inherited a subset of these insertions. The number of insertions transmitted by G0s and the percentage of their transgenic G1 offspring drop with increasing transposon size and with decreasing concentration of the injected donor plasmid and helper mRNA.
16. Establishing stable *Parhyale* transgenic lines by inbreeding is a time-consuming process that takes at least 6 months from the time of injections. An alternative faster approach is to set up cultures with transgenic G1 siblings. However, in this case the *Minos* insertion(s) will be fluctuating in the population, and cultures need to be selected and enriched for *3xP3*-expressing (and/or transgene-expressing) individuals every 6 months.
17. As a faster alternative to the inverse PCR methodology, it may be possible to isolate by standard PCR the intergenic heat-inducible *cis*-regulatory sequences from *hsp70* genes that are physically linked in the genome [40]. In either case, the presence of clusters of putative binding sites (GAANNTTC) for the Heat shock factors (HSFs) in the isolated *cis*-regulatory sequence is a good evidence for its heat responsiveness [14].
18. The subcloning vector pSLfa1180fa contains a super-polylinker (with multiple cloning sites) flanked by two oligos recognized by the rare 8-cutter restriction endonucleases AscI and FseI [46]. Transgene constructs are routinely assembled in pSL-fa1180fa, digested by AscI, gel-purified, and cloned in AscI-digested *Minos* vectors.
19. The expression dynamics of heat-inducible DsRed (or any other transgene fused to a fluorescent protein) can also be imaged live in intact transgenic embryos under a fluorescence microscope, a laser scanning confocal microscope, or a fluorescence light-sheet microscope ([14] see also http://www.cell.com/picture-show/lightsheet2).
20. The two *PhUbx* splice variants exhibited identical expression patterns and their misexpression resulted in similar homeotic transformation. However, expression of *PhUbx-II* was stronger than *PhUbx I*, and the penetrance and severity of induced transformations were much higher with *PhUbx-II* [14]. Interestingly, the evolutionarily conserved NSYF motif required for transcriptional activation [49] is present in *PhUbx-II* and absent in *PhUbx-I* [12].
21. There is a considerable lag between the appearance of PhUbx transcripts and protein [12]; although PhUbx transcripts start

being detected from germband stage onwards (stage 12), the protein comes up about a day later at the early limb bud stage (stage 17).

22. The observed frequencies of homeotic transformations reflect *PhUbx* expression levels required for specification of each appendage type [14]. In particular, ectopic gnathopods that require low *PhUbx* expression are more frequently induced than ectopic walking appendages that require high *PhUbx* expression. The most frequent phenotype of maxilla 2-to-maxilliped transformation is an indirect effect of *PhUbx* misexpression through downregulation of the more anteriorly expressed Hox gene *Sex combs reduced* (*PhScr*). *PhScr* is normally expressed at high levels in maxillae 2 and at lower levels in maxillipeds, and is a sensitive target of *PhUbx*. Even low levels of ectopic *PhUbx* can reduce *PhScr* levels in developing maxillae 2 transforming them into maxillipeds. Antennae develop normally in the absence of any Hox input and acquire thoracic leg identity (gnathopod or pereopod) when they misexpress *PhUbx*.

23. Because of the stochastic and mosaic nature of transient transgenesis, each transient transgenic embryo experiences a unique spatial and temporal pattern of *PhUbx* misexpression. Yet, within each embryo analyzed, there is good association between the pattern and levels of ectopic PhUbx detected and the type of homeotic transformations induced [14]. Furthermore, classification of immunostained embryos based on signal intensity results in class frequencies that are consistent with the frequencies of morphological transformations observed with SEM and cuticle prep analyses (described in **Note 22**). For example, the majority of immunostained embryos express low levels of ectopic PhUbx resulting in the most abundant maxilla 2-to-maxilliped transformation, while few embryos express high levels of ectopic Ubx resulting in rare cases of ectopic walking appendages [14].

24. Homogeneous and prolonged misexpression of *PhUbx* at wild-type levels in stable transgenic embryos subjected to multiple heat-shocks results in embryonic lethality before the effects on appendage morphogenesis can be scored [14]. Misexpression of *PhUbx* at lower levels is tolerated by stable transgenic embryos, but induces only certain phenotypes like maxilla 2-to-maxilliped transformation or antennae-to-thoracic appendage transformations [14]. For this reason, analysis of stable transgenic lines is not sufficient to recover the full spectrum of homeotic transformations induced by *PhUbx*. The full range of homeotic transformations can be recovered with transient transgenics that are genetically mosaic and can express wild-type levels of *PhUbx* locally in the affected appendages and survive to hatching.

Acknowledgements

We dedicate this chapter to the memory of Thanasis Loukeris, whose work paved the way for transgenic approaches in non-model organisms. We are grateful to Frederike Alwes for providing the drawings and photo shown in Fig. 2. Many protocols described in this chapter have been developed in close interaction with our Ph.D. supervisor and mentor Michalis Averof. Z.K. was supported by an EMBO long-term fellowship, and A.P. by a Marie Curie Intra-European fellowship and by the Howard Hughes Medical Institute.

References

1. Brusca RC, Brusca GJ (2003) Invertebrates, 2nd edn. Sinauer Associates, Sunderland, MA
2. Browne WE, Price AL, Gerberding M et al (2005) Stages of embryonic development in the amphipod crustacean, Parhyale hawaiensis. Genesis 42:124–149
3. Rehm EJ, Hannibal RL, Chaw RC et al (2009) The crustacean Parhyale hawaiensis: a new model for arthropod development. Cold Spring Harb Protoc, pdb.emo114
4. Hannibal RL, Price AL, Parchem RJ et al (2012) Analysis of snail genes in the crustacean Parhyale hawaiensis: insight into snail gene family evolution. Dev Genes Evol 222:139–151
5. Prpic NM, Telford MJ (2008) Expression of homothorax and extradenticle mRNA in the legs of the crustacean Parhyale hawaiensis: evidence for a reversal of gene expression regulation in the pancrustacean lineage. Dev Genes Evol 218:333–339
6. Schaeper ND, Pechmann M, Damen WG et al (2010) Evolutionary plasticity of collier function in head development of diverse arthropods. Dev Biol 344:363–376
7. Simanton W, Clark S, Clemons A et al (2009) Conservation of arthropod midline netrin accumulation revealed with a cross-reactive antibody provides evidence for midline cell homology. Evol Dev 11:260–268
8. Vargas-Vila MA, Hannibal RL, Parchem RJ et al (2010) A prominent requirement for single-minded and the ventral midline in patterning the dorsoventral axis of the crustacean Parhyale hawaiensis. Development 137:3469–3476
9. Browne WE, Schmid BG, Wimmer EA et al (2006) Expression of otd orthologs in the amphipod crustacean, Parhyale hawaiensis. Dev Genes Evol 216:581–595
10. Hannibal RL, Price AL, Patel NH (2012) The functional relationship between ectodermal and mesodermal segmentation in the crustacean, Parhyale hawaiensis. Dev Biol 361:427–438
11. Kontarakis Z, Pavlopoulos A, Kiupakis A et al (2011) A versatile strategy for gene trapping and trap conversion in emerging model organisms. Development 138:2625–2630
12. Liubicich DM, Serano JM, Pavlopoulos A et al (2009) Knockdown of Parhyale Ultrabithorax recapitulates evolutionary changes in crustacean appendage morphology. Proc Natl Acad Sci U S A 106:13892–13896
13. Nestorov P, Battke F, Levesque MP et al (2013) The maternal transcriptome of the crustacean Parhyale hawaiensis is inherited asymmetrically to invariant cell lineages of the ectoderm and mesoderm. PLoS One 8:e56049
14. Pavlopoulos A, Kontarakis Z, Liubicich DM et al (2009) Probing the evolution of appendage specialization by Hox gene misexpression in an emerging model crustacean. Proc Natl Acad Sci U S A 106:13897–13902
15. Wolff C, Scholtz G (2002) Cell lineage, axis formation, and the origin of germ layers in the amphipod crustacean Orchestia cavimana. Dev Biol 250:44–58
16. Wolff C, Scholtz G (2008) The clonal composition of biramous and uniramous arthropod limbs. Proc Biol Sci 275:1023–1028
17. Ito A, Aoki MN, Yahata K et al (2011) Embryonic development and expression analysis of Distal-less in Caprella scaura (Crustacea, Amphipoda, Caprellidea). Biol Bull 221:206–214
18. Aspiras AC, Prasad R, Fong DW et al (2012) Parallel reduction in expression of the eye development gene hedgehog in separately

derived cave populations of the amphipod Gammarus minus. J Evol Biol 25:995–1001

19. Alwes F, Hinchen B, Extavour CG (2011) Patterns of cell lineage, movement, and migration from germ layer specification to gastrulation in the amphipod crustacean Parhyale hawaiensis. Dev Biol 359:110–123
20. Gerberding M, Browne WE, Patel NH (2002) Cell lineage analysis of the amphipod crustacean Parhyale hawaiensis reveals an early restriction of cell fates. Development 129:5789–5801
21. Extavour CG (2005) The fate of isolated blastomeres with respect to germ cell formation in the amphipod crustacean Parhyale hawaiensis. Dev Biol 277:387–402
22. Price AL, Modrell MS, Hannibal RL et al (2010) Mesoderm and ectoderm lineages in the crustacean Parhyale hawaiensis display intra-germ layer compensation. Dev Biol 341: 256–266
23. Rehm EJ, Hannibal RL, Chaw RC et al (2009) Injection of Parhyale hawaiensis blastomeres with fluorescently labeled tracers. Cold Spring Harb Protoc, pdb.prot5128
24. Rehm EJ, Hannibal RL, Chaw RC et al (2009) In situ hybridization of labeled RNA probes to fixed Parhyale hawaiensis embryos. Cold Spring Harb Protoc, pdb.prot5130
25. Rehm EJ, Hannibal RL, Chaw RC et al (2009) Antibody staining of Parhyale hawaiensis embryos. Cold Spring Harb Protoc, pdb. prot5129
26. Rehm EJ, Hannibal RL, Chaw RC et al (2009) Fixation and dissection of Parhyale hawaiensis embryos. Cold Spring Harb Protoc, pdb. prot5127
27. Ozhan-Kizil G, Havemann J, Gerberding M (2009) Germ cells in the crustacean Parhyale hawaiensis depend on Vasa protein for their maintenance but not for their formation. Dev Biol 327:230–239
28. Pavlopoulos A, Averof M (2005) Establishing genetic transformation for comparative developmental studies in the crustacean Parhyale hawaiensis. Proc Natl Acad Sci U S A 102: 7888–7893
29. Chaw RC, Patel NH (2012) Independent migration of cell populations in the early gastrulation of the amphipod crustacean Parhyale hawaiensis. Dev Biol 371:94–109
30. Blythe MJ, Malla S, Everall R et al (2012) High through-put sequencing of the Parhyale hawaiensis mRNAs and microRNAs to aid comparative developmental studies. PLoS One 7:e33784
31. Parchem RJ, Poulin F, Stuart AB et al (2010) BAC library for the amphipod crustacean, Parhyale hawaiensis. Genomics 95: 261–267
32. Zeng V, Villanueva KE, Ewen-Campen BS et al (2011) De novo assembly and characterization of a maternal and developmental transcriptome for the emerging model crustacean Parhyale hawaiensis. BMC Genomics 12:581
33. Dohle W, Scholtz G (1988) Clonal analysis of the crustacean segment: the discordance between genealogical and segmental borders. Development 104(Suppl):147–160
34. Scholtz G, Patel NH, Dohle W (1994) Serially homologous engrailed stripes are generated via different cell lineages in the germ band of amphipod crustaceans (Malacostraca, Peracarida). Int J Dev Biol 38:471–478
35. Klinakis AG, Loukeris TG, Pavlopoulos A et al (2000) Mobility assays confirm the broad host-range activity of the Minos transposable element and validate new transformation tools. Insect Mol Biol 9:269–275
36. Pavlopoulos A, Oehler S, Kapetanaki MG et al (2007) The DNA transposon Minos as a tool for transgenesis and functional genomic analysis in vertebrates and invertebrates. Genome Biol 8(Suppl 1):S2
37. Berghammer AJ, Klingler M, Wimmer EA (1999) A universal marker for transgenic insects. Nature 402:370–371
38. Horn C, Schmid BG, Pogoda FS et al (2002) Fluorescent transformation markers for insect transgenesis. Insect Biochem Mol Biol 32: 1221–1235
39. Wimmer EA (2003) Innovations: applications of insect transgenesis. Nat Rev Genet 4: 225–232
40. Pavlopoulos A (2011) Identification of DNA sequences that flank a known region by inverse PCR. Methods Mol Biol 772:267–275
41. Sambrook J, Russell DW (2001) Molecular cloning: a laboratory manual, 3rd edn. Cold Spring Harbor Laboratory Press, Cold Spring Harbor, NY
42. Klinakis AG, Zagoraiou L, Vassilatis DK et al (2000) Genome-wide insertional mutagenesis in human cells by the Drosophila mobile element Minos. EMBO Rep 1:416–421
43. Pavlopoulos A, Berghammer AJ, Averof M et al (2004) Efficient transformation of the beetle Tribolium castaneum using the Minos transposable element: quantitative and qualitative analysis of genomic integration events. Genetics 167:737–746
44. Kapetanaki MG, Loukeris TG, Livadaras I et al (2002) High frequencies of Minos transposon mobilization are obtained in insects by using in vitro synthesized mRNA as a source of transposase. Nucleic Acids Res 30:3333–3340

45. Arca B, Zabalou S, Loukeris TG et al (1997) Mobilization of a Minos transposon in Drosophila melanogaster chromosomes and chromatid repair by heteroduplex formation. Genetics 145:267–279
46. Horn C, Wimmer EA (2000) A versatile vector set for animal transgenesis. Dev Genes Evol 210:630–637
47. Miller DF, Holtzman SL, Kaufman TC (2002) Customized microinjection glass capillary needles for P-element transformations in Drosophila melanogaster. Biotechniques 33:366–367, 369–370, 372 passim
48. Horn C, Jaunich B, Wimmer EA (2000) Highly sensitive, fluorescent transformation marker for Drosophila transgenesis. Dev Genes Evol 210:623–629
49. Tour E, Hittinger CT, McGinnis W (2005) Evolutionarily conserved domains required for activation and repression functions of the Drosophila Hox protein Ultrabithorax. Development 132:5271–5281

Chapter 11

Tissue Specific RNA Isolation in *Drosophila* Embryos: A Strategy to Analyze Context Dependent Transcriptome Landscapes Using FACS

Arnaud Defaye and Laurent Perrin

Abstract

The Hox family of transcription factors defines cell identity along the A/P axis of animal body plan by modulating expression of distinct sets of target genes in a tissue specific manner. Identifying such tissue specific target genes is indispensable if one wants to understand how Hox proteins mediate their context dependent function. Genome wide analysis of transcriptional activity in different tissues and contexts regarding Hox genes activity could help in reaching this goal. Such experiments rely on the possibility to selectively purify the cells of interest from developing embryos and to perform a transcriptomic analysis on such purified cell populations. By combining expression of a fluorescent protein and fluorescent activating cell sorting (FACS) technique, it is possible to obtain highly purified specific cell populations. In this chapter we describe the experimental procedure we have established in *Drosophila*—starting from a genetically marked small cell population (cardiomyocytes, 104 cells)—to dissociate the embryos in order to turn it into a suspension of individual cells, sort cells according to the expression of the introduced genetic marker and purify the total RNA content of the sorted cells. This can be used to analyze the transcriptome landscape of rare cell populations in wild type and mutant contexts. This technique has shown to be useful in the case of cardiac cells but is virtually applicable to any cell type and mutant backgrounds, provided that specific genetic markers are available.

Key words Cell dissociation of *Drosophila* embryo, Fluorescence activated cell sorting, RNA extraction, Transcriptomics, Cell purification, Tissue specific, Context-dependent Hox activity

1 Introduction

Hox genes play a unique function in determining the general morphology of organisms and in making serial structures different. Remarkably, when a Hox gene confers its identity to a given segment, it does it in a variety of tissues, and in a variety of cell types in each tissue [1]. This highlights the importance of the cellular context with respect to Hox function. In addition to this influence of spatial context, Hox function is also largely affected by developmental timing. Indeed, Hox genes are implicated at multiple steps

Yacine Graba and René Rezsohazy (eds.), *Hox Genes: Methods and Protocols*, Methods in Molecular Biology, vol. 1196,
DOI 10.1007/978-1-4939-1242-1_11,

of organogenesis, from cell fate determination to terminal differentiation. For instance, in the developing cardiac system in *Drosophila*, where the Hox genes *Ultrabithorax* (*Ubx*) and *abdominalA* (*abd-A*) play a central role in organogenesis [2–4], both are involved at distinct and successive stages during cardiogenesis, first to drive cell lineage choice and later on to impinge on cell differentiation [5]. In addition, their function is again required in the very same cells to drive the organ remodeling at metamorphosis [6].

The identification of Hox-regulated gene networks is fundamental to understand the developmental processes of morphogenesis and cell differentiation in which they are involved. In this line, comprehensive identification of Hox targets through microarrays [7] gave insightful informations regarding Hox targets in the *Drosophila* embryo. However, a full understanding of how the functional diversification of different organs and tissues is achieved needs approaches using isolated tissues. Such a strategy has recently been carried out in the developing wings disks to analyze the transcriptional landscape of Ubx dependent transformation of wings into halters [8], but, mainly due to the technical challenge that represent tissue specific gene expression profiling in embryos, no such strategy has been applied to developing *Drosophila* embryos so far.

Different methods to analyze tissue specific gene expression have been developed recently. They are either based on tissue specific tagging of RNA for subsequent RNA isolation from whole embryos extracts, or on prior isolation of the cells of interest. The translating ribosome affinity purification (TRAP) technique has recently been adapted in *Drosophila* with the use of the *Gal4/UAS* system [9], making it particularly versatile and applicable to any tissue provided that accurate tissue specific *gal4* driver is available. However, this approach only allows for the purification of mRNA undergoing translation. This caveat is avoided in an alternative method based on the tissue specific incorporation of 4-thiouracil (4-TU) by restricted uracil phosphoribosyltransferase (UPRT) expression [10]. However, 4-TU is best provided by feeding, which makes it hardly applicable to developing embryos.

The INTACT (*i*solation of *n*uclei *ta*gged in specific *c*ell *t*ypes) method has recently been adapted in *Drosophila* [11, 12]. Based on affinity purification of tagged nuclei, it permits the tissue specific analysis of transcriptome landscape since transcriptomes from whole cells and isolated nuclei are highly correlated [11]. This promising method proved efficient in analyzing different cell types. However, it remains to be established whether this is suitable to very small cell populations and how one can discriminate nuclei of different genetic contexts, if one wants to analyze the effect of genes' gain or loss of function.

Fluorescent Activated Cell Sorting (FACS) in this overall context represents a method of choice to purify a specific cell type from complex cell populations. It depends on the fluorescent labeling of the cells to be purified, which in *Drosophila* is easily achieved using tissue specific expression of the Green Fluorescent Protein (GFP) or any other fluorescent protein. When starting from whole embryos, this technic however requires prior mechanical and/or enzymatic dissociation of cells that may affect the whole transcriptome landscape of the targeted cells, and caution should be taken regarding the dissociation and sorting duration. Despite these caveats, FACS sorting is of great promise and has indeed been used successfully in several transcriptomic approaches. One of the major advantages of FACS over other methods, apart from its great specificity (>97 % purity can routinely been achieved), is its versatility: since the sorting strategy can be based on both positive and negatives markers, it allows for sorting very specific cell types in different genetic backgrounds. We and others have previously established a method that allows for the purification of cardiac cells from the developing embryo, which represents a total number of 104 cells in each individual, using GFP, flow cytometry, and FACS [13, 14]. In this chapter, we describe a detailed protocol for dissociation of precisely staged embryos, cell sorting and RNA purification from this very rare cell population. We also propose a way to discriminate between homozygous mutant cells and heterozygous or wild type cells that can be particularly useful when dealing with lethal mutations.

2 Materials

2.1 Egg Collection

1. Small cages of approximately 7.5 cm height, 5 cm wide.
2. Plates for egg deposition. Composition: 43 g/l agar, 43 g/l sucrose, 269 ml/l apple juice, 8 ml/l propionic acid (>99 %). Boil water. Sequentially add agar, sucrose and apple juice, mix well while adding. Wait for temperature to cool down below 60 °C and add propionic acid. Mix well. Distribute in plates (use 5 cm diameter plastic petri dishes): fill half of their height. Wait for the medium to solidify and store at 4 °C. Before using the plates for egg deposition, allow to warm up at room temperature, and spread a paste of alive budding yeast (*Saccharomyces cerevisiae*) over its surface. Cover approximately 2/3 of the area.
3. 0.75 mm and 100 μm nylon sieves.

2.2 Embryo Dissociation

1. Mortar and pestle. Mortar 2 ml volume, borosilicate glass. Smooth surface for the pestle (Teflon). Spacing between pestle and mortar range from 0.1 to 0.15 mm.
2. Screw gun (wireless is more convenient).

3. Bleach (1.3 %) diluted extemporaneously from a 2.6 % stock solution.
4. Precision balance.
5. Thermoregulated centrifuge (4 °C).
6. Ice.
7. Water bath (37 °C).
8. Schneider's medium with calcium (Sigma).
9. Type 2 collagenase (Worthington)—*optional.*
10. Trypsin (2.5 % trypsin, no phenol red, Gibco).
11. Goat serum (20 %).
12. RNase-free DNase Type 1 (Promega RQ1).
13. EDTA (0.5 M).
14. Hemolymph Like: 25 mM KCl, 90 mM NaCl, 4.8 mM $NaHCO_3$, 80 mM d-glucose, 5 mM trehalose, 10 mM Hepes. Adjust pH to 6.9. Aliquots may be stored at −20 °C to avoid microbial contamination. Prepare fresh HL–EDTA (Hemolymph Like supplemented with 0.5 mM EDTA final concentration).
15. Trypan blue solution 0.4 % (Gibco).

2.3 FACS

1. 1.5 ml RNase-free tubes containing 50 μl of TRIzol (Life Technologies).
2. Ice.
3. Dry ice.
4. 30 μm polyamide MACS filters (Miltenyi Biotec).
5. Schneider's medium (Sigma).
6. FACS ARIA II Becton Dickinson equipped with a 488 nm laser (20 mW).
7. BD FACSFlow Sheath Fluid.

2.4 RNA Isolation

1. TRIzol (Life Technologies).
2. Ice.
3. RNase-free water.
4. Linear polyacrylamide (GenElute LPA, SIGMA).
5. Vortex.
6. Chloroform.
7. Phase Lock Gel tubes 1.5 ml Heavy (5Prime).
8. Refrigerated centrifuge (4 °C) up to 14,000 ×*g*.
9. RNase-free tubes.
10. Isopropanol.

11. Ethanol (75 %, dilute absolute EtOH with RNase-free water).
12. Bioanalyzer (Agilent).
13. NanoDrop (Thermo Scientific).

3 Methods

3.1 Embryo Collection

1. Prepare a population cage with approximately 200 mated females. Change food every morning and evening (*see* **Note 1**).
2. Perform three pre-lay collections of 1 h each by changing the plate every hr (*see* **Note 2**).
3. Place the collection plate for the desired period of time.
4. Change the plate and place the egg collection at 25 °C for the desired period of time.
5. Collect the embryos. Put water onto the plate where embryos were deposited. Gently get them into suspension manually using a brush. Position the 0.75 mm sieve above the 100 μm sieve. Filter the embryos through the 0.75 mm sieve and collect them into the second sieve. Proceed to dissociation (*see* **Note 3**).

3.2 Embryo Dissociation (See Note 4)

1. Remove the chorion by bleaching for 2.5 min in 1.3 % bleach at room temperature.
2. Rinse extensively with Milli-Q water. Place embryos on ice (*see* **Notes 4** and **5**).
3. Dry embryos by putting them in between two sheets of paper.
4. Weight the embryos. Use no more than 50 mg per dissociation.
5. Put the embryos in the potter with 1 ml of cold Schneider's solution.
6. Suspend the embryos in the potter using a Pasteur pipette (*see* **Note 6**).
7. Homogenize the embryos: adapt the pestle to the screw gun, make rotations of the pestle at a speed of roughly 50–70 rpm (low speed) while pushing it slowly towards the bottom of the potter. When the pestle has reached the bottom, let it make one complete turn on itself, and then slowly pull it back from the bottom up to the top. Keep the pestle rotate at the same speed all the time, and also when pulling it back from the bottom (*see* **Note 7**).
8. Repeat Subheading 3.2, **step 7** until the solution is homogenous (10–12 times) including manual homogenization of the liquid using a Pasteur pipette in between strokes (*see* **Note 8**).
9. Collect the suspension into a cold 2 ml tube.

10. Optional: add 20 U of type 2 collagenase and incubate at 37 °C for 10 min (*see* **Note 9**).
11. Filter the suspension on a 100 μm nylon sieve using gravity (*see* **Note 10**).
12. Centrifuge for 1 min at 500 × *g* at 4 °C (*see* **Note 11**).
13. Trash the supernatant and gently suspend the pellet with 500 μl of HL–EDTA (*see* **Note 12**).
14. Centrifuge for 1 min at 500 × *g* at 4 °C.
15. Trash the supernatant and suspend the pellet with 450 μl of HL–EDTA + 50 μl trypsin (0.25 % final concentration).
16. Incubate for 10 min at 37 °C (*see* **Note 13**). Gently suspend the cell suspension once during the incubation.
17. Add 450 μl of Schneider's medium, 100 μl of goat serum, and 2 U of DNase (*see* **Note 14**).
18. Incubate for 5 min at 37 °C.
19. Centrifuge for 3 min at 500 × *g* at 4 °C.
20. Trash the supernatant and suspend the cell pellet in 500 μl of Schneider's medium (*see* **Note 15**).
21. Optional: count cells and assess viability using trypan blue solution (*see* **Note 16**).
22. Prepare to go to the FACS with Schneider's medium, dry ice, and RNase-free tubes containing 50 μl of TRIzol.
23. Right before sorting the cells, filter them using 30 μm MACS filters (by gravity) (*see* **Note 17**).

3.3 FACS (See Note 18)

The following settings have been successfully used on a FACS Aria 2 Becton Dickinson controlled by the software FACSDiva version 6.1.2 to sort GFP-expressing cardiac cells using BD FACSFlow sheath fluid. Sort Setup: 70 μm/Precision: 0-24-0/Sheath Pressure: 70.00.

1. Set up the FACS using control (no GFP) samples (*see* **Note 19**).
2. Once gating strategy is defined, start cell sorting.
3. Collect cells into RNase-free tubes containing 50 μl of TRIzol on ice (*see* **Note 20**).
4. Freeze the samples on dry ice and store at −80 °C or process to RNA purification immediately (*see* **Note 21**).

3.4 RNA Isolation (See Notes 22 and 23)

1. Briefly thaw samples on ice.
2. Measure volume. If less than 80 μl, add water up to 80 μl (*see* **Note 24**).
3. Add 250 μl TRIzol (to reach 300 μl final volume of Trizol) and mix thoroughly.

4. Add 5 μg Linear Polyacrylamide—this can be added together with TRIzol if multiple samples are prepared at the same time.
5. Vortex 1 s.
6. Add 60 μl Chloroform.
7. Vortex 30 s.
8. Transfer the material into a Phase Lock Gel tube (pre-spin PLG alone at full speed for 1 min).
9. Spin at full speed (14,000 × *g*) for 5 min at 4 °C.
10. Transfer the aqueous phase (superior) to a clean RNase-free tube.
11. Add 180 μl of isopropanol (0.8 vol).
12. Incubate overnight at −20 °C for RNA precipitation.
13. Spin at full speed (14,000 × *g*) for 30 min at 4 °C.
14. Carefully remove the supernatant. A small whitish pellet should be visible.
15. Wash with 500 μl RNase-free EtOH 75 %.
16. Spin at full speed (14,000 × *g*) for 10 min at 4 °C.
17. Carefully remove the supernatant and make sure to drain the pellet as much as possible.
18. Air-dry (be careful not to over-dry, as it will compromise RNA resuspension efficiency. 1 min is enough if pellet has been drained properly, otherwise 2 min will be required).
19. Dissolve the RNA with 11 μl of RNase-free water. Save 1 μl for qualitative and quantitative analysis of RNA. Store the remaining 10 μl of RNA at −80 °C.
20. Dilute the RNA for qualitative and quantitative analysis in 3 μl (final volume). Use half for qualitative analysis using Bioanalyser (follow supplier instructions) and half for NanoDrop (*see* **Note 25**).

4 Notes

1. This protocol works well when starting the dissociation using 50 mg of embryos. The time required to obtain 50 mg of staged embryos in a cage where flies are laying eggs obviously depends on how many flies are in the cage. Working on a precise temporal window, or even doing an analysis on several successive temporal windows will require adjustments with respect to the number of flies and cages required. We routinely collect 50 mg of embryos in a period of 6 h using 200 females. Note that this will not be true for all experiments, as the genotype and genetic background of the flies may affect their reproductive fitness.

It is also important to let the flies get used to the egg laying chamber, and at least 24 h are required for this purpose. The productivity of the flies increases if they can be placed in an incubator that is dedicated to this experiment. Also, do not forget to prepare a population cage with wild type flies, as nonfluorescent staged matched embryonic cells are required to set up the FACS sorting strategy (*see* Subheading 3.3, **step 1**).

2. *Drosophila* females may keep their egg for several hrs before deposition if the environment is not suitable (egg retention). In order to collect staged embryos it is important to provide a favorable environment to the flies, and refresh it often. We use small cylinder-shaped cages of approximately 7.5 cm high and 5 cm diameter to create an "egg laying chamber." The food for egg laying chambers should be changed regularly, and three pre-egg laying sessions are implemented before the real egg laying session. The aim is to give the flies a good environment to lay the eggs they may have retained, and thus ensure the egg laying session will start with eggs freshly fertilized.
3. Be ready to collect the eggs 30 min before the time at which you want to start the dissociation to make sure you will start the dissociation on time.
4. Keep all materials and solutions on ice (except bleach). Subheading 3.2, **Steps 5–9** should be performed at 4 °C (in a cold room). Try to be fast over the whole process. This is true for several steps of the protocol (Dissociation, FACS, RNA extraction). During the dissociation, the faster the procedure, the better the yield in terms of survival rate in the cell population. The step that requires time is Subheading 3.2, **step 8**, and performing a manual disruption of embryos involves treating samples one by one. It is thus better to handle a small number of samples at a time, to make sure none of them will have to wait a long time before proceeding to the next step. This is also true for the FACS, where samples are once again treated one by one. During RNA extraction and purification, there is no need to treat samples one by one so that they can all be treated in parallel, although it is obvious that the more samples in hands, the longer it takes to go through the protocol. Reducing time of manipulation is a way to reduce risk of RNA degradation. Because of these various constraints, we decided to process no more than four samples at a time.
5. If precise developmental timing is required, this time point may be considered as the most advanced developmental time point.
6. We have been worried of killing cells by making too much of pipetting for homogenizing the mixture of embryos while using the potter and pestle. We have tried to homogenize by

capping and flipping upside down the potter but it turned out that the yield was not as good. This may be due to the fact that even though the pipetting probably kills some cells, it also helps separate cells that are attached in cluster and thus increases the number of cells that are retrieved after the first filtration (Subheading 3.2, **step 11**).

7. The speed at which the pestle is pulled is critical. Indeed a trough may be created at this stage that may damage the cells. To avoid it, push in and pull back the pestle slowly.

8. The number of strokes may vary depending on the stage of the embryos. Example is given here for late (stage 13–16) embryos. It is useful to homogenize the suspension especially at the beginning of the dissociation. It becomes less important when several strokes have been made already. We usually follow this protocol ("H" stands for Homogeneization, "S" for Stroke): H, S, H, S, H, S, S, H, S, S, S, H, S, S, S.

9. The collagenase treatment may facilitate the disruption, especially in old embryos; however, it has not always been useful in our hands. In few cases, it even reduced the yield. This may be due to the fact that it introduces a step at 37 °C and a fraction of the cells may not survive repeated thermal shocks. Note that calcium is required for the activity of the collagenase, so choose the Schneider's medium accordingly. A step for digestion with trypsin is included later in the protocol.

10. We created homemade filters by using 100 μm nylon mesh that were glued at the bottom of a 1.5 ml tube of which the base has been cut. Filtration of the homogenate is done by gravity. Sometimes the filter is saturated with solid material and the filtration stops. In this case, try the following: (1) gently homogenize again the solution in order to remove the materials from the filter, (2) centrifuge at 500 × *g* for 1 min at 4 °C.

11. Centrifugation speed and time may be optimized. These are the best parameters that worked in our hands. Higher speed and longer time will induce more cell death, but will ensure fewer cells are left behind in the supernatant.

12. After centrifugation, pay attention to suspend cells gently. It is useful to cut the extremity of the tip to reduce the damaging of cells due to the shear stress induced by pipetting.

13. Water bath is recommended to ensure optimal thermal exchange with the tube.

14. Addition of goat serum will stop trypsin treatment. The DNase treatment greatly improves the efficiency of the flow cytometry and cell sorting by reducing the viscosity of the suspension.

15. This volume usually makes a cell suspension of very high density. We usually wait for the result of the cell numbering

(Subheading 3.2, **step 21**) to adapt the final volume of Schneider's medium to add to get a cell suspension of approximately 5×10^6 cells/ml.

16. After dissociation, it is interesting to look at the survival rate of the cells. We use trypan blue to perform a dye exclusion test. Mix 45 µl of trypan blue to 5 µl of cells. Viability can be assessed under a standard light microscope: alive cells will not be stained because the trypan blue will not enter into them, while dead cells will be stained blue. We usually obtain more than 90 % viability. If performed on a slide carrying a counting chamber, cell number can be evaluated. We usually obtain 8×10^6 to 16×10^6 alive cells starting from 50 mg of embryos.
17. The filtration aims at removing the doublet cells. Cells tend to attach to each other over time, so it is necessary to filter them just prior starting the sorting. There is no need to filter them a long time in advance because you may end up having to filter them again later.
18. Although we only have experience in sorting GFP positive cells, on the principle any fluorescent marker is suitable, provided that the cell sorter is equipped with the corresponding lasers and filters. It is possible to selectively sort homozygous mutant cells from a fly stock in which the mutation is on a balancer chromosome that ubiquitously expresses a fluorescent maker (provided that this marker is different from the one which is used to sort the cell population of interest). We have used the CyO chromosome which expresses the Cherry fluorescent protein under the control of the *squash* enhancer [15] which is described to produce a signal already at nuclear cycle 4. In that situation, sorted cells are gated in order to be negative for the balancer marker.
19. One needs first to set up the FACS using a control sample. The objective is to define the gating strategy to sort alive, individual cells according to size and granularity of the sorted objects. The next step is then to work with fluorescent expressing cells and define the gate that will sort fluorescent cells out of the nonfluorescent ones. *See* Fig. 1 for details.
20. We collect sorted cells into TRIzol so that cell lysis and inhibition of RNase occurred immediately after cell sorting. Working

Fig. 1 (continued) and their granularity and internal complexity (SSC-A). Small particles and particles that display high granularity or internal complexity are excluded. (**c** and **d**) single cells ("singlets") are selected according to their size. Doublets are producing higher SSC-W and FSC-W values, and are therefore excluded on the basis of these criteria. (**e**) GFP expressing cells are selected according to the signal recovered after excitation using a laser beam at 488 nm. (**f**) Total RNA extracted from sorted cells. RNA quality is assessed on a Bioanalyzer. Typical aspect of non-degraded RNA. Integrity of ribosomal RNA is measured as a proxy for integrity of all RNA molecules. Note that in *Drosophila*, due to the cleavage of 28S rRNA, ribosomal RNAs show two peaks close to one another. The peak at 25 nt is a ladder. *FU* arbitrary unit of fluorescence, *nt* size of RNA molecules in nucleotides

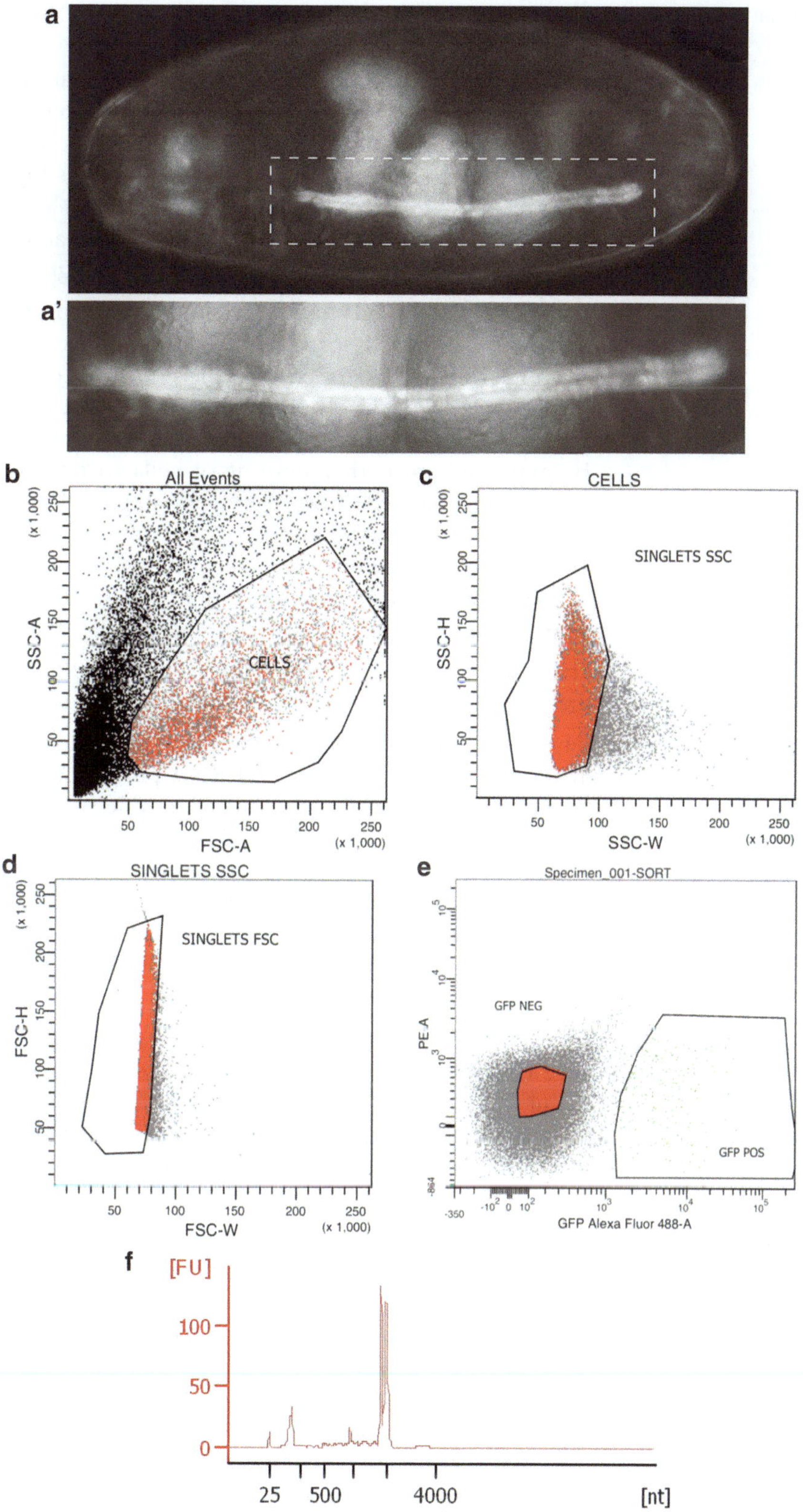

Fig. 1 Gating strategy for cell sorting and RNA quality assessment. An example of the cell sorting strategy performed on dissociated cells from stages 15/16 TinC-Gal4 > UAS-GFP embryos is provided. (**a**) GFP expression pattern of the reporter in the 104 cardiac myocytes. (**b**) Cells are selected according to their size (FSC-A)

this way does not permit the evaluation of cell survival after the sorting procedure; however, the downstream application here is not to perform primary cell culture where cell survival is critical. In a first series of tests, we sorted GFP expressing cells and analyzed the purified cells on the cytometer again a second time. Ninety-nine percent of cells were GFP positive, and more than 90 % showed no signs of cell death (by size and granularity).

21. Avoid too long (>1 h) sorting sessions to prevent RNA degradation. We routinely sort 2×10^4 GFP positive cells in 45 min starting from a cell suspension where positive cells represent less than 0.5 % of the whole population. RNA extraction procedure allows for sufficient amount of RNA (roughly 50 ng) to be extracted from this amount of cells to perform a transcriptomic analysis (most platforms amplify RNA prior to microarray hybridization or library preparation for RNA seq). In addition, it is important to keep the volume of sorted cells (mainly coming from the sheath fluid) below 50 μl, thus preventing the usefulness of collecting more cells.
22. It is recommended to proceed with RNA isolation immediately to maximize the yield and optimize RNA quality.
23. This protocol describes how to extract and purify total RNA. It is a classic protocol for total RNA purification adapted to small cell number and is virtually applicable to any given sample. When handling RNA several precautions are to be taken, the first of which is to manipulate RNA on ice. Make sure to clean everything (bench, pipettes, centrifuge, etc.) with *RNase Away* or any similar anti-RNase solution. Do not hesitate to clean the same object several times if it should get into contact with the RNA several times. Use RNase-free water, tubes, and tips. Wear gloves.
24. The optimal volume of aqueous solution in this protocol is 30 μl (1/10th the volume of Trizol) but the protocol works well up to 50 μl. Aqueous solution comes from the sorted cells which are embedded in sheath fluid droplets and the amount of sorted cells should be set accordingly.
25. The aim of the qualitative analysis is to evaluate the degree of degradation of the RNA, its purity and its concentration. We routinely evaluate degradation by Bioanalyser (Agilent) using RNA Pico Chip which is said to work with as low as 50 pg/μl of total RNA. Integrity of ribosomal RNA is used as a proxy. NanoDrop allows for quantifying total RNA starting from limited amount of material. Ratio of absorbance at 260 and 280 nm (260/280) is indicative of contamination. RNA should give a value close to 2.0, while proteins, phenol, or other contaminants which show absorbance near 280 nm will reduce the ratio below 2.0. EDTA and TRIzol reagents show absorbance at 230 nm, so the ratio 260/230 is indicative of contamination with these compounds.

Acknowledgments

This work was supported by ANR, partner of the ERASysBio+ initiative supported under the EU ERA-NET Plus scheme in FP7.

References

1. Lohmann I, McGinnis W (2002) Hox genes: it's all a matter of context. Curr Biol 12:R514–R516
2. Ponzielli R, Astier M, Chartier A et al (2002) Heart tube patterning in Drosophila requires integration of axial and segmental information provided by the Bithorax Complex genes and hedgehog signaling. Development 129:4509–4521
3. Lo PCH, Skeath JB, Gajewski K et al (2002) Homeotic genes autonomously specify the anteroposterior subdivision of the Drosophila dorsal vessel into aorta and heart. Dev Biol 251:307–319. doi:10.1006/dbio.2002.0839
4. Lovato TL, Nguyen TP, Molina MR, Cripps RM (2002) The Hox gene abdominal-A specifies heart cell fate in the Drosophila dorsal vessel. Development 129:5019–5027
5. Perrin L, Monier B, Ponzielli R et al (2004) Drosophila cardiac tube organogenesis requires multiple phases of Hox activity. Dev Biol 272:419–431. doi:10.1016/j.ydbio.2004.04.036
6. Monier B, Astier M, Sémériva M, Perrin L (2005) Steroid-dependent modification of Hox function drives myocyte reprogramming in the Drosophila heart. Development 132:5283–5293. doi:10.1242/dev.02091
7. Hueber SD, Bezdan D, Henz SR et al (2007) Comparative analysis of Hox downstream genes in Drosophila. Development 134:381–392. doi:10.1242/dev.02746
8. Pavlopoulos A, Akam M (2011) Hox gene Ultrabithorax regulates distinct sets of target genes at successive stages of Drosophila haltere morphogenesis. Proc Natl Acad Sci U S A 108:2855–2860. doi:10.1073/pnas.1015077108
9. Thomas A, Lee P-J, Dalton JE et al (2012) A versatile method for cell-specific profiling of translated mRNAs in Drosophila. PLoS One 7:e40276. doi:10.1371/journal.pone.0040276
10. Miller MR, Robinson KJ, Cleary MD, Doe CQ (2009) TU-tagging: cell type-specific RNA isolation from intact complex tissues. Nat Methods 6:439–441. doi:10.1038/nmeth.1329
11. Steiner FA, Talbert PB, Kasinathan S et al (2012) Cell-type-specific nuclei purification from whole animals for genome-wide expression and chromatin profiling. Genome Res 22:766–777. doi:10.1101/gr.131748.111
12. Henry GL, Davis FP, Picard S, Eddy SR (2012) Cell type-specific genomics of Drosophila neurons. Nucleic Acids Res 40:9691–9704. doi:10.1093/nar/gks671
13. Bryantsev AL, Cripps RM (2012) Purification of cardiac cells from Drosophila embryos. Methods 56:44–49. doi:10.1016/j.ymeth.2011.11.004
14. Salmand P-A, Iché-Torres M, Perrin L (2011) Tissue-specific cell sorting from Drosophila embryos: application to gene expression analysis. Fly (Austin) 5:261–265. doi:10.4161/fly.5.3.16509
15. Abreu-Blanco MT, Verboon JM, Parkhurst SM (2011) Cell wound repair in Drosophila occurs through three distinct phases of membrane and cytoskeletal remodeling. J Cell Biol 193:455–464. doi:10.1083/jcb.201011018

Chapter 12

Hox Transcriptomics in *Drosophila* Embryos

Maria Polychronidou and Ingrid Lohmann

Abstract

Hox proteins are evolutionarily conserved homeodomain containing transcription factors that specify segment identities along the anteroposterior axis of almost all bilaterian animals. They exert their morphogenetic role by transcriptionally regulating a large battery of downstream target genes. Therefore the dissection of transcriptional networks regulated by Hox proteins is an essential step towards a mechanistic understanding of how these transcription factors coordinate multiple developmental and morphogenetic processes. High-throughput techniques allowing whole-transcriptome mRNA expression profiling are powerful tools for the genome-wide identification of Hox downstream target genes in a variety of experimental settings. Here, we describe how to quantitatively identify Hox downstream genes in *Drosophila* embryos by performing a Hox transcriptome analysis using microarrays.

Key words Hox, *Drosophila*, Transcription, Microarray, Transcriptome, Expression profiling

1 Introduction

Hox genes are expressed along the anteroposterior axis of animal embryos where they coordinate morphogenetic processes in a segment-specific manner [1, 2]. The essential role of *Hox* genes in specifying segment identities is evidenced by the "homeotic transformations" that are frequently observed upon disruption of *Hox* function in *Drosophila* [1]. *Hox* genes encode homeodomain transcription factors that exert their regulatory functions in diverse cell and tissue types by activating or repressing downstream target genes [3]. Identification of Hox target genes is crucial for elucidating how Hox proteins carry out their regulatory role in vivo and how distinct morphogenetic programs are executed under the control of Hox proteins in the different embryonic segments. The complexity of Hox regulatory networks in combination with the frequently encountered and degenerated Hox binding sites and the involvement of co-regulatory factors in the transcriptional regulation of Hox target genes present significant limitations for the identification of Hox target genes using in silico or in vitro

Yacine Graba and René Rezsohazy (eds.), *Hox Genes: Methods and Protocols*, Methods in Molecular Biology, vol. 1196,
DOI 10.1007/978-1-4939-1242-1_12,

approaches [4, 5]. Consequently, it is required to employ in vivo strategies in order to identify Hox-responsive gene regulatory networks.

Studies in *Drosophila* have provided a substantial amount of knowledge on the molecular mechanisms underlying Hox protein functions. In recent years, high-throughput analyses of Hox-regulated transcriptional networks using genome-wide methods have shed light on the mechanistic basis of how Hox proteins acquire their high specificity and selectivity (reviewed in refs. 5, 6). A number of studies in which a Hox transcriptome analysis was performed in *Drosophila* allowed the genome-wide identification of Hox downstream genes [7–11]. Importantly, the quantitative identification of target genes for six out of eight *Drosophila* Hox proteins, by performing transcriptome analysis using microarrays and whole-*Drosophila* embryos ubiquitously overexpressing each of the selected Hox proteins [10], demonstrated that Hox proteins control the expression of hundreds of genes and that different Hox proteins, despite their similar DNA-binding properties in vitro, show highly specific effects on the transcriptome (reviewed in refs. 5, 12).

For a variety of reasons, in experiments aiming at target gene identification Hox overexpression is advantageous in comparison to Hox loss of function. First of all, the extensive cross-regulation of Hox genes introduces complications in the interpretation of the results from experiments using mutants [13]. An additional limitation arises from the fact that Hox proteins are only expressed in small subsets of cells [2] and therefore the changes in gene expression taking place in the Hox-deficient cells will be diluted in the heterogeneous RNA mixture isolated from whole embryos. Performing the transcriptome analysis using isolated cells in which the Hox protein of interest is active could solve this problem. Considering that cellular and nuclear mRNA pools are very comparable [14], isolated nuclei are a reliable source of cell-type-specific transcripts. Thus, the recently developed methods allowing isolation of pure populations of nuclei from selected cell types [15, 16] could be employed. Importantly, given the context-dependent activity of Hox proteins, these methods offer the possibility to identify targets of a given Hox protein in different tissue types.

In this chapter we provide the protocol for identification of Hox target genes in *Drosophila* embryos by performing a transcriptome analysis using microarrays. The procedure does not only apply to whole embryos but can also be used with different experimental settings, i.e., when using larval tissues or purified cell populations. Principally, each gene of a given genome is represented on a microarray chip, in the case of Affymetrix Chips by a series of different oligonucleotide probes that serve as unique, sequence-specific detectors. The chip is hybridized with biotinylated cRNA that was generated from oligo-dT-primed cDNA of a given tissue. Each sample (for example, "wild-type" versus "mutant" tissue) is hybridized to a separate array. After hybridization, the chip is stained with a fluorescent molecule (streptavidin-phycoerythrin) that binds to biotin. The staining protocol includes a signal amplification step that employs anti-streptavidin antibody (goat) and biotinylated goat IgG antibody. After washing, transcript levels are calculated by reference to cRNA spikes of known concentration added to the hybridization mixture. Differences in mRNA levels between samples are determined by comparison of any two hybridization patterns produced on separate arrays of the different samples. Here, we describe the procedures for collecting the *Drosophila* embryos, isolating the total RNA, synthesizing the complementary DNA (cDNA) and the complementary RNA (cRNA), controlling the quality of the cRNA probes, performing the hybridization to the microarrays, and scanning the probes.

2 Materials

2.1 Drosophila Embryo Collection

Apple juice agar plates: Add 24 g agar to 740 ml ultrapure water and autoclave. Keep warm so that it does not become solid. Boil 250 ml apple juice with 25 g sugar and add to the autoclaved agar. Stir on a magnetic plate until the mixture had cooled down to approximately 50 °C and pour in petri dishes. The plates can be stored at 4 °C. Fresh yeast paste is spread on the plates directly before use. 100 % bleach is used for dechorionating the embryos.

2.2 Isolation of RNA from Drosophila Embryos

1. RNeasy Mini Kit (QIAGEN).
2. Polypropylene micropestle (Eppendorf).
3. QIAshredder spin columns (QIAGEN).

2.3 Double-Stranded cDNA (ds-cDNA) and cRNA Synthesis and Cleanup, cRNA Fragmentation

MessageAmp™ II-Biotin Enhanced, Single Round aRNA Amplification Kit (Ambion). All reagents required for the synthesis and cleanup of ds-cDNA and biotinylated cRNA are provided by the kit.

2.4 Agarose Gel Analysis

1. Denaturing gel loading buffer: 0.5 μl ethidium bromide (1 mg/ml), 1 μl 10× MOPS buffer, 5 μl formamide, 5 μl formaldehyde (37 %), 3 μl RNA loading dye (50 % glycerol containing 2.5 mg/ml bromophenol blue and xylene cyanol, each).
2. RNA gel: It is not necessary to run a denaturing RNA gel. Prepare a 1 % agarose gel for the unfragmented samples and 2 % agarose gel for the fragmented samples.

2.5 Hybridization, Washing, Staining, and Scanning

Additional arrays and staining solutions can be stored at 4 °C until needed.

1. Arrays: GeneChip Drosophila Genome 2.0 Array (Affymetrix).
2. Streptavidin-phycoerythrin (SAPE) solution (final volume 1,200 µl): 600 µl 2× 2-(*N*-morpholino)ethanesulfonic acid (MES) stain buffer, 120 µl acetylated bovine serum albumin (BSA) (20 mg/ml), 12 µl SAPE, 468 µl water. Split in two aliquots of 600 µl that will both be used for one array (*see* **Note 1**).
3. Antibody solution (final volume 600 µl): 300 µl 2× MES stain buffer, 60 µl acetylated BSA (20 mg/ml), 6 µl goat IgG (10 mg/ml), 3.6 µl biotinylated antibody (0.5 mg/ml), 230.4 µl water.
4. 2× Hybridization buffer: 200 mM MES, 0.5 M NaCl, 10 mM ethylenediaminetetraacetic acid (EDTA), 0.005 % Tween-20.
5. 20× SSPE buffer (1 l): 175.3 g sodium chloride, 27.6 g sodium phosphate monobasic, 9.4 g ethylenediaminetetraacetic acid (EDTA).
6. Wash buffer A: Non-stringent wash buffer consisting of 6× SSPE buffer (use the 20× SSPE stock described in **item 5**) and 0.01 % Tween-20.

3 Methods

3.1 Drosophila Embryo Collection

1. Two days before starting the embryo collection transfer newly hatched adult flies into a collection cage and cover it with an apple juice agar plate with yeast paste (*see* **Note 2**). Collection and aging times depend on the desired embryonic stage(s). When collecting embryos keep in mind that it is recommended to perform microarray experiments using biological triplicates.
2. Add water to the apple juice agar plate and brush the embryos to release them from the agar surface. Transfer the embryos to a sieve, rinse thoroughly with water, and dechorionate in 100 % bleach. Check under the microscope that the chorion has been completely removed. Wash the embryos well with water and remove as much liquid as possible by blotting the sieve with a paper towel. Using a brush, transfer the embryos in a pre-weighed microcentrifuge tube, measure the weight, and snap-freeze in liquid nitrogen. At this point the embryos can be stored at −80 °C.

3.2 Isolation of RNA from Drosophila Embryos

Use of RNeasy Mini Kit (QIAGEN) is recommended for the isolation of total RNA. Make sure that sterile and nuclease-free glassware or plastic are used throughout the procedure (*see* **Note 3**) and follow the instructions for "isolation of total RNA from animal tissues." The following procedure applies to isolation of RNA from whole embryos. The required modifications in cases when tissues (*see* **Note 4**) or sorted cells (*see* **Note 5**) are used are described separately.

1. Remove the frozen embryos from storage. Do not use more than 30 mg of embryos per sample. Do not allow the frozen embryos to thaw and proceed to **step 2** as fast as possible.
2. Before use, ensure that β-mercaptoethanol is added to Buffer RLT (RNA easy mini kit, QIAGEN). Add 400 μl Buffer RLT to the embryos and lyse them by homogenizing in the microcentrifuge tube using a polypropylene pestle. Alternatively the embryos can be lysed in a Dounce homogenizer using a tight pestle.
3. Transfer the lysate to a new labeled tube.
4. Rinse both pestle and grinder with 200 μl RLT and combine the solution with the lysate of the previous step.
5. Incubate at RT for 5 min.
6. Centrifuge the lysate at maximum speed in a microcentrifuge for 3 min.
7. Carefully transfer the supernatant (cleared lysate) to a new microcentrifuge tube.
8. Add 600 μl of 70 % ethanol (prepared with diethylpyrocarbonate (DEPC)-treated water) to the cleared lysate and mix immediately by pipetting.
9. Transfer up to 700 μl of the sample, including any precipitate that may have formed, to an RNeasy spin column placed in a 2 ml collection tube. Centrifuge for 15 s at 8,000 × *g* and discard the flow-through. (If the sample volume exceeds 700 μl, centrifuge successive aliquots in the spin column.)
10. Add 700 μl Buffer RW1 (QIAshredder spin columns, QIAGEN) to the RNeasy spin column. Centrifuge for 15 s at 8,000 × *g* and discard the flow-through.
11. Before use, make sure that ethanol is added to Buffer RPE (RNA easy mini kit, QIAGEN). Add 500 μl Buffer RPE to the RNeasy spin column. Centrifuge for 15 s at 8,000 × *g* and discard the flow-through.
12. Repeat **step** 7 and centrifuge for 2 min at 8,000 × *g*.
13. Transfer the RNeasy spin column in a new 2 ml collection tube and centrifuge at full speed for 1 min to ensure no carryover of ethanol.
14. Place the RNeasy spin column in a new 1.5 ml microcentrifuge tube and add 30 μl RNase-free water directly to the membrane. Centrifuge for 1 min at 8,000 × *g*.

15. Optional: Repeat **step 10** using another 30–50 µl RNase-free water. Do not mix the two eluates before measuring the RNA concentration.
16. Measure RNA concentration and 260 nm/280 nm absorbance ratio (as a control of RNA purity) using Nanodrop. At this step the two eluates can be pooled together depending on their concentration. The RNA can be stored at −80 °C.

3.3 ds-cDNA Synthesis

Use the "MessageAmpTm II-Biotin Enhanced, Single Round aRNA Amplification Kit" (Ambion).

3.3.1 First-Strand cDNA Synthesis

1. Before you start prepare spike controls (*see* **Note 6**) according to the following dilution table. The first dilution can be stored at −20 °C and reused up to three times.

Total RNA (µg)	Spike control dilution			
	First dilution	Second dilution	Third dilution	Final dilution
1	1:20	1:50	1:25	1:25,000
5	1:20	1:50	1:5	1:5,000
10	1:20	1:50	1:2.5	1:2,500

2. Prepare the following mix and keep the amount of RNA constant for all samples:
   10 µl total RNA (1 µg in 10 µl water is recommended when using the Ambion kit), 1 µl T7 Oligo(dT) primer, 1 µl spike controls (according to the dilution table).
3. Incubate RNA at 70 °C for 10 min.
4. Chill rapidly on ice.
5. For each sample, prepare a master mix containing the following reagents: 2 µl 10× first-strand buffer, 4 µl dNTP Mix, 1 µl RNase inhibitor, and 1 µl ArrayScript Reverse Transcriptase.
6. Add 8 µl master mix to each RNA sample (final volume 20 µl) and mix gently. Spin down briefly if necessary.
7. Incubate at 42 °C for 2 h.

3.3.2 Second-Strand cDNA Synthesis

1. Place the first-strand cDNA synthesized in Subheading 3.3.1 on ice.
2. For each sample, prepare a master mix containing the following reagents: 63 µl RNase-free water, 10 µl 10× second-strand buffer, 4 µl dNTP mix, 2 µl DNA polymerase, 1 µl RNase H.
3. Mix gently and spin down briefly if necessary.

4. Add 80 μl master mix to each sample. Mix thoroughly by pipetting up and down two to three times. Spin down briefly if necessary.
5. Incubate at 16 °C for 2 h. Take care not to let the temperature of the samples rise above 16 °C.

3.3.3 ds-cDNA Cleanup

1. Before you begin preheat the nuclease-free water at 50–55 °C and make sure that ethanol has been added to the bottle of wash buffer.
2. Add 250 μl cDNA-binding buffer to each sample and mix thoroughly by pipetting up and down two to three times. Spin down briefly if necessary.
3. Apply the sample mix to a cDNA Cleanup Spin Column placed in a 2 ml collection tube.
4. Spin at 8,000 × *g* for 1 min and discard flow-through.
5. Apply 500 μl cDNA wash buffer to each spin column.
6. Spin at 8,000 × *g* for 1 min and discard flow-through.
7. Spin once again at 8,000 × *g* for 1 min in order to remove trace amounts of wash buffer.
8. Transfer the spin column into a new 1.5 ml collection tube and pipet 10 μl preheated nuclease-free water directly onto the center of the membrane.
9. Incubate for 2 min.
10. Spin at 8,000 × *g* for 1 min.
11. Pipet 12 μl preheated nuclease-free water directly onto the center of the membrane and incubate for 2 min.
12. Spin at 8,000 × *g* for 1 min.
13. The total elution volume will be 20 μl. At this point the ds-cDNA can be stored at −80 °C.

3.4 cRNA Synthesis

Use the "MessageAmpTm II-Biotin Enhanced, Single Round aRNA Amplification Kit" (Ambion).

3.4.1 In Vitro Transcription

1. Thaw ds-cDNA on ice and prepare the following in vitro transcription master mix for each sample: 12 μl biotin-NTP mix, 4 μl T7 10× reaction buffer, 4 μl T7 enzyme mix.
2. Add 20 μl in vitro transcription master mix to 20 μl ds-cDNA (final volume 40 μl) and mix gently. Spin down briefly if necessary.
3. Incubate at 37 °C overnight (approximately 16 h).
4. Add 60 μl nuclease-free water to each cRNA sample (final volume 100 μl). At this point the cRNA can be stored at −80 °C.

3.4.2 cRNA Cleanup

1. Before you begin preheat nuclease-free water to 50–60 °C and assemble cRNA filter cartridges and tubes.
2. Add 350 μl cRNA-binding buffer to the cRNA and vortex.
3. Add 250 μl ethanol (>99 %) to each sample and mix thoroughly by pipetting the mixture up and down two to three times. Do not vortex and do not centrifuge the samples.
4. Apply sample mix (final volume 700 μl) to a cRNA Cleanup Spin Column placed in a 2 ml collection tube.
5. Spin at 8,000 × *g* for 1 min and discard flow-through.
6. Add 650 μl wash buffer (supplemented with ethanol) to each sample.
7. Spin at 8,000 × *g* for 1 min and discard flow-through.
8. Spin at 8,000 × *g* for 1 min to remove trace amounts of wash buffer.
9. Transfer spin column into new 1.5 ml collection tube and pipet 40 μl preheated nuclease-free water directly onto the membrane.
10. Incubate for 2 min.
11. Spin at 8,000 × *g* for 1 min to elute cRNA.
12. Transfer spin column into new 1.5 ml collection tube and repeat elution with an additional 60 μl preheated nuclease-free water.
13. Use 1 μl of the first eluate to measure cRNA concentration and 260/280 ratio.

    Optional steps, in case the cRNA concentration is lower than 1.0 μg/μl:
14. Combine the two cRNA eluates (final volume 100 μl).
15. Add 10 μl 5 M ammonium acetate and 275 μl 100 % ethanol.
16. Mix well and incubate at −20 °C for at least 30 min.
17. Centrifuge at full speed for 15 min at 4 °C and discard supernatant.
18. Wash pellet with 500 μl cold 70 % ethanol.
19. Centrifuge at full speed for 15 min at 4 °C or room temperature.
20. Remove 70 % ethanol.
21. Spin briefly, remove any residual fluid, and air-dry the pellet.
22. Resuspend the cRNA in an appropriate volume of nuclease-free water.
23. Use 1 μl to measure cRNA concentration and 260 nm/280 nm absorbance ratio.

3.5 cRNA Fragmentation

1. Set up the fragmentation reaction (final volume: 20 μl) by mixing the following: 15 μg cRNA in a total volume of 16 μl nuclease-free water and 4 μl 5× fragmentation buffer.
2. Incubate at 94 °C for 35 min.
3. Chill rapidly on ice.

3.6 Agarose Gel Analysis

It is recommended to check the quality of the synthesized cRNA and the fragmentation on an agarose gel in order to confirm the quality of the probe before starting the hybridization procedure.

1. Add 5 μl gel loading buffer to the following samples:
   (a) 2 μl unfragmented cRNA.
   (b) 2 μl fragmented cRNA.
2. Incubate at 65 °C for 10 min and chill on ice.
3. Run the unfragmented samples on a 1 % agarose gel and the fragmented samples on a 2 % agarose gel.

3.7 Preparation of Hybridization Cocktail and Hybridization

1. For each sample prepare a master mix containing the following reagents: 4.25 μl control oligo B2, 12.5 μl 20× eukaryotic hybridization controls, 2.5 μl salmon sperm DNA, 6.25 μl BSA (20 mg/ml), 125.0 μl 2× hybridization buffer, 25.0 μl dimethyl sulfoxide (DMSO), 56.5 μl water.
2. Add 232 μl master mix to the remaining 18 μl fragmented cRNA from each sample (final volume 250 μl). At this point the samples can be stored at −80 °C.
3. Before hybridization, denature the probe for 5 min at 95 °C, followed by 5 min at 45 °C.
4. Prehybridize the probe array with 250 μl 1× hybridization buffer at 45 °C for at least 10 min.
5. Hybridize the probe array for 16 h at 45 °C while shaking at 60 rpm.

3.8 Washing, Staining, and Scanning of the Arrays

Before you start enter experiment description into Affymetrix workstation (GCOS). Unless the experiments have been defined first, you will not be able to wash, stain, and scan your arrays. Wash and stain the probe array using the Affymentrix Fluidics Station 450, which is operated using GCOS/Microarray Suite. Throughout the procedure, follow the instructions in the LCD window on the fluidics station.

1. Prime fluidics station with wash buffers in order to make sure that the lines of the fluidics station are filled with the appropriate buffers.
2. Remove probe from arrays and fill arrays with wash buffer A. Take care not to introduce any bubbles. If necessary, the probes can be frozen at −80 °C at this point (*see* **Note 7**).

3. Select the correct experiment name and the appropriate antibody amplification protocol to control the washing and staining of the probe array, from the drop-down Experiment list.
4. Load the three experiment microcentrifuge vials into the sample holders on the fluidics station as follows: Place one vial containing 600 μl of SAPE solution in sample holder 1, one vial containing 600 μl of antibody solution in sample holder 2, and one vial containing 600 μl of SAPE solution in sample holder 3.
5. Start the run. The Fluidics Station dialog box displays the status of the washing and staining during the procedure. After the run has finished, if you do not plan to scan the arrays immediately, keep them at 4 °C in the dark until ready for scanning.
6. Clean excess fluid from around septa on the back of the probe array cartridge.
7. Carefully apply one Tough-Spots to each of the two septa (*see* **Note 8**).
8. Insert the cartridge into the scanner and ensure that the spots lie flat.
9. Select the correct experiment name, insert the probe array into the holder, and start the scan.
10. After completion of the scan, the data is ready for analysis.

4 Notes

1. SAPE should be stored in the dark at 4 °C and should not be frozen. Before preparing staining solution take SAPE out of the refrigerator and mix well. The SAPE solution should always be prepared fresh.
2. The apple juice agar plates and yeast paste must be fresh and pre-warmed before use in order to maximize the yield of the collections.
3. To provide RNase-free conditions, bake all the glassware for 2 h at 300 °C, use DEPC-H_2O for preparing the solutions, and wear gloves and a lab coat when working with RNA samples.
4. A few parameters need to be considered when isolating RNA from dissected tissues (i.e., imaginal discs or other larval tissues). During dissection, the tissues of interest should be kept on ice at all times so that the RNA does not degrade. Snap-freeze the tissue in liquid nitrogen as soon a possible. If the amount of dissected tissue is smaller than 20 mg, use 300 μl of RLT buffer.
5. When using isolated cells (i.e., isolated by FACS sorting), follow the protocol "Purification of Total RNA from Animal Cells"

provided in the RNeasy Mini Kit (QIAGEN) and adjust the volume of RLT according to the number of cells as described in the manual.

6. The use of external RNA controls ("spikes") allows assessing the performance of the microarray experiment. The probes that correspond to the spikes are informative of the fidelity of the microarray technology in determining their presence. It is recommended to spike external controls in an early step of sample processing in order to monitor several steps.
7. If the probes were frozen, equilibrate them to room temperature.
8. Press to ensure that the spots remain flat. If the Tough-Spots do not apply smoothly and you observe bumps, bubbles, tears, or curled edges, remove the spot and apply a new one. Do not attempt to smoothen the spot.

Acknowledgements

We would like to thank Jan U. Lohmann and Markus Schmid for sharing protocols with us.

References

1. Lewis EB (1978) A gene complex controlling segmentation in *Drosophila*. Nature 276: 565–570
2. McGinnis W, Krumlauf R (1992) Homeobox genes and axial patterning. Cell 68:283–302
3. Hombría JC-G, Lovegrove B (2003) Beyond homeosis—HOX function in morphogenesis and organogenesis. Differentiation 71:461–476. doi:10.1046/j.1432-0436.2003.7108004.x
4. Pearson JC, Lemons D, McGinnis W (2005) Modulating *Hox* gene functions during animal body patterning. Nat Rev Genet 6:893–904. doi:10.1038/nrg1726
5. Hueber SD, Lohmann I (2008) Shaping segments: *Hox* gene function in the genomic age. Bioessays 30:965–979. doi:10.1002/bies.20823
6. Polychronidou M, Lohmann I (2013) Cell-type specific cis-regulatory networks: insights from Hox transcription factors. Fly (Austin) 7:13–17. doi:10.4161/fly.22939
7. Leemans R, Loop T, Egger B et al (2001) Identification of candidate downstream genes for the homeodomain transcription factor Labial in *Drosophila* through oligonucleotide-array transcript imaging. Genome Biol 2:RESEARCH0015
8. Mohit P, Makhijani K, Madhavi MB et al (2006) Modulation of AP and DV signaling pathways by the homeotic gene *Ultrabithorax* during haltere development in *Drosophila*. Dev Biol 291:356–367. doi:10.1016/j.ydbio.2005.12.022
9. Hersh BM, Nelson CE, Stoll SJ et al (2007) The UBX-regulated network in the haltere imaginal disc of *D. melanogaster*. Dev Biol 302:717–727. doi:10.1016/j.ydbio.2006.11.011
10. Hueber SD, Bezdan D, Henz SR et al (2007) Comparative analysis of Hox downstream genes in *Drosophila*. Development 134:381–392. doi:10.1242/dev.02746
11. Pavlopoulos A, Akam M (2011) *Hox* gene *Ultrabithorax* regulates distinct sets of target genes at successive stages of *Drosophila* haltere morphogenesis. Proc Natl Acad Sci U S A 108:2855–2860. doi:10.1073/pnas.1015077108
12. Pavlopoulos A, Akam M (2007) Hox go omics: insights from *Drosophila* into Hox gene targets. Genome Biol 8:208. doi:10.1186/gb-2007-8-3-208
13. Miller DF, Rogers BT, Kalkbrenner A et al (2001) Cross-regulation of *Hox* genes in the *Drosophila melanogaster* embryo. Mech Dev 102:3–16
14. Barthelson RA, Lambert GM, Vanier C et al (2007) Comparison of the contributions of

the nuclear and cytoplasmic compartments to global gene expression in human cells. BMC Genomics 8:340. doi:10.1186/1471-2164-8-340

15. Bonn S, Zinzen RP, Perez-Gonzalez A et al (2012) Cell type-specific chromatin immunoprecipitation from multicellular complex samples using BiTS-ChIP. Nat Protoc 7:978–994. doi:10.1038/nprot.2012.049
16. Steiner FA, Talbert PB, Kasinathan S et al (2012) Cell-type-specific nuclei purification from whole animals for genome-wide expression and chromatin profiling. Genome Res 22:766–777. doi:10.1101/gr.131748.111

Part II

Hox Proteins: Mode of Action and Biomedical Applications

Chapter 13

Measuring Hox-DNA Binding by Electrophoretic Mobility Shift Analysis

Kelly Churion, Ying Liu, Hao-Ching Hsiao, Kathleen S. Matthews, and Sarah E. Bondos

Abstract

Understanding gene regulation by Hox transcription factors requires understanding the forces that underlie DNA binding by these proteins. Electrophoretic mobility shift analysis (EMSA) not only allows measurement of protein affinity and cooperativity but also permits visualization of differently migrating protein-DNA complexes, including complexes with different compositions or complexes with identical compositions yet assembled in different geometries. Furthermore, protein activity can be measured, allowing correction of binding constants for the percentage of protein that is properly folded and capable of binding DNA. Protocols for measuring protein activity and the equilibrium DNA-binding dissociation constant (K_d) are provided. This versatile assay system can be adjusted based on specific needs to measure other parameters, including the kinetic association and dissociation constants (k_a and k_d) and the formation of heterologous protein-protein interactions.

Key words Hox, Gel retardation, Gel shift, Electrophoretic mobility shift analysis, EMSA, DNA binding, Activity, Affinity, Cooperativity

1 Introduction

DNA binding is an important, yet poorly understood, aspect of Hox transcription factor function. The "Hox paradox" states a long-standing question in the field: How do Hox proteins, which have very similar DNA-binding homeodomains, recognize different target DNA sites in vivo and thus regulate different genes? An extension of the Hox paradox is also problematic: How does a single Hox protein sense the identity of the tissue in which it is expressed and respond by binding the appropriate subset of its target genes? Understanding the forces that regulate DNA binding in full-length Hox proteins is necessary to answer these questions.

Electrophoretic mobility shift analysis (EMSA) is a simple, inexpensive, rapid, and versatile technique that is compatible with a wide variety of protein and DNA sizes. EMSA can be applied in

Yacine Graba and René Rezsohazy (eds.), *Hox Genes: Methods and Protocols*, Methods in Molecular Biology, vol. 1196, DOI 10.1007/978-1-4939-1242-1_13, © Springer Science+Business Media New York 2014

a qualitative manner to identify a protein in a mixture that binds a specific DNA sequence or to locate a DNA sequence in a mixture of DNAs that bind a specific protein [1]. Quantitative EMSA is capable of measuring DNA-binding affinity, cooperativity, and stoichiometry by directly visualizing all differentially migrating species, including complexes with different components and, in some cases, complexes with the same components arranged in different geometries [1, 2]. This aspect is particularly important for Hox proteins, which often bind DNA as homomeric or heteromeric multimers [3–5]. Common difficulties with Hox proteins can also be diagnosed with EMSA. Misfolded Hox proteins can be detected using EMSA activity assays, which quantify the percentage of protein capable of binding DNA. Hox proteins can also be prone to aggregation [6], and the intrinsically disordered regions of Hox proteins render them susceptible to proteolysis. These last two common problems can be easily diagnosed from the banding patterns observed.

The migration distance of free DNA in EMSA is dependent on DNA length and curvature [7]. Although we have only used DNA sequences 16–200 bp in length, others have successfully used DNA up to 500 bp in length [7]. Protein binding to DNA generally elicits a reduction in mobility in native gels, creating a second band "shifted" from the free DNA band. The size of proteins that can be used in these experiments ranges from peptides up to 1,000 kDa, allowing separation and visualization of higher order Hox protein complexes [8]. Radioactive labeling of DNA using end labeling or primer extension allows detection of both the bound and free species. The ability to detect protein binding by EMSA depends on the resolution of bound DNA and free DNA bands, the stability of the protein-DNA complex, the salt concentration of the gel running buffer, the temperature of the gel, and the gel run time [7, 9]. Although the duration of a typical EMSA experiment is typically longer than the half-life of protein-DNA complexes, the confinement of the complex within a gel pore generally promotes the rebinding of dissociated complexes [7–10]. The separation of the bands, and hence the resolution of the technique, depends on the size and charge of the protein(s), the structure and geometry of each migrating species, and the pore size of the gel [7]. Although DNA binding measured by EMSA generally correlates with measurements by other assays, validation by an alternate method (e.g., nitrocellulose filter binding [9]) is strongly recommended.

The ability of EMSA to separate and detect protein-DNA complexes with different stoichiometries offers a more in-depth view of cooperative DNA binding [3, 4]. Likewise, "supershift" experiments allow analysis of the impact of heterologous protein-protein interactions on DNA binding [3, 4, 11–13]. In addition, competition experiments can probe DNA-binding kinetics [4]. These and other useful variations of EMSA experiments are described in detail elsewhere [7–10]. For simplicity, this chapter focuses on measuring Hox activity and affinity for DNA sequences harboring a single DNA-binding site. Activity assays measure the fraction of protein that is capable of binding DNA. In any protein purification, some protein molecules will misfold or aggregate. Consequently protein affinity measurements must be corrected for the activity of the protein in that purification. In activity assays, the concentration of DNA must be at least tenfold greater than K_d, and protein concentrations span four orders of magnitude, with the mid-range reaction containing equimolar protein and DNA; hence, activity assays are sometimes called stoichiometry assays. These conditions force protein binding to DNA:

$$\frac{[\mathrm{DNA}]}{K_\mathrm{d}} = \frac{[\mathrm{P} * \mathrm{D}]}{[\mathrm{P}]_\mathrm{F}} > 10 \tag{1}$$

in which K_d is the equilibrium dissociation constant, [DNA] is the concentration of DNA, $[P]_F$ is the concentration of free protein, and [P*D] is the concentration of the bound complex. Under these conditions, the extent of protein binding is linearly dependent on protein concentration until the DNA-binding site is saturated. Quantitation of the density of the bound bands reveals a sloped line at protein concentrations below the saturation point, and a horizontal line at protein concentrations above the saturation point. These lines intersect at the point where the concentration of DNA equals the concentration of active protein capable of binding DNA (*see* **Note 1**). Division of the concentration of DNA by the total concentration of protein at this intersection yields the protein activity. To ensure that all measurements rely on high-quality protein samples, we recommend measuring activity each day that affinity is measured and rejecting the binding affinity data if the activity falls below 80 %.

In contrast, affinity (nonstoichiometric) assays are used to measure K_d and cooperativity. In affinity assays, the concentration of DNA must be at least tenfold lower than K_d. Under these conditions, the concentration of free protein greatly exceeds that of bound protein (Eq. 2), and thus the concentration of free protein, which is not measured, can be reasonably approximated by the concentration of total protein, which is known. This approximation simplifies data analysis [14]. Although some protein-DNA systems use ethidium bromide or fluorescent labels to detect DNA, the extremely high affinity of Hox homeodomains forces the use of picomolar DNA concentrations, which are best detected using radiolabels:

$$\frac{K_\mathrm{d}}{[\mathrm{DNA}]} = \frac{[\mathrm{P}]_\mathrm{F}}{[\mathrm{P} * \mathrm{D}]} > 10 \tag{2}$$

The use of Hox proteins in EMSA experiments creates specific challenges. As mentioned above, full-length Hox proteins are highly prone to aggregation. Aggregation can be minimized by expressing proteins from weaker promoters or at lower temperatures to reduce accumulation of unfolded or partly folded proteins [15]. We have developed an assay that rapidly identifies buffer additives that maintain proteins in a soluble, monomeric, and active state [6, 15, 16]. The aggregation status of purified Hox proteins should be tested before beginning gel shift experiments, and only active, monomeric protein should be used in DNA-binding assays. Because Hox proteins contain intrinsically disordered regions, they are especially susceptible to proteolysis during expression and purification. This issue can be generally solved by expressing recombinant Hox proteins for a short period of time and/or by purifying the products as rapidly as possible. Finally, buffer conditions, tight binding to columns, and the use of purification tags can potentially cause Hox proteins to misfold. These conditions can either result in the loss of activity or the loss of inhibitory autoregulatory interactions with the folded homeodomain and thus a paradoxical increase in the apparent DNA-binding affinity. Despite these difficulties, the DNA complexes of many Hox proteins can be accurately detected by EMSA [3, 5, 17–20].

2 Materials

2.1 DNA Annealing, DNA Labeling, and EMSA Equipment (Fig. 1)

1. Forward- and reverse-strand DNA oligos, HPLC purified.
2. 4 l beaker.
3. Stirring hot plate.
4. Foam floater to support microcentrifuge tubes in a water bath.
5. Disposable UV cuvettes.
6. Spectrophotometer.
7. Sand bath preheated to 37 °C (in radiation area).
8. Cover for tubes in sand bath that cannot be penetrated by ^{32}P radiation.
9. Vortex mixer (in radiation area).
10. Microcentrifuge (in radiation area).
11. Nick columns (GE Healthcare), a gel filtration column prepacked with Sephadex G-50 and used to separate labeled DNA from unincorporated ^{32}P-ATP.
12. Scintillation counter.
13. Gel boxes capable of running gels 20 cm long × 20 cm wide and open at the top to allow loading while the current is running.

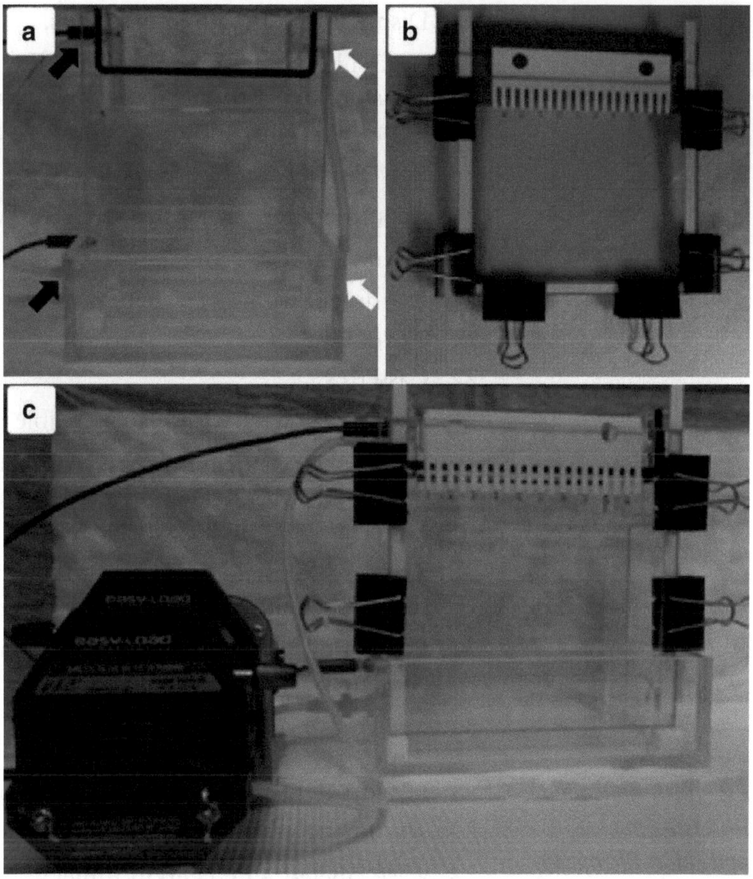

Fig. 1 EMSA apparatus assembly. (**a**) A custom-made gel box used for EMSA and designed to fit 20 cm × 20 cm glass gel plates. Ports leading to (bottom) and from (top) the pump are marked with *white arrows*, and the ports connecting the top and bottom chambers are marked with *black arrows*. (**b**) Assembly of plates for pouring gels. (**c**) Assembly of the EMSA apparatus. The bottom of the gel should be below the liquid in the bottom reservoir. The square glass plate should face outward, such that the top of the gel is exposed to buffer in the top gel reservoir. Finally, the port connecting the top buffer reservoir with the lower reservoir should be positioned above the top of the gel so that the top of the gel will remain submerged while the recirculation pump runs

Ports should be available at the top and bottom of the gel boxes to allow recirculation of the running buffer (Fig. 1).

14. Glass plates, one square and one toothed as pictured in Fig. 1.
15. Kimwipes.
16. Combs (20 lanes, 2 mm thick).
17. Spacers (2 mm thick).
18. Binder clips, 2 in., 8 counts per gel.
19. Syringe with needle narrower than the gel spacers.
20. Plastic disposable transfer pipets.

21. Disposable gel-loading pipet tips.
22. Recirculation pumps (MasterFlex L/S model 7518-60) (MasterFlex silicone tubing (platinum)) with tubing that fits both the pumps and the gel box ports.
23. Power supply and leads.
24. Cellulose filter paper (3MM CHR, Whatman).
25. Saran Wrap (*see* **Note 2**).
26. Gel dryer (we use Drygel Sr. model SE1160; other types are acceptable).
27. Cassettes.
28. Phosphorimager with plates capable of detecting ^{32}P.
29. Purified Hox protein at micromolar concentration, free of proteolytic products and tested for aggregation. Aggregation assays are described elsewhere [6, 15].
30. Software for data analysis: IgorPro version 6, SigmaPlot, or similar that allows you to write equations for data fitting.
31. Radiation shielding and shielded microcentrifuge tube containers.
32. Microwave oven.

2.2 DNA Annealing, DNA Labeling, and EMSA Chemicals and Reagents

1. H_2O in all protocols refers to Millipore-filtered ddH_2O.
2. 10× PCR buffer (Invitrogen).
3. γ-^{32}P labeled ATP (Perkin Elmer).
4. $MgCl_2$.
5. T4-PNK enzyme, including 10× PNK buffer (NEB).
6. EDTA.
7. Tris base (hydroxymethyl)aminomethane (Tris base).
8. Sigmacote (Sigma).
9. Fresh TEMED (Sigma).
10. Ammonium Persulfate (APS) (BioRad).
11. 6× Blue/Orange Loading Dye (Promega).
12. HCl and NaOH for adjusting pH of buffers.
13. Agarose.
14. Dithiothreitol (DTT) (Research Products International).
15. Ethanol.
16. Boric acid.
17. Potassium chloride (KCl).
18. Glycerol.
19. Bovine serum albumin (BSA) (Sigma).
20. Polyacrylamide (40 %, 37.5:1) (Fisher).

2.3 Preparation of 0.5× TBE, the EMSA Running Buffer

1. Prepare 0.5 M EDTA: Mix 9.3 g EDTA with 40 ml H_2O and adjust pH to 8.0 with NaOH to dissolve EDTA. Bring the final volume to 50 ml and store at room temperature.
2. Prepare 1× TBE buffer: Mix 10.9 g Tris base, 5.56 g boric acid, and 4 ml of 0.5 M EDTA stock with 500 ml of H_2O and stir. This typically results in a solution close to pH 8.3. Adjust pH if necessary with NaOH. Bring the final volume to 1 l with H_2O.
3. Prepare 0.5× TBE buffer: Mix 1 l of 1× TBE buffer with 1 l of H_2O.

2.4 Preparation of 0.1 M DTT

Prepare a 0.1 M DTT stock: Dissolve 15 mg DTT in 1 ml H_2O. Gently vortex and store at 4 °C. Use that same day.

2.5 Preparation of TE Buffer (for the Nick Column)

Dissolve 0.121 g Tris base and 37.2 mg EDTA in 50 ml H_2O and stir. Adjust pH to 7.5 with HCl, and dilute with H_2O to a final volume of 100 ml. Final concentrations are 10 mM Tris base and 1 mM EDTA. Store at room temperature.

2.6 Preparation of DNA-Binding Buffer (See Note 3)

1. Prepare a 2 M KCl stock: Dissolve 37.28 g of KCl in 250 ml of H_2O and store at room temperature.
2. Prepare a 0.4 M Tris base stock: Weigh 12.11 g Tris base and mix in a beaker containing 100 ml of H_2O. Adjust pH to 7.5 and then dilute to 250 ml final volume. Store this 0.4 M Tris base stock at room temperature.
3. Prepare an 80 % glycerol stock: In a 100 ml graduated cylinder measure 80 ml of glycerol. Fill to 100 ml with H_2O. Mix thoroughly and store at room temperature.
4. Prepare a 2 mg/ml BSA stock: Weigh 20 mg of BSA, gently mix into 10 ml H_2O, and store at 4 °C.
5. DNA-binding buffer: Add 5 ml H_2O to a 15 ml sterile disposable conical centrifuge tube. Add 500 μl from the 2 M KCl stock, 500 μl Tris base from the 0.4 M stock, 500 μl from the 2 mg/ml BSA stock, 1.25 ml from the 80 % glycerol stock, and 100 μl of 0.1 M DTT. Adjust pH to 7.5 with HCl. Bring the final volume to 10 ml, gently vortex solution, and store at 4 °C until use on the same day. Final buffer composition: 0.1 M KCl, 20 mM Tris base, 100 μg/ml BSA, 10 % (v/v) glycerol, and 1 mM DTT, pH 7.5.

2.7 Preparation of Ammonium Persulfate (APS) Solution

1. Prepare a 10 % stock solution: Dissolve 0.1 g of ammonium persulfate in 1 ml H_2O. Vortex gently and immediately store at 4 °C.
2. This APS stock should be replaced after 2 weeks.

2.8 Preparation of 1 % Agarose

1. Prepare a 1 % stock solution: Weigh 1 g of agarose and mix with 99 ml of H_2O in a 250 ml Erlenmeyer flask. Gently swirl

and place in microwave for 1–2 min until the solution begins to boil. Remove carefully (wearing protection) and allow to cool to room temperature. Use the solution before it solidifies.

2. Solidified agarose can be stored at room temperature. Reheat the agarose to liquefy and reuse the solution.

3 Methods

3.1 Annealing DNA Oligonucleotides (Oligos) (See Note 4)

1. Dilute each oligo in sterile water to a concentration of 100 μM.
2. Anneal DNA: Mix 10 μl of 10× PCR buffer, 20 μl of each oligo, 3 μl 50 mM $MgCl_2$, and 47 μl of H_2O for a total volume of 100 μl. Be certain that the final concentration of both oligos is the same. The final buffer composition is 1× PCR buffer and 1.5 μM $MgCl_2$.
3. Heat a 4 l glass beaker filled with approximately 3 l H_2O to 95 °C.
4. Using the floater, incubate the annealing reaction in the water bath at 95 °C for 10 min.
5. Turn off the heat and let the temperature fall to less than 40 °C (~3 h). Store the annealed DNA at −20 °C until needed.
6. Calculate or use UV spectroscopy to measure the DNA concentration of each annealed oligo. Use 60 μl of H_2O in a cuvette to measure the baseline spectrum. Add 2 μl of your DNA to the H_2O, and mix by pipetting. Calculate the concentration of DNA based on the absorbance of DNA at 260 nm using the following equation:

$$[\text{DNA}]\text{in } \mu g / \mu l = \frac{((A_{260})(\text{dilution factor})(50))}{1{,}000} \tag{3}$$

in which the dilution factor is 31 (2 μl into 62 μl), and A_{260} is the absorbance at 260 nm. If you are using a particularly long DNA sequence, additional dilution may be required for accurate measurement.

3.2 Labeling DNA Oligos

1. Use proper procedures and equipment for handling all radiolabeled materials, including shielding, double gloves, lab coats, and monitoring devices.
2. Thaw the γ^{32}P-labeled ATP at room temperature for ~30 min or until liquid. ATP should be less than 30 days old.
3. Assemble the kinase reaction: Mix in a microcentrifuge tube 1 μl 0.1 M DTT, 1 μl annealed DNA (3×10^{-7} M), 1.5 μl 10× PNK buffer, 10.5 μl γ^{32}P ATP, and 2 μl T4-PNK enzyme (add this component last) for a total volume of 16 μl.

4. Mix all the components very well by gently vortexing.
5. Incubate the reaction at 37 °C for 1.5 h in the sand bath. Keep the reactions covered to prevent evaporation/condensation and reduce radiation exposure.
6. Stop the reaction by adding 2 μl of 0.5 M EDTA pH 8.0.
7. Add 32 μl H_2O for a final volume of 50 μl and vortex gently.
8. At this point, most of the radioactivity is due to unincorporated ^{32}P-ATP. To decrease radiation exposure and improve band visibility, unincorporated nucleotide must be removed by gel filtration chromatography. We find Nick columns, which are pre-packed with Sephadex G-50 and disposable, to be a convenient approach. Attach a Nick column to a support. Allow the liquid in the column to flow through the resin, leaving the frit at the top almost dry. Equilibrate the column by adding 4 ml of 1× TE buffer and letting it flow through the column. Be ready to add the completed labeling reaction; the column should not be allowed to dry out.
9. Load the labeling reaction onto the column, and as soon as it is absorbed into the column add 400 μl of 1× TE buffer. Collect the wash in a 1.5 ml microcentrifuge tube. This first wash does not contain your radiolabeled dsDNA.
10. As soon as all of the first TE buffer aliquot from **step 9** enters the column, add another 400 μl of 1× TE buffer to the column. Collect the eluent in another 1.5 ml Falcon tube. This second wash will contain the radiolabeled dsDNA carefully mix so that the DNA concentration is homogeneous.
11. Dilute 1 μl of labeled oligo into 5 ml scintillation fluid and measure radiation levels in a scintillation counter. The cpm/μl should be ~15,000.
12. Aliquot 40 μl into microcentrifuge tubes and freeze in a shielded container in a freezer approved for use with radiation.

3.3 Pouring the Gel

1. To help the gel release easily from the glass plates, the inner face of each place must be siliconized. Hold a pad of KimWipes across the top of an open bottle of Sigmacote and tilt to partially wet the KimWipes. Rub the side of the glass plates that will face the gel with the Sigmacote-soaked wipes, and allow the surface to dry (~10 min). To rinse, gently apply H_2O to the treated surface, and wipe dry with KimWipes. This procedure will prevent the gel from sticking to the plate.
2. Clean the spacers and comb with water and ethanol.
3. Place the spacers between the plates on the vertical edges of the gel and at the bottom of the gel. Make sure that there are no gaps where the side spacers meet the bottom (Fig. 1). Use two binder clips to hold each of the sides and bottom of the assembly together. Lay this apparatus horizontally on its side.

4. To prevent any leaks, melt a 1 % agarose solution in the microwave and carefully seal the edges of the glass plates by pipetting the agarose down the sides of the glass plate and the spacers. Although agarose will seep between the spacers and the glass plates, no agarose should accumulate in the gel bed area.
5. Once the agarose solidifies, mix 21.75 ml H_2O, 2.25 ml 80 % glycerol, 30 ml 1× TBE, and 6 ml 40 % 37.5:1 acrylamide for a final concentration of 4 % acrylamide. This recipe is sufficient to pour one gel (*see* **Note 5**).
6. When you are ready to pour the gel, add 200 μl APS and 100 μl TEMED and swirl to mix. Hold the plates so that the notched plate is on top and the notched (unclipped) end is slanted up (Fig. 1b). Pour the gel so that the gel solution overflows slightly. Avoid trapping any bubbles. If bubbles form in the gel, gently tap the glass until they rise to the top and burst. Insert the comb and use an additional 2 in. binder clip to hold the comb between both glass plates. The binder clip should not touch the gel or the inner face of the square glass plate.
7. The gel will require ~30 min to polymerize. Monitor the gel assembly to ensure that there are no leaks. After 30 min, carefully hold the gel up to a light without spilling any unpolymerized acrylamide. The presence of a line just under the meniscus and around the comb indicates that some of the acrylamide remains unpolymerized. Do not proceed until the gel is completely polymerized.
8. Attach the tubing from the pump to the gel box. Prepare 2 l 0.5× TBE and use it to fill the bottom of the gel box about ¾ full (above the outflow port).
9. Remove all clips and the bottom spacer from the gel. Use the four side clips to attach the plates with the polymerized gel to the gel box (Fig. 1c). If you see bubbles along the bottom of the gel between the gel plates, bend the needle of a syringe into a U shape and use it to carefully and gently rinse the bottom with 0.5× TBE, thus pushing the bubbles out of the cavity.
10. Pour 0.5× TBE into the top reservoir of the gel box, making sure that the top of the gel is under the liquid.
11. Mark the position of the bottom edge of the comb teeth on the glass plate with a Sharpie pen and number the lanes.
12. Turn on the pump to generate vigorous recirculation without overflowing the buffer in the top chamber. Make sure that the top reservoir is draining effectively.
13. Remove the comb and gently rinse the wells with 0.5× TBE using the straight needle syringe to remove any unpolymerized acrylamide. This step will be repeated immediately before

loading. Remove any gel that polymerized above the notched glass plate.

14. Turn on the power supply to 115 V constant current (about 15 mA) and pre-run the gel for at least 30 min. Re-rinse the wells with 0.5× TBE before loading.

3.4 Diluting Protein and DNA Stocks

1. Measure the initial concentration of cold DNA, and dilute to yield 100 μl of 60 nM cold DNA. This stock will be named "C1."
2. Calculate the concentration of [^{32}P]-labeled DNA based on the labeling protocol. Assume 10 % loss due to DNA either remaining on the column or eluting prior to the collected fraction. Based on this protocol, you should have 1 nM hot DNA. Dilute the labeled stock to yield 60 pM hot DNA (termed "H1") and dilute this 1:10 to yield 6 pM hot DNA ("H2").
3. Measure the protein concentration (*see* **Note 6**), and dilute the protein stock to yield 100 μl of 3×10^{-7} M protein (P2). Serially dilute the P2 stock 1:10 to create 100 μl protein stocks at concentrations ranging from 3×10^{-8} M (P3) to 3×10^{-11} M (P6) for use in activity assays (Table 1) and affinity assays (Table 2).

3.5 Mixing the Binding Reactions

1. Label 20 microcentrifuge tubes 1–20.
2. Binding reactions should be mixed according to Table 1 (activity assay) or Table 2 (affinity assay). To each microcentrifuge tube, add buffer first, then protein, and finally DNA to start the binding reaction.
3. Allow the reaction to incubate at room temperature for 15 min (*see* **Note 7**).

3.6 Loading and Running the Gel

1. Rinse the wells with 0.5× TBE just prior to loading.
2. Increase the current to 300 V, which should be approximately 65–75 mA.
3. Starting with lane 1, quickly but carefully load 15 μl of each sample. Ejecting the samples with too much force will mix and dilute the reaction with the running buffer. After lane 20 is loaded, the free DNA sample in lane 1 should have entered the gel, and ~5 μl of 6× Blue/Orange Loading Dye can be added into lane 1 to track the progress of the gel without altering migration of the free DNA band.
4. Continue to run the gel at 300 V for ~5 min to ensure that all of the bound complexes have entered the gel (*see* **Note 8**). For a single-site DNA, run at high voltage until the dye in lane 1 has separated and the cyan dye has completely entered the gel. When the reactions contain DNA with multiple binding sites, this time interval may need to be longer. All of the bound complexes must enter the gel before the voltage is reduced.

Table 1
Activity assay reaction components for a sample experiment

Lane	[Protein]	Protein stock	Binding buffer (μl)	Protein (μl)	DNA (μl)
1	0		20	0	10
2	2.50E−11	P5	17.5	2.5	10
3	4.00E−11	P5	16	4	10
4	6.30E−11	P5	13.7	6.3	10
5	1.00E−10	P5	10	10	10
6	1.60E−10	P5	4	16	10
7	2.50E−10	P4	17.5	2.5	10
8	4.00E−10	P4	16	4	10
9	6.30E−10	P4	13.7	6.3	10
10	1.00E−09	P4	10	10	10
11	1.60E−09	P4	4	16	10
12	2.50E−10	P3	17.5	2.5	10
13	4.00E−09	P3	16	4	10
14	6.30E−09	P3	13.7	6.3	10
15	1.00E−08	P3	10	10	10
16	1.60E−08	P3	4	16	10
17	2.50E−08	P2	17.5	2.5	10
18	4.00E−08	P2	16	4	10
19	6.30E−08	P2	13.7	6.3	10
20	1.00E−07	P2	10	10	10

The concentration of DNA should always be at least tenfold ***higher*** than K_d, and the protein in lane 11 should be approximately the same concentration as the DNA. In this example, these parameters are correct for the *Drosophila melanogaster* Ultrabithorax, for which the K_d is 160 pM [3, 17]. Therefore, the DNA concentration and the protein concentration in lane 11 should be 1.6 nM. Both protein and DNA concentrations must be adjusted based on the K_d of the Hox protein being measured

5. Reduce the voltage to 115 V (15–20 mA) and allow the gel to run until all complexes have separated. For single-site DNA with a full-length Hox protein, running the gel until the leading edge of the yellow dye is 8 cm from the bottom of the wells is generally sufficient. Separating multi-protein complexes may require 12–16 cm.

3.7 Drying the Gel and Phosphorimaging

1. Turn off the power supply and pumps.
2. Remove the clips holding the gel plates to the gel box and carefully remove the gel sandwich.

Table 2
Affinity assay reaction components for a sample experiment

Lane	[Protein]	Protein stock	Binding buffer (μl)	Protein (μl)	DNA (μl)
1	0		20	0	10
2	2.50E−12	P6	17.5	2.5	10
3	4.00E−12	P6	16	4	10
4	6.30E−12	P6	13.7	6.3	10
5	1.00E−11	P6	10	10	10
6	1.60E−11	P6	4	16	10
7	2.50E−11	P5	17.5	2.5	10
8	4.00E−11	P5	16	4	10
9	6.30E−11	P5	13.7	6.3	10
10	1.00E−10	P5	10	10	10
11	1.60E−10	P5	4	16	10
12	2.50E−10	P4	17.5	2.5	10
13	4.00E−10	P4	16	4	10
14	6.30E−10	P4	13.7	6.3	10
15	1.00E−09	P4	10	10	10
16	1.60E−09	P4	4	16	10
17	2.50E−09	P3	17.5	2.5	10
18	4.00E−09	P3	16	4	10
19	6.30E−09	P3	13.7	6.3	10
20	1.00E−08	P3	10	10	10

In an affinity assay, the concentration of DNA should always be at least tenfold *lower* than K_d, and the protein concentration in lane 11 should be approximately the protein K_d. Both protein and DNA concentrations should be adjusted as needed based on the K_d of the Hox protein being measured. In this example, these values are correct for Ultrabithorax, for which the K_d is 160 pM [3, 17]. We typically use a DNA concentration of 6 pM for Ultrabithorax affinity assays

3. Lay the gel on the lab bench with square glass plate on the bottom. Using a spacer, pry apart the glass plates and carefully lift the notched plate such that the gel remains affixed to the square plate.
4. Use the edge of the spacer to trim off the top half of the wells (leave the well base). Also use the spacer to cut horizontally across the bottom of the gel ~3 cm below the yellow dye front in lane 1. Discard the bottom piece and the well fragments.

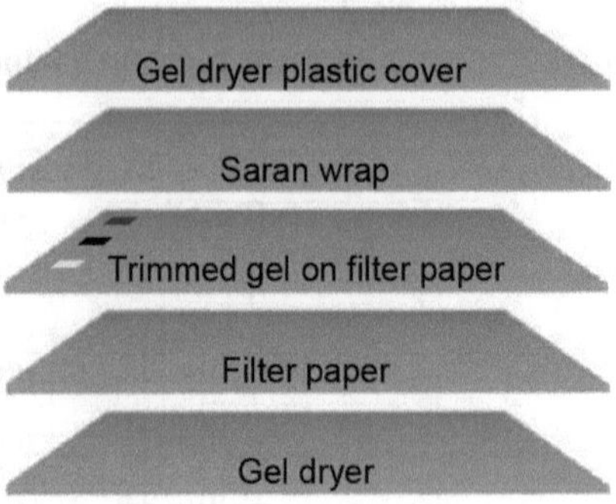

Fig. 2 Stacking components for drying gels in a gel dryer. At this point the only visible marks on the gel will be the separated dye colors in lane 1, drawn in grayscale here

5. Cut a piece of filter paper to be ~1 cm larger than the remaining gel on all sides. Gently press the filter paper onto the gel such that the gel evenly adheres and can be peeled off the glass plate.
6. On the gel dryer bed, stack a large piece of filter paper, the gel/filter paper from **step 5**, a piece of Saran Wrap, and the gel dryer plastic cover (Fig. 2).
7. Run the gel dryer at 90 °C for 45–90 min or until completely dry.
8. While the gel is drying, erase a phosphorimaging plate.
9. Wrap the gel, dried onto the small piece of filter paper, in Saran Wrap. Be careful to have only one wrinkle-free layer of Saran Wrap on the gel side. Avoid creating lumps on the back side of the gel or you will damage your phosphorimaging plate.
10. Place the wrapped gel in a cassette gel side up, and place a phosphorimaging plate sensitive to ^{32}P face (white side) down on top of the gel. Close the cassette and expose the plate overnight.
11. Image the phosphorimaging plate as per phosphorimager instructions. The data should look like Fig. 3a.
12. Use the imaging software to draw boxes around the bound and free bands to quantitate band density. All band density should be completely encased in the box, and all boxes for a single species (e.g., free DNA) should be exactly the same size as shown in Fig. 3b. Common problems when measuring binding by Hox proteins with EMSA are depicted in Fig. 4 and described in **Note 9**.

3.8 Data Analysis for Activity Assays

1. Using Microsoft Excel, plot the density of the bound band (*y*-axis) versus the concentration of the Hox protein (*x*-axis).
2. By visual inspection, identify data points that clearly fall on either the DNA-binding line (slanted) or the saturation line

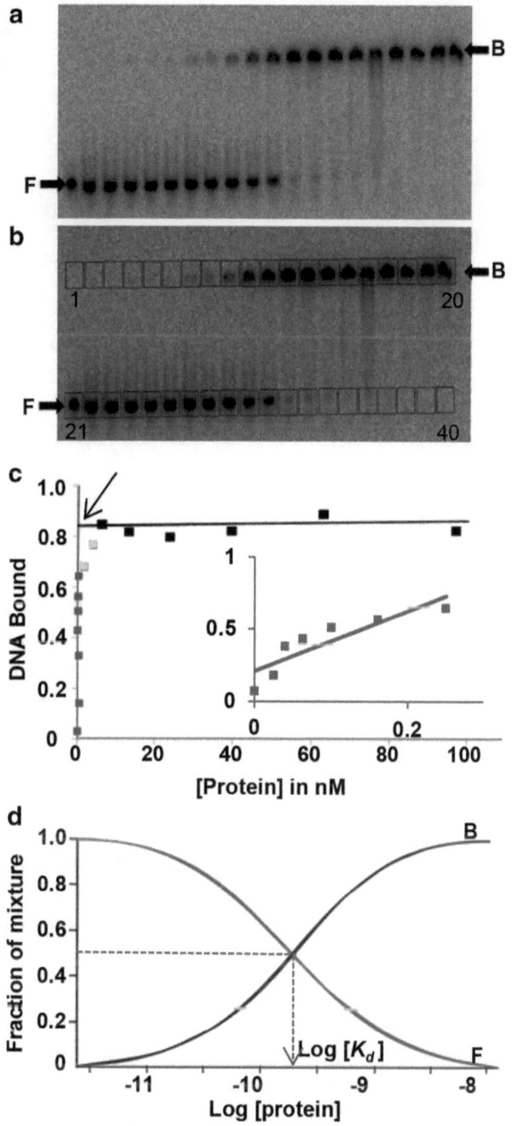

Fig. 3 EMSA data analysis. (**a**) Autoradiograph of an EMSA gel. F indicates the free DNA band and B indicates the bound DNA band. (**b**) *Boxes* have been defined that encompass all band density. Importantly, all *boxes* in the same row are the same size. (**c**) An activity assay. Data used to fit the binding line are in *grey*, and data used to fit the saturation line are in *black*. The *inset* shows the binding line on a smaller scale. The two *light grey data points* near the intersection show normal levels of deviation from both lines and were not included in either fit. Using even higher DNA concentrations in the assay would reduce this deviation. (**d**) A theoretical graph based on Eq. 4 for an affinity assay, showing the loss of the free DNA band (F, *light grey*) and the appearance of the bound DNA band (B, *dark grey*). These lines intersect where half of the DNA is bound and half is free, a point corresponding to the log of K_d on the *x*-axis

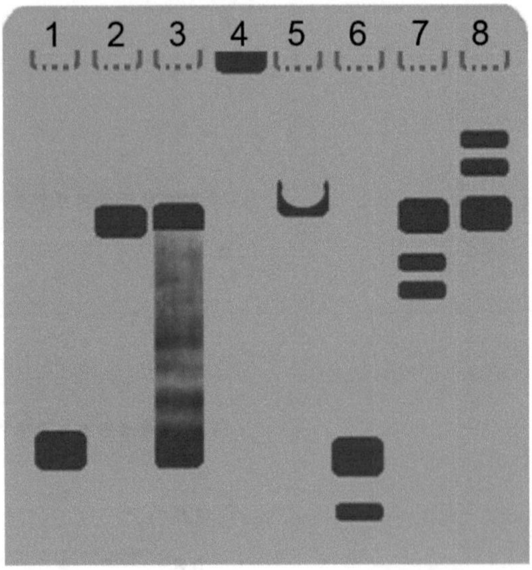

Fig. 4 Cartoon of common problems with gel shifts. Common problems for gel shift are described previously [1]. The sensitive nature of Hox proteins can generate additional problems which are diagrammed in this cartoon, in which the *dashed lines* represent the wells. A smear between the bound and free bands (*lane 3*) indicates that the bound complex is dissociating while the gel loads and/or runs. To address dissociation while loading, *see* **Note 8**. Dissociation while running may be fixed by reducing the percentage of acrylamide or the cross-linking density in the gel, running the gel at lower current, or using fans or a lower room temperature to keep the gel cooler during electrophoresis. The bound DNA can appear trapped in a well (*lane 4*). This problem can be caused by unpolymerized acrylamide that was not sufficiently rinsed from the wells, old acrylamide, old APS, old TEMED, or aggregated Hox protein. If rinsing the wells more thoroughly and replacing reagents do not solve the problem, the buffers that maintain the Hox protein in a soluble, monomeric state need to be identified as described elsewhere [6, 15, 16]. If the bound band is U shaped instead of flat (*lane 5*), then either a supershift band is failing to completely form (solve as for the smear in *lane 3*) or the purified Hox protein is aging and a new purification is needed. An unexpected band below the free DNA (*lane 6*) can indicate ssDNA due to incomplete annealing, which can be solved as described in **Note 4**, or DNA degradation. Unexpected bands between the bound and free bands indicate proteolysis of the Hox protein (*lane 7*). Unexpected bands above the bound band suggest that additional Hox·Hox or Hox·DNA interactions have formed (*lane 8*)

(close to horizontal). If there is significant curvature rather than intersecting lines, repeat the experiment using higher protein and DNA concentrations to minimize curvature (*see* **Note 1**).

3. Determine the percent active protein using the following equation:

$$\%\text{Active protein} = \frac{[\text{DNA}]}{[\text{P}]_{\text{int}}} \tag{4}$$

where [DNA] is the concentration of DNA in the activity assay and $[P]_{int}$ is the concentration of protein at the intersection point.

3.9 Data Analysis for Affinity Assays

1. Raw data for DNA binding are obtained from scanning phosphorimaging plates using a phosphorimaging system. Determine the relative amount of radioactivity within each bound and free species on the gel. Initial data set can be exported to Microsoft Excel for further analysis using any equivalent software that allows data fitting to equations entered manually. Software-specific instructions in the next steps refer to IgorPro, version 6.0.
2. Analysis of the raw data:
   (a) Quantify the density of the free band and import the data into the software program.
   (b) Determine the fraction of bound DNA and the fraction of free DNA in each lane. The sample in lane 1 consisted only of DNA. Therefore the intensity of the "bound box" (Box 1 in Fig. 3b) is the background intensity and should be subtracted from all other data points. The intensity of the free DNA in lane 1 (Box 21 in Fig. 3b) is equal to the total possible signal for either free DNA or bound protein.
   (c) The raw data can be fit to the following equation:

$$R = \Upsilon_{\max} \frac{[P]^n}{\left(K_d^n + [P]^n\right)} \qquad (5)$$

   where R is the fractional saturation, K_d is the equilibrium dissociation constant, n is the Hill coefficient, and [P] is the total concentration of protein in each well (*see* **Note 10**). The indicated equation represents the simplest binding process with no linked equilibria, such as protein interactions that impact DNA binding. This equation should be generally applicable and is sufficient to fit the data for the Hox protein Ultrabithorax [21]. However, data is more commonly displayed and fit using a logarithmic scale for the *x*-axis, which enables display of data over a large range of protein concentrations required to measure binding [1, 14]. Using this approach, the binding data will appear sigmoidal (Fig. 3d).
   (d) To generate logarithmic binding curves, use IgorPro to fit the data to "sigmoidal formula parameters" using the default equation:

$$R = \text{base} + \frac{\max}{1 + \exp^{\left(\frac{\text{Xhalf} - x]}{\text{rate}}\right)}} \qquad (6)$$

where xhalf is the K_d, x is the total concentration of protein, and max is the maximum value of the fractional saturation.

(e) After data have been fitted, multiply the K_d by the percent activity to correct for the fraction of protein that is inactive and thus incapable of binding DNA.

4 Notes

1. The closer the DNA concentration is to the K_d, the more the binding diverges from two intersecting lines and instead forms a curve.
2. Saran Wrap brand plastic wrap is required. Other brands can leak any residual water in the gel and ruin the expensive phosphorimaging plate.
3. This DNA-binding buffer has been optimized for *Drosophila melanogaster* Ultrabithorax. Other Hox proteins may require slight adjustments to maintain protein activity and enhance band resolution.
4. ssDNA oligos with significant structure may need to be cooled more slowly by covering the water bath with Styrofoam box and cooling overnight. In this case, dsDNA should be separated from ssDNA by polyacrylamide gel electrophoresis and gel purified as described elsewhere [17].
5. This gel recipe creates a gel with loose, open pores that is easily distorted if not handled carefully. This open matrix is required to observe clear bands for experiments involving the full-length *Drosophila melanogaster* Hox protein Ultrabithorax. If you are using a smaller, less disordered Hox protein or a fragment of a Hox protein, then a tighter, more solid gel is needed to resolve bound and free bands. For instance, when we measure binding by the isolated Ubx homeodomain, our gels contain 10 % 19:1 polyacrylamide [3, 17]. Measuring supershifts using a full-length Hox protein may require a more porous gel matrix. These gels can be physically supported by using a mixture of acrylamide (to define pore size) and agarose (for structural integrity) as described previously [3].
6. Because Hox proteins generally contain intrinsically disordered regions that hamper standard protein concentration assays, any equilibrium or kinetic measurements of DNA-binding affinity require nontraditional methods to determine protein concentration [3, 17].
7. The binding reaction must come to equilibrium during the incubation time. A typical incubation time is 15 min [1, 3, 15]. Hox proteins with different kinetic association and dissociation times than *Drosophila melanogaster* Ultrabithorax may

require more time to equilibrate. To test this possibility, compare results from experiments using 15- and 30-min incubation times. If the K_d is unchanged, then a 15-min incubation can be utilized.

8. During sample entry into the gel, the binding reactions will have the opportunity to re-equilibrate. Depending on the extent of this process, a smear of free DNA may appear between the bound and free bands. Loading at higher voltage or leaving the gels at high voltage for a longer period of time after loading can mitigate this problem.
9. Common problems encountered in EMSA experiments are previously described [1]. In addition, the sensitive nature of Hox proteins and the repetitive sequences in Hox target DNAs can lead to additional challenges. One of the advantages of the EMSA technique is that these problems are easily diagnosed by the banding pattern on the gel (Fig. 4).
10. This simplified equation (Eq. 4) can only be applied when the concentration of total protein ($[P]_T$ in this equation) is a reasonable approximation for the concentration of free protein. This condition is met if the concentration of DNA is more than tenfold less than K_d (Eq. 2).

Acknowledgements

This work was supported by an RDEAP grant from the Texas A&M Health Science Center to S.E.B. and a Robert A. Welch Foundation grant (C-576) to K.S.M.

References

1. Carey J (1991) Gel retardation. Methods Enzymol 208:103–117
2. Senear DF, Brenowitz M (1991) Determination of binding constants for cooperative site-specific protein-DNA interactions using the gel mobility-shift assay. J Biol Chem 266: 13661–13671
3. Liu Y, Matthews KS, Bondos SE (2009) Internal regulatory interactions determine DNA binding specificity by a Hox transcription factor. J Mol Biol 390:760–771
4. Beachy PA, Varkey J, Young KE et al (1993) Cooperative binding of an Ultrabithorax homeodomain protein to nearby and distant DNA sites. Mol Cell Biol 13:6941–6956
5. Joshi R, Sun L, Mann R (2010) Dissecting the functional specificities of two Hox proteins. Genes Dev 24:1533–1545
6. Bondos SE, Bicknell AA (2003) Detection and prevention of protein aggregation before, during, and after purification. Anal Biochem 316:223–231
7. Lane D, Prentki P, Chandler M (1992) Use of gel retardation to analyze protein-nucleic acid interactions. Microbiol Rev 56: 509–528
8. Hellman LM, Fried MG (2007) Electrophoretic mobility shift assay (EMSA) for detecting protein-nucleic acid interactions. Nat Protoc 2:1849–1861
9. Garner MM, Revzin A (1981) A gel electrophoresis method for quantifying the binding of proteins to specific DNA regions: application to components of the *Escherichia coli* lactose operon regulatory system. Nucleic Acids Res 9:6505–6525

10. Gerstle JT, Fried MG (1991) Measurement of binding-kinetics using the gel-electrophoresis mobility shift assay. Electrophoresis 14: 725–731
11. Bondos SE, Catanese DJ Jr, Tan XX et al (2004) Hox transcription factor Ultrabithorax physically and genetically interacts with Disconnected Interacting Protein 1, a double-stranded RNA-binding protein. J Biol Chem 279:26433–26444
12. Merabet S, Saadaoui M, Sambrani N et al (2007) A unique Extradenticle recruitment mode in the *Drosophila* Hox protein Ultrabithorax. Proc Natl Acad Sci U S A 104:16946–16951
13. Chan SK, Wang XA, Mak SS et al (1994) The DNA binding specificity of Ultrabithorax is modulated by cooperative interactions with extradenticle, another homeoprotein. Cell 78:603–615
14. Rippe K (1997) Analysis of protein-DNA binding at equilibrium. B I F Futura 12:20–26
15. Bondos SE (2006) Methods for measuring protein aggregation. Curr Anal Chem 2: 157–170
16. Churion KA, Bondos SE (2012) Identifying solubility-promoting buffers for intrinsically disordered proteins prior to purification. Methods Mol Biol 896:415–427
17. Liu Y, Matthews KS, Bondos SE (2008) Multiple intrinsically disordered sequences alter DNA binding by the homeodomain of the Drosophila Hox protein Ultrabithorax. J Biol Chem 283:20874–20887
18. Wong EYM, Wang XA, Mak SS et al (2011) Hoxb3 negatively regulates Hoxb1 expression in mouse hindbrain patterning. Dev Biol 352:382–392
19. Shen WF, Rozenfeld S, Kwong A et al (1999) HOXA9 forms triple complexes with PBX2 and MEIS1 in myeloid cells. Mol Cell Biol 19:3051–3061
20. Galant R, Walsh CM, Carroll SB (2002) Hox repression of a target gene: extradenticle-independent, additive action through multiple monomer binding sites. Development 129: 3115–3126
21. Li L, von Kessler D, Beachy PA et al (1996) pH-dependent enhancement of DNA binding by the Ultrabithorax homeodomain. Biochemistry 35:9832–9839

Chapter 14

Chromatin Immunoprecipitation and Chromatin Immunoprecipitation with Massively Parallel Sequencing on Mouse Embryonic Tissue

Shilu Amin and Nicoletta Bobola

Abstract

Regulation of gene expression must be tightly controlled during embryonic development. A central mechanism to control gene expression is the binding of sequence-specific transcription factors to *cis*-regulatory elements in the genome. Chromatin immunoprecipitation (ChIP) is a widely used technique to analyze binding of transcription factors and histone modifications on chromatin; however, it is limited to looking at a small number of genes. ChIP with massively parallel sequencing (ChIP-seq) is a recently developed powerful tool to analyze genome-wide binding of transcription factors and histone modifications and provides a vast amount of information into the regulation of gene expression. This chapter describes how ChIP and ChIP-seq are performed on mouse embryonic tissue.

Key words ChIP, ChIP-seq, Embryo, Mouse, Transcription factor, DNA

1 Introduction

ChIP and on a genome-wide scale ChIP-seq are powerful tools for mapping whole-genome histone modifications and transcription factors occupancy in vivo. In this technique, proteins are cross-linked to chromatin, and chromatin is sheared into small fragments. These are used as starting material for immunoprecipitation (IP) with an antibody against an epitope of interest (a transcription factor or a histone modification). Following several washes and the reversal of cross-links, IP DNA is purified and PCR is performed on specific loci to assess their enrichment relative to background (input).

The advent of next-generation sequencing platforms has led to the development of high-throughput ChIP techniques. ChIP-seq utilizes IP DNA and performs massively parallel sequencing to produce a high-resolution genome-wide map of protein binding and histone modifications [1, 2]. ChIP-seq was first developed in

Yacine Graba and René Rezsohazy (eds.), *Hox Genes: Methods and Protocols*, Methods in Molecular Biology, vol. 1196, DOI 10.1007/978-1-4939-1242-1_14, human T cells [3] and mouse ES cells [4] to analyze histone modifications and in cell lines to analyze genome-wide STAT1 [5] and NRSF [6] binding. Other useful applications of ChIP-seq include nucleosome positioning to assess chromatin structure and accessibility, chromosome conformation capture to assess long-range chromatin interactions, and association of histone mark signatures to key regulatory regions. Analysis of ChIP-seq data can reveal frequently occurring sequence motifs (to identify transcription factor-binding sites and/or potential binding partners) and provide information on the functional relevance of binding through gene ontology annotation, in particular associated biological processes and molecular function [6–8]. Recent modifications of ChIP-seq include ChIP-exo, where an exonuclease trims ChIP DNA to a precise distance from the cross-linking site and provides single base pair accuracy [9] and nano-ChIP-seq, a tailored procedure for generating high-throughput sequencing libraries from scarce amounts of ChIP DNA [10].

ChIP has commonly utilized cell lines as starting material. We have successfully performed ChIP on first and second branchial arches (IBA and IIBA, respectively) isolated from E11.5 mouse embryos [11] and ChIP-seq on the IIBA from mouse embryos to analyze genome-wide binding of Hoxa2 [12]. ChIP-seq becomes crucial for identifying binding locations of Hox, which cannot be predicted based on their short recognition motif (TAAT). Here we describe both ChIP and ChIP-seq methods on mouse embryonic tissue.

2 Materials

Prepare all solutions with nuclease-free water and use molecular biology-grade reagents. ChIP-seq-specific materials are highlighted in italics.

2.1 Mouse Embryo Dissection and Cross-Linking

1. Dissection tools.
2. Ice-cold sterile PBS.
3. 10 cm Petri dish.
4. Sterile 50 ml Falcon tube.
5. 1.5 ml clear Eppendorf tubes.
6. 37 % Formaldehyde.
7. 1 M Glycine (freshly prepared).
8. Pipettes (1 ml and 200 μl) and tips.
9. Dry ice.

2.2 Chromatin Extraction

1. 25G needle and syringe.
2. L1 buffer (freshly prepared): 50 mM Tris–HCl pH 8.1, 2 mM EDTA pH 8.0, 0.1 % NP-40, 10 % glycerol. Add before use 1 mM PMSF and protease inhibitors (Roche, cOmplete, Mini, EDTA-free tablets).
3. SDS lysis buffer: 50 mM Tris–HCl pH 8.1, 10 mM EDTA pH 8.0, 1 % SDS. Add before use 1 mM PMSF and protease inhibitors (Roche, cOmplete, Mini, EDTA-free tablets).

2.3 Immuno-precipitation

1. ChIP dilution buffer: 0.5 % NP-40, 5 mM EDTA pH 8.0, 200 mM NaCl, 50 mM Tris–HCl pH 8.1. Add before use 1 mM PMSF.
2. Wash buffer: 0.1 % SDS, 1 % NP-40, 2 mM EDTA pH 8.0, 500 mM NaCl, 20 mM Tris–HCl pH 8.1. Add before use 1 mM PMSF.
3. LiCl buffer: 0.1 % SDS, 1 % NP-40, 2 mM EDTA pH 8.0, 500 mM LiCl, 20 mM 1 M Tris–HCl pH 8.1. Add before use 1 mM PMSF.
4. TE buffer: 1 mM EDTA pH 8.0, 10 mM Tris–HCl pH 8.1.
5. 1.5 and 2 ml clear Eppendorf tubes.
6. Dynabeads Protein A/Protein G magnetic beads (Life Technologies) (*see* **Note 1**).
7. Antibody (*see* **Note 2**) and isotype control IgG.
8. Magnetic rack suitable for 1.5 and 2 ml Eppendorfs.

2.4 DNA Extraction

1. Proteinase K.
2. RNase A.
3. PCR purification kit (Qiagen).
4. *ChIP-seq: Minelute purification kit (Qiagen).*

2.5 Equipment

1. Microscope.
2. Microcentrifuge.
3. Vortex mixer.
4. Rotating wheel (store in cold room).
5. Agarose gel electrophoresis equipment.
6. Sonicator and probe: Ultrasonic Processor 130 W (Sonics & Materials, VCX-130) equipped with 2 mm probe (Fisher, 630-0423).
7. Thermomixer.
8. PCR machine.

3 Methods

ChIP-seq-specific protocols are highlighted in italics.

3.1 Mouse Tissue Dissection and Cross-Linking

The following procedure must be carried out in accordance with relevant animal ethics approval and regulations.

1. Dissect uterus into 50 ml Falcon tube with sterile PBS and place embryos into a 10 cm Petri dish containing fresh PBS on ice. Dissect the tissue of interest under the microscope in a 10 cm Petri dish containing PBS (this protocol has been optimized on branchial arches).
2. Transfer tissue isolated from 10 to 12 embryos into an Eppendorf with PBS (keep on ice). Once all dissected, replace with ice-cold PBS and adjust the volume to 1 ml (if using a larger area of the embryo it may be necessary to split the material into more tubes).
3. Cross-linking: Add 27 μl of 37 % formaldehyde (final concentration 1 %). Incubate for approximately 20 min at 4 °C rotating (*see* **Note 3**).
4. Stop the reaction by adding 142 μl 1 M glycine (freshly prepared) to reach a final concentration of 0.125 M, and incubate for 5 min at 4 °C rotating.
5. Remove glycine, and rinse once with 1 ml PBS (*see* **Note 4**).
6. Wash twice for 5 min with PBS at 4 °C rotating (*see* **Note 5**).
7. Remove as much of PBS as possible, and freeze immediately on dry ice. Store at −80 °C (*see* **Note 6**).

Pooling of tissues is generally required at mid-stage embryogenesis in order to perform ChIP. For instance, we pool 20–25 pairs of IIBA or 10–12 pairs of IBA from mouse embryos at stage E11.5 to perform two ChIPs (antibody of interest and control IgG).

ChIP-seq: Scale up the ChIP protocol for ChIP-seq to obtain enough chromatin for sequencing. Collect 100–110 pairs of IIBA or 65–75 pairs of IBA from mouse embryos at stage E11.5 to perform eight ChIPs (i.e., antibody of interest for seven ChIPs and control IgG for one ChIP). Ideally collect arches over a short period of time to avoid long-term storage (over 2 months).

3.2 Chromatin Extraction and Sonication

1. Defreeze cross-linked tissue quickly on ice. Add 300 μl of PBS.
2. Disintegrate the tissue by passing it 25 times through a 25G needle until no clumps are visible (*see* **Note 7**). Wash the needle and syringe two to three times by passing through 300 μl PBS and assemble all in one tube. *ChIP-seq: Process first four tubes (IIBA) or two tubes (IBA) and keep the remaining tubes at −20°C. Process the remaining tubes whilst first tubes rest on ice.*

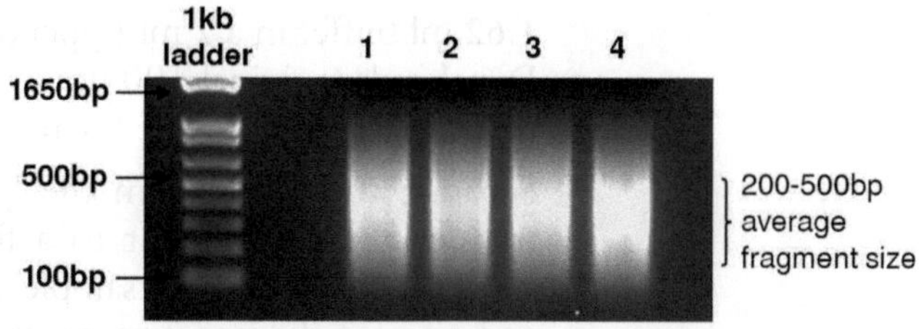

Fig. 1 Chromatin from branchial arches of E11.5 mouse embryos sonicated for 10 × 10 s pulses at 50 % amplitude. The four separate tubes (*1–4*) show similar sonication pattern by gel electrophoresis, with the majority of fragment size between 200 and 500 bp suitable for ChIP-seq experiments

3. Centrifuge for 5 min at 2,000 × *g* in a microcentrifuge at 4 °C.
4. Resuspend the cell pellet in 500 μl L1 buffer containing protease inhibitors and PMSF, and incubate for 5 min on ice.
5. Centrifuge for 5 min at 800 × *g*, 8 °C, to pellet the nuclei.
6. Resuspend the nuclei pellet in 450 μl SDS lysis buffer containing protease inhibitors and PMSF (*see* **Note 8**).
7. Transfer the lysate to a fresh tube. Sonicate to obtain the fragments between 200 and 1,000 bp. Using the tip sonicator, sonicate 7 × 10 s on, and 45 s off (on ice) with 40 % amplitude (*see* **Note 9**). *ChIP-seq: Using the tip sonicator, sonicate 10× 10 s pulses without interruption*, and *50% amplitude to obtain fragments between 200* and *500 bp* (*ideally average fragment size should be 300–350 bp*).
8. Centrifuge for 10 min at 15,700 × *g*, 8 °C. Transfer the supernatant to a fresh tube (*see* **Note 10**).
9. In order to evaluate the size of the sheared chromatin, add 80 μl of TE and 5 μl 5 M NaCl to 20 μl aliquot of the supernatant (*see* **Note 11**). Reverse cross-links overnight at 65 °C. *ChIP-seq: Take an aliquot from each tube and check the size of sheared chromatin before pooling samples.*
10. Add 1 μl (1 mg/ml) RNase A to the aliquot of sheared chromatin and incubate for 30 min at 37 °C. Purify using the PCR purification kit from Qiagen. Combine 50 μl eluate with 10 μl 6× loading buffer and run 12 μl on a 1.5 % agarose gel (Fig. 1).

3.3 Immunoprecipitation and DNA Extraction

1. Prepare beads (Dynabeads Protein A or G, depending on the species of the antibody host, refer to the manufacturer's instructions for choice of A vs. G beads). Use 100 μl beads per sample. Wash three times with 1 ml each of ChIP dilution buffer and resuspend in 50 μl ChIP dilution buffer.
2. Dilute the sonicated lysate 1:10 in ChIP dilution buffer containing PMSF. Use 180 μl lysate per reaction (one for the specific antibody and one for the negative control antibody) and add

1.62 ml buffer in a 2 ml Eppendorf tube. Add 50 μl of washed Dynabeads (original 100 μl volume). Incubate for 1 h at 4 °C rotating to pre-clear the lysate.

3. Separate the beads from the lysate on a magnetic rack and transfer the supernatant to a fresh tube. Take 50 μl of the supernatant from each sample (one for the specific antibody and one for the negative control antibody) and pool. This is the input DNA and should be kept separate from the IP DNA.
4. Add the antibody for the protein to be tested to the 1.75 ml diluted lysate. Use between 1 and 5 μg antibody. To the remaining 1.75 ml sample, add the same amount of unspecific IgG (this will be the negative control). Incubate overnight at 4 °C, rotating (*see* **Note 12**). *ChIP-seq: Add antibody to seven tubes and unspecific IgG control to one tube only.*
5. To the 100 μl of the diluted lysate from **step 3** (input DNA), add 5 μl 5 M NaCl and reverse cross-links overnight at 65 °C. Digest with proteinase K (2 μg for 1 h at 45 °C) and purify using PCR purification kit from Qiagen. Elute the DNA in 50 μl EB buffer. Store at −20 °C.
6. The following day, prepare the Dynabeads Protein A/G as in **step 1**, but instead use 50 μl per sample and resuspend in 50 μl ChIP dilution buffer after the washes. Add the 50 μl of beads to each of the 1.75 ml of antibody and negative control reactions. Incubate for 30 min at 4 °C, rotating.
7. Separate on a magnetic rack and rinse once with wash buffer. At this point, transfer the beads in their wash buffer to 1.5 ml Eppendorf.
8. Wash the beads for 5 min on ice five times using 800 μl wash buffer containing PMSF, each time separating on a magnetic rack (*see* **Note 13**).
9. Wash three times using 800 μl LiCl buffer containing PMSF as above.
10. Wash three times using 800 μl TE as above.
11. Elute by adding 50 μl 2 % SDS in 1× TE freshly made to the beads, and mix by tipping the bottom of the tube. Shake in a thermomixer for 15 min at 25 °C, 1,400 rpm. Repeat once again at 65 °C, 1,400 rpm. Assemble the eluates.
12. Add 5 μl 5 M NaCl to each 100 μl eluate (*see* **Note 11**). Reverse cross-links overnight at 65 °C.
13. Purify the DNA using Qiagen columns (*see* **Note 14**). Elute in 50 μl EB and transfer the eluate to a fresh tube. Keep at 4 °C, do not freeze, and for best results use within 2 days.
14. Use approximately one-tenth of eluted IP DNA per qPCR reaction and a corresponding volume of diluted input (dilution 1:300). We use 4 μl of IP DNA and diluted input for SYBR

green qPCR in a total volume of 50 μl per PCR reaction. Perform each reaction in duplicate (*see* **Note 15**).

ChIP-seq: Pool the content of the seven tubes and divide again into identical aliquots (95 μl each). ChIP validation: Use one aliquot along with the negative IgG control to test if the ChIP has worked by SYBR green qPCR. For the remaining six aliquots, purify using Minelute kit (Qiagen) and elute in a total volume of 20 μl (to achieve this perform some of the elutions using the eluates instead of elution buffer). IP DNA can then be sent for sequencing.

4 Notes

1. Magnetic beads do not require blocking and washing steps are quicker and more efficient compared to agarose beads.
2. The choice of antibody is crucial; not all antibodies will work in ChIP. To test if your antibody is ChIP-grade, follow the ChIP protocol until Subheading 3.3, **step 6**. Incubate each antibody and negative control reactions with beads for 1 h and proceed with all washes. At this point, instead of eluting DNA and reversing cross-links (to analyze by PCR), the beads are boiled in 2× SDS loading buffer (100 mM Tris–HCl pH 6.8, 4 % (w/v) SDS, 0.2 % (w/v) bromophenol blue, 20 % (v/v) glycerol, 200 mM DTT) for 30 min and a western blot is performed. A longer boiling time is necessary to reverse cross-links.
3. During cross-linking, proteins will become covalently attached to DNA. Optimization of cross-linking time is required. Perform a time course of cross-linking to identify the optimal length of time for the embryonic tissue of choice (increasing cross-linking time will eventually interfere with DNA fragmentation).
4. There is no need to centrifuge as the tissue will deposit on the bottom of the tube.
5. Tissue may need brief spin.
6. Avoid long-term storage. Ideally use material in the next 2 weeks but can be kept up to 2 months. Store in dry ice at -80 °C.
7. A small aliquot can be checked under the microscope; most of the cells should be "single."
8. First resuspend the pellet in 200 μl by pipetting through the narrower 200 μl tip and then add the additional 250 μl after. The lysate should be relatively clear; no clumps should be visible at this point. If clumps are visible, it means that either the tissue was not disintegrated properly or not enough buffer was used to resuspend the pellet (too much tissue). The lysis can be

completed by additional pipetting and longer incubation times at 4 °C (also by adding more buffer).

9. Prior to sonication, optimize conditions to obtain 200–1,000 bp fragments (or 200–500 bp fragments for ChIP-seq). Insert the tip around 0.5 cm below the liquid level and hold the tube firmly (avoid moving the tube when sonicating to prevent foaming). If foaming occurs, centrifuge for 3 min at 2,300 × *g*, 8 °C (do not count the round because the efficiency of sonication is dramatically reduced when the sample foams). Mix the sample by pipetting a few times. If you have to spin down the tube more than once due to the foam, add a small amount (40 μl) of the lysis buffer to return to the original volume of 450 μl (this helps to prevent additional foaming). Keep the samples waiting for sonication in the fridge rather than on ice to prevent the SDS from precipitating. If the SDS precipitates, warm up the tube in your hands before proceeding to the sonication or spinning. For the same reason, spin the samples at 8 °C rather than at 4 °C.
10. No pellet should be visible at this point. If a larger pellet is visible, the sonication did not work (possibly too much tissue was used).
11. At high concentrations of NaCl, SDS precipitates. Vortex the sample to resuspend it.
12. Secure tubes with parafilm.
13. From this point on, be very careful. Use clean set of pipets, fresh tubes, and filter tips to avoid contamination.
14. Keep a separate kit for the purification of ChIP IP DNA samples to avoid the contamination with input DNA.
15. SYBR green qPCR should be performed on the IP DNA of both specific antibody and control IgG samples to ensure that there is no or minimal background. Additional controls such as a no-antibody sample can also be included. It is essential to test known binding regions as a positive control for the PCR and at least one region where no binding is known as a negative control. Percent input and/or fold enrichment methods can be used to analyze results (Fig. 2).

Acknowledgements

The authors thank Eva Kutejova for her invaluable input in setting up the ChIP protocol. S.A. is supported by the Biotechnology and Biological Sciences Research Council BB/H018123/2 to N.B.

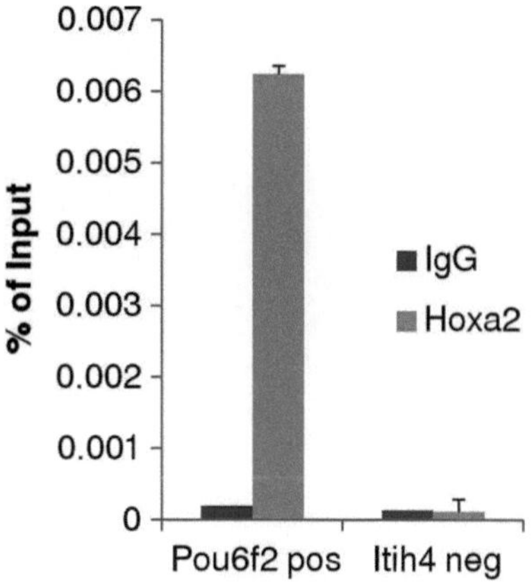

Fig. 2 ChIP experiment on IIBA of E11.5 embryos. SYBR green qPCR on chromatin IP using Hoxa2 antibody and a control IgG. Enrichment is calculated as percent input or fold enrichment (for instance see ChIP analysis section of Life Technologies online manual). Significant enrichment as a percentage of the input is observed at the Pou6f2 locus (positive control), whereas there is no enrichment at the Itih4 locus (negative control). These genes are subsequently used to test further Hoxa2 ChIP/ChIP-seq experiments. Values represent the average of duplicate samples and the standard error of the mean is indicated by error bars for each sample

References

1. Barski A, Zhao K (2009) Genomic location analysis by ChIP-Seq. J Cell Biochem 107: 11–18
2. Farnham PJ (2009) Insights from genomic profiling of transcription factors. Nat Rev Genet 10:605–616
3. Barski A, Cuddapah S, Cui K et al (2007) High-resolution profiling of histone methylations in the human genome. Cell 129:823–837
4. Mikkelsen TS, Ku M, Jaffe DB et al (2007) Genome-wide maps of chromatin state in pluripotent and lineage-committed cells. Nature 448:553–560
5. Robertson G, Hirst M, Bainbridge M et al (2007) Genome-wide profiles of STAT1 DNA association using chromatin immunoprecipitation and massively parallel sequencing. Nat Methods 4:651–657
6. Johnson DS, Mortazavi A, Myers RM et al (2007) Genome-wide mapping of in vivo protein-DNA interactions. Science 316: 1497–1502
7. Jiang C, Pugh BF (2009) Nucleosome positioning and gene regulation: advances through genomics. Nat Rev Genet 10:161–172
8. Schones DE, Zhao K (2008) Genome-wide approaches to studying chromatin modifications. Nat Rev Genet 9:179–191
9. Rhee HS, Pugh BF (2011) Comprehensive genome-wide protein-DNA interactions detected at single-nucleotide resolution. Cell 147:1408–1419
10. Adli M, Bernstein BE (2011) Whole-genome chromatin profiling from limited numbers of cells using nano-ChIP-seq. Nat Protoc 6: 1656–1668
11. Kutejova E, Engist B, Self M et al (2008) Six2 functions redundantly immediately downstream of Hoxa2. Development 135:1463–1470
12. Donaldson IJ, Amin S, Hensman JJ et al (2012) Genome-wide occupancy links Hoxa2 to Wnt-β-catenin signaling in mouse embryonic development. Nucleic Acids Res 40: 3990–4001

Fig. 2 [illegible]

Chapter 15

ChIP for Hox Proteins from *Drosophila* Imaginal Discs

Pavan Agrawal and L.S. Shashidhara

Abstract

Chromatin immunoprecipitation (ChIP) is a technique that reveals in vivo location of a protein bound to DNA. ChIP coupled with DNA microarrays (ChIP-chip) or next-generation sequencing (ChIP-seq) allows for identification of binding sites of transcription factors on a global scale. Here we describe a protocol for ChIP to identify binding of the Ultrabithorax (Ubx) Hox transcription factors from imaginal discs of *Drosophila* larvae. The protocol can be extended to other model organisms and transcription factors.

Key words *Drosophila*, *Ubx*, Hox, GAGA factor, Chromatin immunoprecipitation, Wing, Haltere, Imaginal discs

1 Introduction

DNA in the nucleus is transcribed by its interaction with proteins. How these interactions lead to transcription of genes has been a long-standing question in biology. Chromatin immunoprecipitation (ChIP) is a technique that reveals location of a protein bound to DNA in the cells. When ChIP is coupled with microarray hybridization (ChIP-chip) or DNA sequencing (ChIP-seq) it can reveal DNA-protein interaction on a global scale [1–4]. In a typical ChIP experiment, a chemical cross-linker is used to "stitch" proteins interacting with DNA. This chromatin is fragmented by sonication and precipitated using immune-affinity method. The DNA fragments bound to the protein are enriched and detected by various methods such as quantitative PCR (qPCR), ChIP-chip, or ChIP-seq.

Since ChIP was first used to identify binding of yeast transcription factors on a global scale [1, 2], various modifications have improved it considerably. As a variation of the technique, post-immunoprecipitation, the complex can be resolved on an SDS-PAGE gel to visualize enrichment of pulled-down protein over control IP by Western-blot hybridization (ChIP-Western). If standardized appropriately, ChIP-western can be utilized to detect

Yacine Graba and René Rezsohazy (eds.), *Hox Genes: Methods and Protocols*, Methods in Molecular Biology, vol. 1196,
DOI 10.1007/978-1-4939-1242-1_15, © Springer Science+Business Media New York 2014

proteins that bind in the vicinity of the transcription factor of interest. Using ChIP-western and other methods we could identify that GAGA-associated factor (GAF) shares targets with Ubx in imaginal discs [5].

Our aim is to equip the readers to understand the basic principles of ChIP, share our experience to avoid possible pitfalls, and provide methodology that can be used to perform ChIP in *Drosophila* for any transcription factor with a special emphasis on identifying targets of a Hox protein.

2 Materials

Prepare all solutions using nuclease-free water and sterilize by passing through a 0.22 μm filter. Store all buffers at 4 °C unless mentioned otherwise. If precipitate forms in buffers due to cold storage, warm the solution to room temperature for few minutes and gently mix till it dissolves. Add protease inhibitor cocktails (*see* **Note 1**) in all the buffers before use.

2.1 ChIP Buffers

1. Nuclear Lysis buffer (*see* **Note 2**): SDS 1 %, EDTA 10 mM, Tris–HCl 50 mM, pH 8.
2. ChIP dilution buffer: 0.01 % SDS, 1.1 % TritonX-100, 1.2 mM EDTA, 16.7 mM Tris–HCl, pH 8, 167 mM NaCl.
3. Low-salt buffer: 0.1 % SDS, 1 % TritonX-100, 2 mM EDTA, 20 mM Tris–HCl, pH 8, 150 mM NaCl.
4. High-salt buffer: 0.1 % SDS, 1 % TritonX-100, 2 mM EDTA, 20 mM Tris–HCl, pH 8, 0.5 M NaCl.
5. LiCl buffer: 0.25 M LiCl, 1 % NP-40, 1 % NaDOC, 1 mM EDTA, 10 mM Tris–HCl, pH 8.
6. TE: 10 mM Tris–HCl, pH 8, 1 mM EDTA.
7. 5× PK buffer: 50 mM Tris–HCl, pH 8, 25 mM $EDTA.Na_2$, pH 8, 1.25 % SDS.
8. ChIP elution buffer (*see* **Note 3**): 50 mM $NaHCO_3$, 1 % SDS.
9. Phosphate-buffered saline (PBS) solution: Prepare 1× solution by adding nuclease-free water to 10× solution, add 2× protease inhibitor cocktail, and store at 4 °C for up to 2 weeks (*see* **Note 1**).
10. 1 % Formaldehyde solution: Dilute 37 % formaldehyde with 1× PBS to get 1 % formaldehyde. Store in brown-colored bottle. Use caution while handling this chemical (*see* **Note 4**). Fixation should be performed at room temperature (*see* **Note 5**).
11. Antibodies for ChIP: Polyclonal sera against homeodomain-deleted version of Ubx was used for ChIP. Please *see* Subheading 3.6 for guidelines to use antibodies for different Hox proteins.

12. Protease inhibitor cocktail: EDTA-free protease inhibitor cocktail tablets (Roche) (*see* **Note 1**).
13. ChIP DNA Clean and Concentrator™ kit (Zymo Research): The columns in the kit can tolerate most of the buffers described in this protocol. We observed better DNA yields than many other column-based kits.

2.2 Protein-A Sepharose Slurry

Protein-A Sepharose binds to Fc fragment of antibodies used in pull down (*see* **Notes 6** and **7**). Follow the instruction below to prepare the slurry for pull down.

1. Take 100 mg Protein-A Sepharose beads supplied as lyophilized powder (CL-4B, GE Healthcare) in a DNA Lo-bind tube (Eppendorf) and add 800 μl nuclease-free water.
2. Swell the beads by gently inverting the microfuge tube few times and keep the tube on ice for 10 min (*see* **Note 7**).
3. Centrifuge at low speed (less than 500 rcf) to settle the beads.
4. Remove supernatant and repeat **steps 2** and **3** three times to wash the beads.
5. Keep for another 10 min on ice.
6. Gently spin, remove water, add Tris–HCl pH 7.4 to make up the volume to 1,000 μl, and keep on ice.
7. Repeat **step 6** with Tris–HCl pH 7.4, centrifuge, discard the supernatant, and block the beads by adding 0.5 mg/ml BSA, 200 μl Salmon Sperm DNA (Life Technologies, Invitrogen), and 0.05 % sodium azide (*see* **Note 4**). Make up the final volume to 1 ml by adding TE.
8. Store overnight at 4 °C for blocking before first use. Slurry can be stored at 4–8 °C for future use.
9. Alternatively, readymade Protein A Agarose/Salmon Sperm DNA from Millipore can be used.

2.3 Consumable and Equipment

1. DNA Lo-bind tube (Eppendorf)—Recommended for better yields of DNA throughout the protocol.
2. Bioruptor, Diagenode, Model UCD-200 TO: Bioruptor should be precooled by adding ice for 30 min. Before sonication, add ice-cold water up to the "fill line" in the water bath and add a thin layer of ice. Do not allow samples to overheat during sonication, use appropriate time of "on" and "off" cycles for sonication, and replace fresh water-ice mixture in between on cycles, if needed.
3. Qubit 2.0 (Life Technology) along with DNA HS assay kit (Life Technology).
4. 2100 Bioanalyzer, Agilent Technologies (*see* **Note 8**).

3 Methods

The entire protocol takes 4 days to perform. Since it is a complex protocol, we include suggestions along with all the steps that require standardization in order to perform the assay.

3.1 Experimental Controls

ChIP detects relative enrichment of targets over a reference sample. It is not an absolute measurement of binding events; therefore choice of reference sample (control) determines the outcome. While unenriched fragmented genomic DNA (input) is a most common control used in ChIP, a good control should allow detecting subtle changes in binding due to different variables introduced during the experiment. Therefore, we highly recommend using a control in parallel to the experiment with specific antibody. The control can include immunoprecipitation in the absence of antibody (Mock IP) to identify nonspecific binding of beads to chromatin or pre-immune sera (nonspecific IgG) to differentiate binding events due to nonspecificity of antibodies.

3.2 Tissue/Cell Isolation

The capacity to interpret ChIP data largely depends on the cellular homogeneity of the starting material. Thus it is important to begin with the most homogenous population of cells possible. There have been various attempts to use whole-*Drosophila* embryo or larvae to identify binding of transcription factors which are expressed only in a subset of cells [6, 7], which can lead to increase in noise and loss of binding information. Therefore, attempts were made to FACS-sort specific cells from *Drosophila* embryo to obtain purified population of cells, in a technique named BiTS-ChIP [8] resulting in better signal to noise.

In developing *Drosophila* larvae, imaginal discs provide an attractive alternative to FACS sorting since the discs can be dissected out manually and they provide relatively homogenous population of cells expressing specific Hox proteins. We have used *Drosophila* wing and haltere tissue to identify binding of the Hox transcription factor Ubx [5]. The protocol described here uses ~10^6 cells per IP reaction. It might be possible to further modify the protocol to detect binding events in ~10,000 cells [9, 10] albeit with higher noise than a ChIP experiment with 100-fold higher cell numbers.

Imaginal discs can be isolated from wandering third instar larvae to obtain purified population of cells. We identified targets of Hox protein Ubx by using *Cbx*$^{Hm/+}$ wing discs. *Cbx*$^{Hm/+}$ wing discs ectopically express Ubx in the pouch region which helped us focus on suppression of wing fate by Ubx without mixing with the Ubx targets involved in notum specification [5]. We also used the protocol mentioned here to identify targets of a trithorax group protein GAGA-associated factor (GAF) from *Cbx*$^{Hm/+}$ wing discs [5] and targets of Ubx from haltere imaginal discs (P.A., unpublished).

1. From a bottle of appropriate genotype flies, take out wandering third-instar larvae using a paintbrush. Take the larvae in the batches of 10–15 and wash in a glass well dish kept on ice with cold PBS.
2. Dissect larvae inside out in cold PBS with 2× protease inhibitor cocktail and discard the gut material.
3. Wash once with PBS, with 2× protease inhibitor cocktail kept at room temperature.

3.3 Chromatin Cross-Link

An ideal cross-linking agent should diffuse rapidly inside the tissues or the cells and cross-link proteins interacting with DNA and these cross links should be reversible to be able to extract the DNA. Formaldehyde is a reversible cross-linker that covalently links proteins to DNA when they are within few angstrom distance to each other [11]. When used at lower concentrations (typically 1 % formaldehyde), the cross-links can be reversed by heating the chromatin for 4–5 h. The ideal cross-linking time for ChIP depends on tissue/cell type, cell number, sonication method being used, and transcription factor under study and hence should be determined empirically. The protocol described here should serve as a general guideline for cross-linking time. However, if it is much easier or difficult to obtain sonicated fragments of specific size it can be due to under- or over-cross-linking, respectively. Typically, cross-linking times required to capture transient binding events of a Hox transcription factors are higher than Pol II or a stable histone modification in the same tissue.

1. From **step 3** (Subheading 3.2), fix inside-out larval heads in 1 % formaldehyde solution (*see* **Note 4**) at room temperature for 20 min or a specific time. Please see steps listed in Subheading 3.4 and also *see* **Note 9** to standardize the time required for fixation.
2. Stop the cross-linking reaction by adding glycine solution (final concentration of 0.125 M).
3. Wash fixed heads twice with cold PBS solution with 2× protease inhibitor cocktail.
4. Isolate appropriate imaginal discs in cold PBS with 2× protease inhibitor cocktail.
5. Use pipette tip rinsed with 0.1 % SDS solution to pool 100–120 discs for each immunoprecipitation (IP) reaction in a DNA Lo-bind tube kept on ice and proceed for nuclear lysis.

3.4 Nuclear Lysis and Sonication

Sonication is required to shear the chromatin and also solubilize it for immunoprecipitation. It is the most variable step in ChIP, which depends on multiple factors including number of cells, tissue type, degree of cross-linking and type of sonicator, and volume used for sonication. If the chromatin is under-sonicated, it will yield larger fragments and this will increase the detection of fragments away from binding (noise). On the other hand, if the samples are over-sonicated, they will yield smaller DNA fragments resulting in loss of detection of specific binding events or, in case of genome-wide analysis, noise in detection. In a typical ChIP reaction, DNA fragments used should be around 200–1,000 bp in length with the majority of DNA fragments around 500 bp in length.

1. From **step 5** (Subheading 3.3), remove excess PBS from the discs, add 400 μl nuclear lysis buffer with protease inhibitor cocktail, mix by pipetting up and down few times, and incubate on ice for 20 min to allow lysis (*see* **Note 10**).
2. Put the tubes inside a precooled Bioruptor and proceed for sonication.
3. Sonicate the sample as follows: amplitude 200 W, and two pulses of 30 s on and off. Sonication conditions need to be adjusted if using a different device than Bioruptor (*see* **Note 11**).
4. After sonication keep the sample on ice for 2 min. Centrifuge at 14,000 rcf for 10 min at 4 °C.
5. Carefully transfer supernatant to a fresh Lo-bind tube.
6. Save about 10 % sonicated chromatin, i.e., 30 μl to obtain input DNA and DNA for analyzing the fragment size. At this step remaining chromatin can be flash frozen in liquid nitrogen and stored at −80 °C till use (up to few months).

3.5 Analyzing the Size of Sheared Chromatin

Figure 1 shows attempts to standardize sonication for imaginal discs cross-linked with 1 % formaldehyde for 20 min. At 300 μl chromatin volume in a 1.5 ml microfuge tube, the spread of DNA was in broader range (Fig. 1a), while increasing the chromatin volume to 400 μl resulted in decrease in the spread of DNA fragments (Fig. 1b). In our experience, two pulses of sonication in 400 μl chromatin volume resulted in the distribution of majority of IPed fragments in the ideal range (as detected by LM-PCR amplification on Agilent Bioanalyzer 2100, Fig. 2).

1. From **step 6** (Subheading 3.4), take ~30 μl chromatin and add 150 μl nuclear lysis buffer and 2 μl RNase (10 mg/ml stock).
2. Incubate at 65 °C for 5 h to overnight to reverse the cross-links.
3. Add 3 μl Proteinase K (Roche 3115 887) and incubate at 45 °C for 2 h.
4. Use Zymo ChIP DNA clean and concentrator kit (Cat #. D5205) to purify DNA.
5. Quantitate DNA using Nanodrop and save about 10–20 ng DNA as input. Resolve about 50 ng or more DNA on 0.8 %

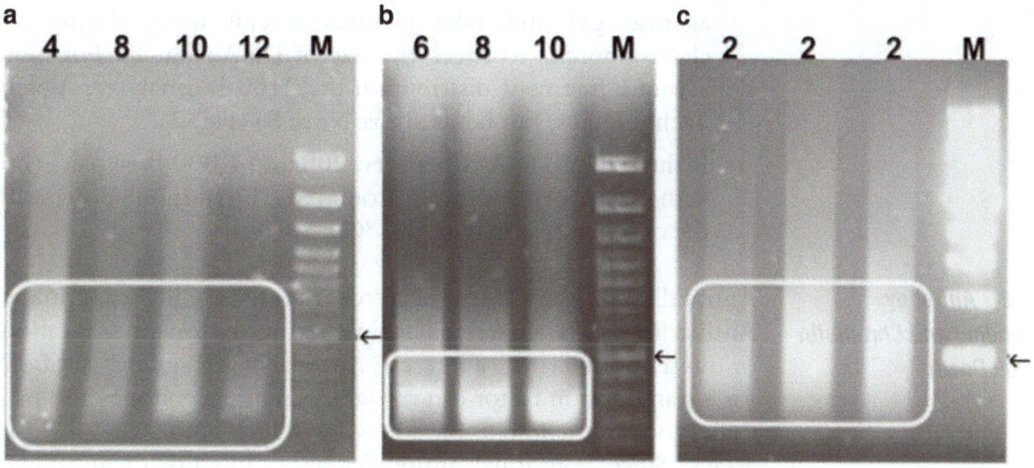

Fig. 1 Standardization of chromatin shearing using sonication. Diagenode Bioruptor UCD-200 was used to sonicate chromatin from 40 to 50 imaginal discs in each tube. Tubes were removed after different number of sonication pulses from the sonicator. Reverse cross-linked DNA was resolved on 1 % agarose gel. (**a** and **b**) Show chromatin fragmented in 300 μl and 400 μl volume, respectively. Note that the size of DNA fragment reduces due to increase in sonication volume, but the DNA does not peak around 500 bp. (**c**) When sonication volume was kept at 400 μl and number of sonication pulses was decreased to two pulses, majority of sheared DNA was found to be around 500 bp. This image shows three different samples treated the same way, indicating reproducibility. Numbers above different lanes show number of pulses used for chromatin sonication, M denotes the molecular weight marker, and *arrow* denotes the 500 bp size DNA fragment in it

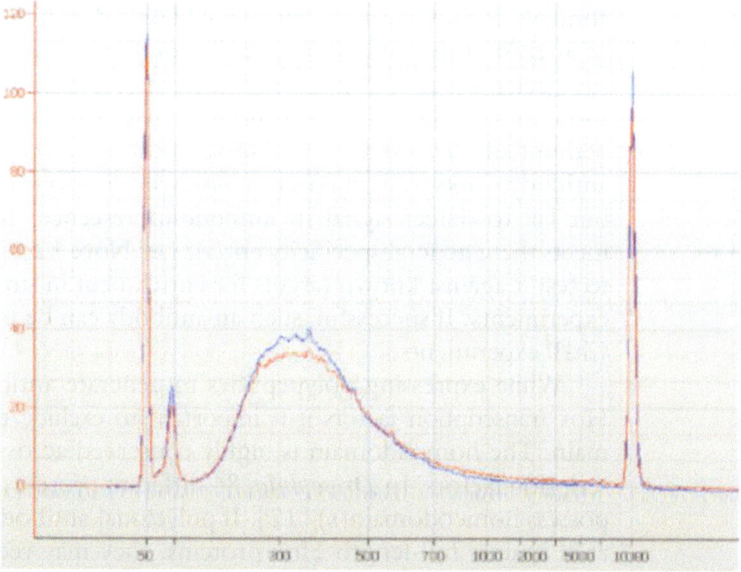

Fig. 2 Bioanalyzer profile of LM-PCR-amplified ChIP samples. Amplified DNA shows distribution between 200 and 500 bp size with a peak at around 300–500 bp, similar to un-amplified template. Numbers on *X*-axis represent DNA size in bp and *Y*-axis represent absorbance in fluorescence units (FU)

agarose gel and take pictures several times during gel electrophoresis to check the size of DNA (Fig. 1). For more accurate fragment distribution the 2100 Bioanalyzer, Agilent Technologies, can be used (*see* **Note 8**) (Fig. 2).

6. If the size of DNA is between 200 and 1,000 base pairs (bp) with majority of fragments around 500 bp start pull down with the chromatin stored at −80 °C.

3.6 Antibody Binding and Chromatin Pull Down

Antibodies are the most crucial reagents for a ChIP experiment. Antibodies can be generated against various histone modifications or transcription factor of interest. An ideal "ChIP-grade" antibody for a transcription factor can recognize multiple epitopes and produce good enrichment for known binding sequences in ChIP assays. Since polyclonal antibodies are a mixture of antibodies against multiple epitopes of a protein, at least few epitopes would be available for binding during ChIP that enables efficient pull down of the antigen of interest that is, in turn, bound to the DNA.

It is important to keep in mind that an antibody that can specifically recognize a protein on Western blot and immune fluorescence might still not enrich targets in ChIP experiments; hence it should be tested in small-scale ChIP experiments before one sets out to use them for global binding analysis. If starting with a new antibody against a Hox transcription factor, it should be tested by Western blot hybridization for specificity using purified proteins and tissue lysates expressing Hox protein of interest. Purified non-target Hox proteins and nontarget tissues can be used as negative controls. Ideally the antibodies should only detect the Hox protein of interest in the target tissues and should not cross-react with tissues lacking that Hox protein. Furthermore, immunofluorescence should reveal the expected expression pattern of the Hox protein without any cross-detection. However there are few cases when an antibody works well on Western blot hybridization and ChIP assays but fail to detect signal in immunofluorescence. If an antibody meets these general specificity criteria (*see* **Note 12**), it is important to test it against known targets for enrichment in small-scale ChIP experiments. If successful, such an antibody can be used for global ChIP experiments.

While expressing polypeptides to generate antibodies against Hox transcription factors it is important to exclude the homeodomain. The homeodomain is highly conserved across several transcription factors. In *Drosophila*, 84 different proteins are known to possess homeodomain(s) [12]. If polyclonal antibodies are generated against full-length Hox proteins, they may recognize other homeodomain-containing transcription factors expressed in that tissue. For example, a study using antibodies against full-length Ubx protein to detect its targets in ChIP-chip resulted in much higher number of targets [13] than other studies that investigated binding of a tagged version of Ubx protein [14] or used antibodies raised against truncated Ubx protein lacking the conserved homeodomain [5].

The concentration of antibody required for ChIP depends on target abundance, cell/tissue type, and downstream processing and has to be determined empirically. In general, small-scale ChIP experiments can be performed with multiple antibody dilutions including the one that works for the immunofluorescence in the target tissue and enrichment can be compared with different dilutions of antibodies. For example our Ubx antibodies work in immunofluorescence at 1:500 dilution; therefore, we tested it in ChIP experiments at 1:250, 1:500, and 1:1,000 dilutions. As a readout we compared the enrichment of Ubx targets *knot*, *Spalt-major*, and *CG13222* [15–17] and found that 1:500 dilution gave us best signal to noise for these targets.

1. If starting with chromatin stored at −80 °C allow slow thawing of the chromatin on ice. If starting with 400 μl sonication volume, use 170–180 μl of chromatin per IP tube and add ChIP dilution buffer to make up the final volume to 300 μl. A typical pull-down tube contains chromatin 180 μl + protease inhibitor cocktail (Roche) 2× + polyclonal anti-Ubx sera (1:500 final concentration), or appropriate ChIP-grade antibodies + ChIP dilution buffer to make up the final volume to 300 μl. Label the tube as Test-IP. Mock IP tube contains abovementioned ingredients except ChIP antibodies. Alternatively, pre-immune sera in the same amount as ChIP antibody can be added to Mock IP tube.
2. Incubate on rotating wheel at 20 rpm at 4 °C overnight to allow binding of antibodies to cross-linked chromatin.
3. Add 20 μl Protein-A Sepharose 50 % slurry in each IP reaction tube and incubate at 4 °C on rotating wheel for 3 h.
4. To pull down chromatin-antibody-matrix complex centrifuge tubes at 500 rcf for 1 min (*see* **Note 7**) and carefully remove the supernatant which will have unbound nonspecific DNA.
5. Wash Protein-A Sepharose beads with different ChIP buffers in 1 ml volumes, as described below. After adding each wash buffer gently swirl the complex for 1 min and then centrifuge at 500 rcf for 1 min to remove the supernatant. The buffers should be used in the sequence given below:
   (a) Low-salt immune complex wash buffer—One wash.
   (b) High-salt immune complex wash buffer—One wash.
   (c) LiCl immune complex wash buffer—One wash.
   (d) TE buffer—Two washes.
6. Complex is ready for elution at this stage; use freshly made elution buffer to elute DNA out.

7. Resuspend pellet in 150 μl ChIP elution buffer (*see* **Note 3**) and vortex gently for 15 min at room temperature.
8. Centrifuge for 1 min at 500 rcf and transfer supernatant to a fresh tube.
9. Repeat **steps** 7 and **8** one more time and pool the eluates.
10. Centrifuge eluate for 1 min at 3,000 rcf and transfer to a fresh tube if there are any residual beads.

3.7 Reversal of Cross-Links and DNA Purification

1. Add 1 μl RNase (stock 10 mg/ml, in water) and 18 μl of 5 M NaCl per IP tube.
2. Incubate samples at 67 °C for 4–5 h for reversal of cross-links and digestion of RNA simultaneously (*see* **Note 13**).
3. Cool tubes for 1–2 min on ice. Add 800 μl ice-cold >99 % ethanol (i.e., 2.5 times of present volume), mix by inversion, and store at −20 °C overnight for precipitating pulled-down DNA.
4. Centrifuge at 4 °C for 20–30 min at 14,000 rcf to precipitate DNA and remove and discard supernatant with a pipette tip.
5. Centrifuge at 4 °C for 1–2 min at 14,000 rcf and remove the remaining residual liquid.
6. Air-dry DNA pellet for ~1 h. The pellet will appear white due to the presence of salt.
7. Resuspend DNA pellet in 100 μl TE and proceed for protein digestion.
8. Add 25 μl 5× PK buffer and 1.5 μl proteinase K. Incubate at 45 °C for 2 h.
9. Elute DNA using Zymo ChIP DNA clean and concentrator Kit (Cat# D5205).
10. The amount of DNA eluted at this stage is usually lower than detectable range of most spectrophotometers, e.g., Nanodrop (ND-2000, Thermo-Fisher). Use fluorometer, e.g., Qubit 2.0 (Life Technology, Cat #. Q32866 along with the kit Q32851), to estimate the amount of total pulled-down DNA.
11. Store the eluted DNA at −20 °C for qPCR, sequencing, and other downstream applications.

3.8 Detection of Target Enrichment

Eluted DNA (**step 11**, Subheading 3.7) is used to detect binding on a localized or a global scale. If using qPCR to detect enrichment it is important to use appropriate negative control where you expect no binding (regions of DNA where the protein of interest does not bind, for example, coding region of a housekeeping gene) and a known target as a positive control. Readers are advised to follow MIQE guidelines [18] while designing and publishing qPCR experiments.

4 Notes

1. Protease inhibitor cocktail tablets without EDTA should be added (twice the recommended amount) to all the buffers before use. Keep the solutions at 4 °C until use. The solution loses protease inhibitor activity if stored for longer than 2 weeks at 4 °C.
2. Prepare nuclear lysis buffer fresh for each use. SDS in the buffer can precipitate at low temperatures; bring the buffer to room temperature and gently mix to dissolve the precipitate.
3. Prepare elution buffer fresh for each use. 1 M $NaHCO_3$ solution can be prepared and stored in aliquots at −20 °C. Thaw an aliquot to prepare elution buffer.
4. Caution: The chemical is extremely harmful/carcinogenic/toxic. Use appropriate personal protective equipment and follow the manufacturer's instruction for safe use and local safety guidelines for disposal.
5. Paraformaldehyde is a polymer of formaldehyde and sold as a powder. When dissolved in the presence of alkali at room temperature it produces hydrated formaldehyde monomers which are utilized for cross-linking [19]. Since monomerization of this polymer depends on temperature and pH leading to high variability in fixation, we do not recommend using paraformaldehyde powder to prepare fixative. Instead liquid formaldehyde solution (usually 37–41 %) can be diluted in water to desired concentration and used at room temperature to obtain reproducible tissue fixation.
6. Protein A has better affinity to Fc fragments of rabbit polyclonal antibodies, whereas protein G binds with better affinity to mouse antibodies.
7. Do not vortex Protein-A Sepharose slurry as it may fracture the beads. Mix beads by gentle inversion and when required, settle by slow centrifugation (less than 500 rcf) to avoid damaging the beads. Once resuspended the beads should not be allowed to dry.
8. 2100 Agilent Bioanalyzer uses capillary-based electrophoresis to separate DNA, RNA, and proteins to provide sizing, quantitation, and quality control. It is a more robust and sensitive method for separation and visualization of nucleic acids than routine agarose gel electrophoresis. Please follow the manufacturer's instructions for use.
9. Chromatin size is crucial to a ChIP experiment as the resolution of binding events depends on the size of the pulled-down fragment. If chromatin size is very large, resolution of true binding would be reduced and true binding event would be masked due to noise from the background. On the other hand,

trying to generate too small fragments during sonication would lead to disruption of nucleosomes and also difficulty in detecting binding events by PCR or qPCR. Typically 200–1,000 bp size range with a peak of ~300–500 bp is ideal for chromatin used in ChIP reaction.

10. Resuspend pellet after nuclear lysis in PBS solution with DAPI and put a drop of this solution on a glass slide to observe under microscope. If the lysis was effective very few intact nuclei will be detected.
11. We have observed a tighter distribution of DNA fragments using water bath sonicators (Bioruptor, UCD-200) than probe-based sonicators. Sonication time and amplitude need to be adjusted for different models. We have observed variations in DNA fragment sizes for same power settings on different machines sourced from a single manufacturer (Bioruptor, UCD-200 and Bioruptor-XL). However, once standardized, fragment sizes are highly reproducible for a given machine.
12. ChIP-grade antibodies are usually affinity purified although the use of sera is also common. In our hands we detected loss of specific signal by affinity purification of Ubx antibodies and therefore used the sera that passed our quality control (QC) procedure mentioned above and produced good signal in all the assays.
13. While for detection of certain targets by ChIP-qPCR overnight de-cross-linking can be performed, Lee et al. [20] have observed increased noise in global ChIP experiments due to overnight de-cross-linking. Therefore, we recommend 4–5-h de-cross-linking time for ChIP-chip and ChIP-seq assays.

Acknowledgements

The authors thank R Mishra, R White, Y Graba, and members of LSS and R Mishra lab for help with experiments and discussion. We thank Council of Scientific and Industrial Research, India, for a fellowship to PA and Department of Atomic Energy and Department of Science and Technology (Government of India) for research grants to LSS.

References

1. Ren B, Robert F, Wyrick JJ et al (2000) Genome-wide location and function of DNA binding proteins. Science 290:2306–2309. doi:10.1126/science.290.5500.2306
2. Iyer VR, Horak CE, Scafe CS et al (2001) Genomic binding sites of the yeast cell-cycle transcription factors SBF and MBF. Nature 409:533–538. doi:10.1038/35054095
3. Wu J, Smith LT, Plass C, Huang TH-M (2006) ChIP-chip comes of age for genome-wide functional analysis. Cancer Res 66:6899–6902. doi:10.1158/0008-5472.CAN-06-0276

4. Park PJ (2009) ChIP-seq: advantages and challenges of a maturing technology. Nat Rev Genet 10:669–680. doi:10.1038/nrg2641
5. Agrawal P, Habib F, Yelagandula R, Shashidhara LS (2011) Genome-level identification of targets of Hox protein Ultrabithorax in Drosophila: novel mechanisms for target selection. Sci Rep 1:1–10. doi:10.1038/srep00205
6. Li X, MacArthur S, Bourgon R et al (2008) Transcription factors bind thousands of active and inactive regions in the Drosophila blastoderm. PLoS Biol 6:e27. doi:10.1371/journal.pbio.0060027
7. Negre N, Brown CD, Ma L et al (2011) A cis-regulatory map of the Drosophila genome. Nature 471:527–531. doi:10.1038/nature09990
8. Bonn S, Zinzen RP, Perez-Gonzalez A et al (2012) Cell type-specific chromatin immunoprecipitation from multicellular complex samples using BiTS-ChIP. Nat Protoc 7:978–994. doi:10.1038/nprot.2012.049
9. Shankaranarayanan P, Walia M, Wang L et al (2011) Single-tube linear DNA amplification (LinDA) for robust ChiP-seq. Nat Methods 8:565–568. doi:10.1038/nmeth.1626
10. Ng J-H, Kumar V, Muratani M et al (2013) In vivo epigenomic profiling of germ cells reveals germ cell molecular signatures. Dev Cell 24:324–333. doi:10.1016/j.devcel.2012.12.011
11. Collas P (2009) Chromatin immunoprecipitation assays. In: Collas P (ed) Methods in molecular biology, 567th edn. Humana, Totowa, NJ, pp 1–25
12. Noyes MB, Christensen RG, Wakabayashi A et al (2008) Analysis of homeodomain specificities allows the family-wide prediction of preferred recognition sites. Cell 133:1277–1289, doi: http://dx.doi.org/10.1016/j.cell.2008.05.023
13. Slattery M, Ma L, Négre N et al (2011) Genome-wide tissue-specific occupancy of the Hox protein Ultrabithorax and Hox cofactor homothorax in Drosophila. PLoS One 6:e14686. doi:10.1371/journal.pone.0014686
14. Choo SW, White R, Russell S (2011) Genome-wide analysis of the binding of the Hox protein Ultrabithorax and the Hox cofactor homothorax in Drosophila. PLoS One 6:e14778. doi:10.1371/journal.pone.0014778
15. Hersh BM, Carroll SB (2005) Direct regulation of knot gene expression by Ultrabithorax and the evolution of cis-regulatory elements in Drosophila. Development 132:1567–1577. doi:10.1242/dev.01737
16. Hersh BM, Nelson CE, Stoll SJ et al (2007) The UBX-regulated network in the haltere imaginal disc of D. melanogaster. Dev Biol 302:717–727. doi:10.1016/j.ydbio.2006.11.011
17. Makhijani K, Kalyani C, Srividya T, Shashidhara LS (2007) Modulation of Decapentaplegic gradient during haltere specification in Drosophila. Dev Biol 302:243–255. doi:10.1016/j.ydbio.2006.09.029
18. Bustin SA, Benes V, Garson JA et al (2009) The MIQE guidelines: minimum information for publication of quantitative real-time PCR experiments. Clin Chem 55:611–622. doi:10.1373/clinchem.2008.112797
19. Manoonkitiwongsa PS, Schultz RL (2003) Proper nomenclature of formaldehyde and paraformaldehyde fixatives for histochemistry. Histochem J 34:365–367. doi:10.1023/A:1023342929105
20. Lee TI, Johnstone SE, Young RA (2006) Chromatin immunoprecipitation and microarray-based analysis of protein location. Nat Protoc 1:729–748. doi:10.1038/nprot2006.98

Chapter 16

SELEX-seq: A Method for Characterizing the Complete Repertoire of Binding Site Preferences for Transcription Factor Complexes

Todd R. Riley*, Matthew Slattery*, Namiko Abe, Chaitanya Rastogi, Dahong Liu, Richard S. Mann, and Harmen J. Bussemaker

Abstract

The closely related members of the Hox family of homeodomain transcription factors have similar DNA binding preferences as monomers, yet carry out distinct functions in vivo. Transcription factors often bind DNA as multiprotein complexes, raising the possibility that complex formation might modify their DNA-binding specificities. To test this hypothesis we developed a new experimental and computational platform, termed SELEX-seq, to characterize DNA-binding specificities of Hox-based multiprotein complexes. We found that complex formation with the same cofactor reveals latent specificities that are not observed for monomeric Hox factors. The findings from this in vitro platform are consistent with in vivo data, and the "latent specificity" concept serves as a precedent for how the specificities of similar transcription factors might be distinguished in vivo. Importantly, the SELEX-seq platform is flexible and can be used to determine the relative affinities to any DNA sequence for any transcription factor or multiprotein complex.

Key words Hox proteins, Transcription factor specificity, Extradenticle, Pbx, SELEX, Next-generation sequencing, Computational analysis

1 Introduction

Members of the homeodomain-containing Hox family of transcription factors (TFs) play integral roles in many aspects of metazoan development, from anterior-posterior patterning during embryogenesis to organogenesis and stem cell maintenance [1–3]. As with most TFs, the DNA-binding activity of Hox proteins is integral to their function—at least with regard to their control of developmental processes [4]. Still, one of the more puzzling aspects of the Hox factors is that they direct exquisitely specific functions in vivo yet have very little specificity when binding to DNA in vitro [4].

* Author contributed equally with all other contributors.

Yacine Graba and René Rezsohazy (eds.), *Hox Genes: Methods and Protocols*, Methods in Molecular Biology, vol. 1196,
DOI 10.1007/978-1-4939-1242-1_16,

Importantly, this quandary is not limited to the Hox family of TFs; a similar disconnect between DNA-binding specificity in vitro and functional specificity in vivo is observed for many TF families, including those encompassing the T-box, Ets, and bHLH factors [5–7]. Given the importance of TF-DNA interactions in gene regulation, understanding the DNA-binding specificities is one of the critical steps in deciphering the function of noncoding, regulatory DNA, and it is clear that monomeric DNA-binding properties cannot fully explain the specific in vivo functions of many TFs.

For the Hox proteins, one potential solution to this specificity paradox lies in the finding that Hox proteins can bind DNA in vivo in conjunction with cofactors. The most critical cofactors known to perform this function are the PBC (pre-B cell) homeodomain TFs: Exd (Extradenticle) in *Drosophila* and Pbx (pre-B cell leukemia homeobox) in vertebrates. Both Exd and Pbx bind in a highly cooperative manner with Hox proteins to composite Hox-Exd-binding sites [4]. Importantly, beyond increasing Hox DNA-binding affinities, PBC proteins enhance the DNA-binding specificities of Hox proteins. That is, PBC-Hox heterodimers are more selective in their DNA-binding preferences than Hox monomers. For the *Drosophila* Hox TF Scr (Sex combs reduced), structural studies demonstrated that interaction with Exd allows the Scr protein to recognize the unique minor groove structure of an Exd-Scr-specific DNA motif [8]. Thus, in this case, interaction with the PBC protein Exd reveals a latent specificity that is intrinsic to the Hox protein Scr. This latent specificity example, combined with the fact that PBC proteins can heterodimerize with all Hox family members, suggested that the formation of Hox-PBC complexes might have a widespread impact on Hox DNA-binding specificity [8].

The Homothorax (Hth)-Meis family of homeodomain proteins can also influence PBC-Hox DNA binding [4]. Interaction with Hth-Meis family members stabilizes PBC proteins and can promote nuclear localization of PBC proteins. Further, Hth-Meis factors also promote cooperative PBC-Hox binding on certain sequences. In *Drosophila*, an Hth isoform that lacks a DNA-binding domain but contains the Exd-interacting domain—the "Hth-Meis" (HM) domain—is sufficient for Exd nuclear localization and stability, and can carry out most Hox-related functions of *hth* [9, 4].

We recently described a novel high-throughput method, termed SELEX-seq, that we used to systematically characterize the DNA-binding specificities of all *Drosophila* Hox-Exd-HM complexes (hereafter referred to as Hox-Exd complexes for simplicity) [10]. This approach combines the classical method of SELEX (Systematic Evolution of Ligands by Exponential Enrichment; also known as in vitro selection) [11, 12] with the power of next-generation sequencing technology, and is ideally suited for exploring the DNA-binding preferences of multiprotein complexes. As with traditional SELEX, an oligonucleotide containing a randomized region that is flanked by defined primer docking sites is used to bind the Hox-Exd complex of interest. DNA bound by the complex is then separated from unbound DNA, in this case using EMSA (electrophoretic mobility shift assay), though immunopurification-based assays can also be employed, and the bound DNA is then amplified by PCR and used for subsequent rounds of DNA binding and selection.

SELEX-seq differs from traditional SELEX in two respects: the number of selected (bound) DNA oligos characterized and the number of rounds of selection performed. Unlike traditional SELEX, where on the order of 10^2 selected DNA oligos are identified at the very end of the reiterative selection process, SELEX-seq leverages the depth of next-generation sequencing to characterize 10^7 or more selected DNA molecules at each round of selection. Additionally, whereas traditional SELEX requires many rounds of selection, the greater sequencing depth of SELEX-seq allows for identification of relevant binding sites after only one to two rounds of selection. Using a biophysical model of the SELEX procedure, relative affinities for selected sequences are then obtained by comparing the sequence composition of later rounds to that of the unselected DNA library. The combined impact of these improvements—greater coverage of selected DNA, fewer rounds of selection, and biophysical sequence-to-affinity model—is that SELEX-seq captures more than just high-affinity binding sites, and thus provides a more complete view of the binding preferences for a TF or TF complex. And because moderate-affinity binding sites are just as likely to be relevant in vivo, techniques that reveal the entire binding site repertoire for a TF or a TF complex are essential.

Applying the SELEX-seq strategy to all eight *Drosophila* Hox-Exd heterodimers demonstrated that complex formation with the same cofactor reveals latent specificities that are not observed for monomeric Hox factors [10], and the findings from SELEX-seq are consistent with in vivo data [13]. Thus, the latent specificity phenomenon extends well beyond the Hox protein Scr, and serves as a precedent for how the specificities of other transcription factor families might be distinguished in vivo. Therefore, SELEX-seq now serves as a platform for studying numerous TFs and multiprotein TF complexes [14]. In this chapter, we describe the detailed procedures for all of the wet lab and computational steps for executing SELEX-seq.

2 Materials

2.1 Preparation of SELEX Library and Control EMSA Probe

1. Oligonucleotides for SELEX library (*see* **Note 1**):

   SELEX_16mer_Multiplex1: 5′GTTCAGAGTTCTACAGTC CGACGATCTGG-[N_{16}]-CCA**gc**T**g**TCGTATGC CGTCTTCTGCTTG3′.

SELEX_16mer_Multiplex2: 5′GTTCAGAGTTCTACAGTCCGACGATCTGG-[N_{16}]-CCA**cg**T**c**TCGTATGCCGTCTTCTGCTTG3′.

SELEX_SR1: 5′ CAAGCAGAAGACGGCATACGA 3′.

2. Oligonucleotides for control EMSA probe (*see* **Note 2**):

fkhCON_Tracking_Fwd: 5′GCTATACTGTGCTATCCACAGTTCAGAGTCGTCAAGATTTATGGCCTGCTGGTCACTGGTCGTTTCCCTCTT3′.

fkhCON_Tracking_Rev: 5′AAGAGGGAAACGACCAGTGACCAGCAGGCCATAAATCTTGACGACTCTGAACTGTGGATAGCACAGTATAGC3′.

3. 10× STE buffer: 10 mM Tris pH 8, 1 M NaCl, 1 mM EDTA pH 8.
4. DNA Polymerase I, Large (Klenow) Fragment and 10× NEB Buffer 2 (New England BioLabs).
5. T4 Polynucleotide Kinase and 10× polynucleotide kinase reaction buffer.
6. 10 mM dNTP mix.
7. ATP [γ-^{32}P] (6,000 Ci/mmol 10 mCi/ml; PerkinElmer) (*see* **Note 3**).
8. TE stop buffer: 10 mM Tris pH 8, 6.6 mM EDTA pH 8.
9. MinElute PCR Purification Kit (Qiagen).
10. 5 % TBE Acrylamide Gel (BioRad) and 5× TBE.

2.2 DNA-Binding Reaction and EMSA

1. Purified DNA-binding protein of interest (*see* **Note 4**).
2. 5× Binding buffer: 50 mM Tris pH 7.5, 250 mM NaCl, 5 mM MgCl2, 20 % glycerol, 2.5 mM DTT, 2.5 mM EDTA pH 8, 250 μg/ml polydI-dC, 1 mg/ml BSA.
3. 80 % Glycerol.
4. 40 % Acrylamide solution.
5. 30 % Acrylamide/bis-acrylamide solution, 37.5:1 (BioRad).
6. 10 % Ammonium persulfate.
7. TEMED (Tetramethylethylenediamine).
8. 5× TBE.
9. Gel dryer and phosphorimager cassette.

2.3 Isolation and Elution of Bound DNA

1. Elution buffer: 0.5 M NH_4OAc, 1 mM EDTA, 0.1 % SDS.
2. Phenol:Chloroform:Isoamyl Alcohol (25:24:1, v/v) (Invitrogen/Life Technologies).
3. Ethanol.
4. 3 M Sodium acetate, pH 5.2.
5. TE: 10 mM Tris pH 8, 1 mM EDTA pH 8.

2.4 DNA Amplification and Preparation of Sequencing Library

1. Oligonucleotides for amplification of SELEX library:
   SELEX_SR0: 5′GTTCAGAGTTCTACAGTCCGA3′.
   SELEX_SR1: 5′CAAGCAGAAGACGGCATACGA3′.
2. Oligonucleotides for preparation of sequencing library:
   SELEX_SR1: 5′CAAGCAGAAGACGGCATACGA3′.
   SELEX_SR2: 5′AATGATACGGCGACCACCGACAGGTTC AGAGTTCTACAGTCCGA3′.
3. *Taq* DNA Polymerase and 10× *Taq* Reaction Buffer (New England BioLabs).
4. Phusion High-Fidelity DNA Polymerase and 5× Phusion HF Buffer (New England BioLabs).
5. 10 mM dNTP mix.
6. MinElute PCR Purification Kit (Qiagen).
7. 5 % TBE Acrylamide Gel (BioRad) and 5× TBE.
8. 10× NEB Buffer 2 (New England BioLabs).
9. Corning Costar Spin-X centrifuge tube filters (Sigma).
10. Ethanol.
11. 3 M Sodium acetate, pH 5.2.
12. 20 mg/ml Glycogen (Roche).

2.5 Computational Analysis of the Data

1. A personal computer, or an account on a shared computer system (a minimum of 4 GB random access memory and 256 MB hard drive storage recommended, but more may be required depending on the data and analysis options).
2. SELEX analysis software (available for download as R package at bussemakerlab.org/software/SELEX/).
3. R version 2.14.0 or newer.
4. Java runtime environment version 1.6.0 or newer.

3 Methods

3.1 Preparation of SELEX Library and Control EMSA Probe

The success of a SELEX-seq experiment is primarily driven by two components: the DNA-binding protein(s) of interest and the randomized DNA SELEX library. We have had much success using bacterially expressed (*E. coli*), purified His-tagged Hox and HM-Exd proteins, and the methods described here assume access to purified protein preparations. However, there is much flexibility in the methods of protein expression and purification that are suitable for SELEX-seq [10, 14], so this facet of the experimental design must be optimized on a case-by-case basis.

SELEX

```
SELEX_16mer_Multplex1
5' GTTCAGAGTTCTACAGT CCGACGATCTGG [NNN...] CCAgcTgTCGTATGCCGTCT TCTGCTTG 3'
3' ----------------- ------------ [------] -------------------- -------- 5'

SELEX_SR1
5' ----------------- ------------ [------] -------------------- -------- 3'
3' ----------------- ------------ [------] -------AGCATACGGCAGA AGACGAAC 5'

SELEX_16mer_Multplex1 + SELEX_SR1
5' GTTCAGAGTTCTACAGT CCGACGATCTGG [NNN...] CCAgcTgTCGTATGCCGTCT TCTGCTTG 3'
3' ----------------- ------------ [------] -------AGCATACGGCAGA AGACGAAC 5'

SELEX Library
5' GTTCAGAGTTCTACAGT CCGACGATCTGG [NNN...] CCAgcTgTCGTATGCCGTCT TCTGCTTG 3'
3' CAAGTCTCAAGATGTCA GGCTGCTAGACC [NNN...] GGTcgAcAGCATACGGCAGA AGACGAAC 5'

SELEX_SR0
5' GTTCAGAGTTCTACAGT CCGA-------- [------] -------------------- -------- 3'
3' ----------------- ------------ [------] -------------------- -------- 5'
```

Sequencing

```
SELEX_SR2
5' AATGATACGGCGACCACCGA CAGGTTCAGAGTTCTACAGT CCGA-------- [------] -------------------- -------- 3'
3' -------------------- -------------------- ------------ [------] -------------------- -------- 5'

SELEX Sequencing Library
5' AATGATACGGCGACCACCGA CAGGTTCAGAGTTCTACAGT CCGACGATCTGG [NNN...] CCAgcTgTCGTATGCCGTCT TCTGCTTG 3'
3' TTACTATGCCGCTGGTGGCT GTCCAAGTCTCAAGATGTCA GGCTGCTAGACC [NNN...] GGTcgAcAGCATACGGCAGA AGACGAAC 5'

Sequencing Primer**
5' -----------------CGA CAGGTTCAGAGTTCTACAGT CCGACGATC--- [------] -------------------- -------- 3'
3' -------------------- -------------------- ------------ [------] -------------------- -------- 5'
```

Fig. 1 SELEX-seq oligonucleotide sequences. Sequences of oligonucleotides used for SELEX library generation, DNA amplification, and Illumina sequencing

Beyond protein preparation, multiple variables must be considered when it comes to generation of the randomized SELEX library. The randomized region must be large enough to encompass the expected core DNA-binding motif and also allow for capturing information regarding sequences immediately flanking the core motif; for our exploration of Hox-Exd binding we used a 16 bp randomized region because all previously characterized Hox-Exd-binding sites were 10–11 bp long. In addition to the randomized region, the SELEX library must eventually carry adapter sequences that are compatible with Illumina sequencing. Sequence information for various Illumina-compatible adapters can be accessed via Illumina's website (www.illumina.com) and the oligonucleotides used for characterizing Hox-Exd binding are shown in Fig. 1. In this case, the adapter regions are exact matches to the sequences used for the Illumina small RNA sequencing library, with the exception of the TGG/CCA sequences immediately flanking the random region (TGG[N_{16}]CCA) and the 4 bp barcode region (gcTg)

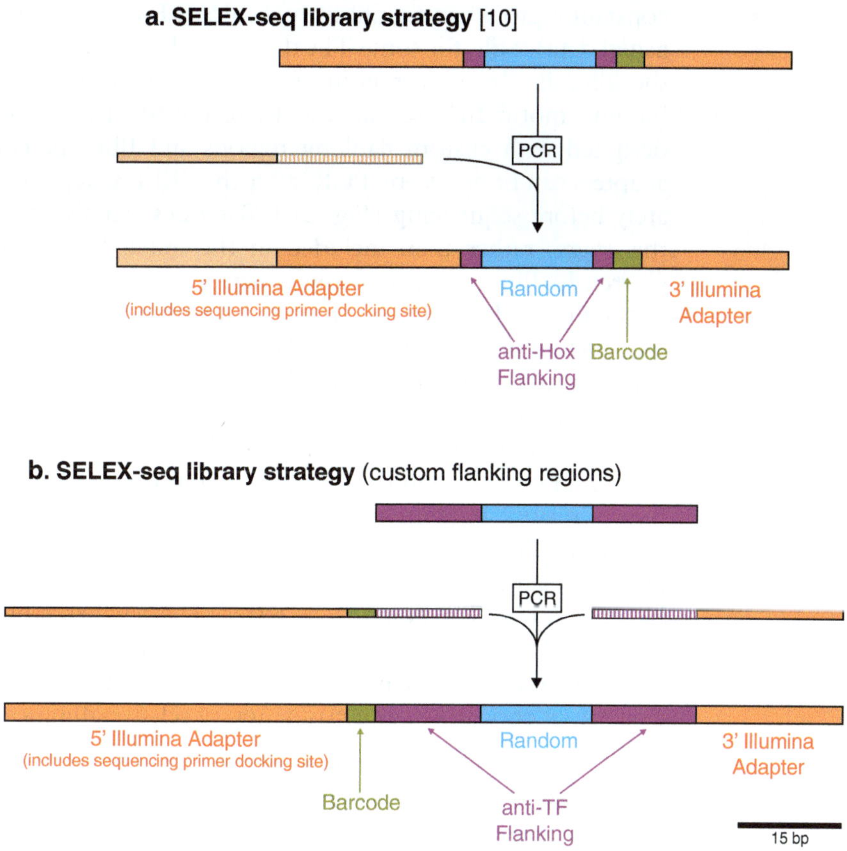

Fig. 2 Strategies for SELEX-seq library design. (**a**) SELEX-seq library strategy used for the Hox-Exd study described here and in [10] (**b**) SELEX-seq strategy in which Illumina adapters are not added until after selection

just 3′ of the CCA (Fig. 1). The TGG/CCA sequences flanking the randomized oligonucleotides act as "anti-Hox" flanking regions, meant to discourage Hox-Exd binding that overlaps the constant region. The 4 bp barcode region allows for multiplexing of multiple SELEX-seq libraries in the same Illumina sequencing lane; two multiplex barcoded libraries ("SELEX_16mer_Multiplex1" and "SELEX_16mer_Multiplex2") are described here, though additional barcodes can be used to increase multiplexing depth (i.e., running >2 samples per sequencing lane). The 5′ and 3′ adapter regions used for Illumina sequencing differ significantly in length (Fig. 1), so the SELEX library is designed to be near symmetrical in length—29 and 28 bp constant regions flanking the randomized region—with the additional 5′ adapter added via PCR immediately before sequencing (Figs. 1 and 2a).

As described above, the "anti-Hox" sequences were sufficient to prevent Hox-Exd binding to the constant adapter sequences. This is important because TF binding within or overlapping the constant regions can significantly skew SELEX-seq results (unpublished data). Still, for some TFs there may be sequences integral to the Illumina adapter regions that also match the TF's DNA-binding motif. In cases such as these the SELEX library can be designed with custom flanking regions and Illumina-compatible adapters can be added by PCR, after the SELEX step and immediately before sequencing (Fig. 2b). Barcodes can also be added at this stage, rather than included in the initial library (Fig. 2b). Currently, Illumina sequencing is performed in 50 or 100 cycles, so the full randomized region must fall within 50 or 100 bp of the sequencing primer (of course 100 cycles of sequencing costs approximately twice as much as 50 cycles).

Finally, one of the key features of SELEX-seq is that it allows for calculation of motif enrichment after selection relative to the initial SELEX library (Round 0, or R0), which in turn allows for estimation of relative binding affinities. Because this requires sequencing data from R0, it is best to make enough double-stranded SELEX library for an entire experiment, especially if different TFs or combinations of TFs are going to be tested. If a new batch of the double-stranded SELEX library is generated in the middle of an experiment, re-sequencing of R0 may be necessary for accurate calculation of enrichment and estimation of relative affinities.

1. Anneal SELEX library template (SELEX_16mer_Multiplex1 or SELEX_16mer_Multiplex2) and SELEX_SR1. Standard annealing mix consists of 3 μl of 100 μM SELEX_16mer_Multiplex1 or _Multplex2 template, 6 μl 100 μM SELEX_SR1, 3 μl of 10× STE buffer, and 18 μl of water. Boil the mixture for 5 min and allow to cool slowly back to room temperature; this is most easily performed by floating the tube for 5 min in 500–700 ml boiling water in a 1 l beaker, then removing the whole beaker from heat, and letting it cool down to room temperature overnight. 30 μl of 10 μM annealed product will be generated.
2. Extend the primer to generate the double-stranded SELEX library. The reaction described here makes use of one-third of the annealed product from **step 1**, but can be scaled up to use all 30 μl of the annealed solution, if necessary. Mix 10 μl of annealed product from **step 1** with 8 μl water, 2.5 μl 10× NEB buffer 2, 2 μl 10 mM dNTPs, and 2.5 μl DNA Polymerase I, Large (Klenow) Fragment. Allow extension reaction to proceed for at least 30 min at room temperature, and add 1 μl of 0.5 M EDTA to stop the reaction.
3. Purify the double-stranded SELEX library (hereafter referred to as "SELEX library") using the MinElute PCR purification kit, using the standard protocol provided by Qiagen. Measure the concentration of the SELEX library using a NanoDrop spectrophotometer and adjust to a convenient concentration for SELEX (3 μM will be used for the protocol described here).

Also check that the purified SELEX library is primarily a single band of the expected size by gel electrophoresis on a 5 % TBE Acrylamide Gel (BioRad). SELEX library can be stored at −20 °C.

4. Generation of a control EMSA probe—which is used for tracking protein-DNA complexes—does not require a Klenow extension reaction. Full-sense and antisense oligonucleotides can be used (because control probes do not have a random region). Anneal 2 μl of 50 μM fkhCON_Tracking_Fwd and 2 μl of 50 μM fkhCON_Tracking_Rev in 86 μl water and 10 μl 10× STE buffer. Anneal by boiling for 5 min and slowly cooling to room temperature as described above in **step 1**. The annealed control probe can then be labeled with ^{32}P in the following reaction: 3.5 μl annealed probe (~1 μM, from annealing reaction), 3.5 μl water, 1 μl 10× polynucleotide kinase reaction buffer, 1 μl T4 Polynucleotide Kinase, and 1 μl ATP [γ-^{32}P]. Allow reaction to proceed for 10 min at 37 °C, and then stop the reaction by adding 90 μl TE stop buffer, for a final ^{32}P-labeled probe concentration of 35 nM.

3.2 DNA-Binding Reaction and EMSA

When running the SELEX EMSA the control binding lanes serve as tracking lanes to monitor the mobility of the protein-DNA complex. It is best to set up two tracking lanes, run on each side of the SELEX lane, to facilitate accurate tracking and isolation of the appropriate region in the SELEX lane. In addition to the tracking lanes, a "no-protein" lane containing the radiolabeled control probe and no TF(s) should also be included to monitor the mobility of the free probe; this lane should also include a loading dye such as bromophenol blue for monitoring progression of the gel. When working with a multiprotein complex one should also set up lanes containing individual proteins to monitor the mobility of monomeric protein-DNA binding. For simplicity, only the SELEX and control/tracking lanes are described here.

1. Prepare a gel for EMSA by adding 3.5 ml 5× TBE, 3.1 ml 30 % acrylamide/bis-acrylamide solution (37.5:1), 2.33 ml 40 % acrylamide solution, and 1.1 ml 80 % glycerol to 24.25 ml water. Mix well, without generating bubbles, and then add 262.5 μl 10 % ammonium persulfate and 17.5 μl TEMED to catalyze polymerization. Mix briefly, pour gel, and allow gel to solidify for 1 h. After gel has solidified, be sure to flush the wells with 0.5× TBE to remove unpolymerized acrylamide and pre-run gel for 20–30 min (150 V) in 0.5× TBE buffer. Run gel in cold room at 4 °C (*see* **Note 4**).
2. Set up the 30 μl control binding reactions as follows: 1 μl of 35 nM labeled control probe, 6 μl 5× binding buffer, 6 μl of 500 nM Exd-HM, 6 μl of 1 μM Hox, 11 μl of water (*see* **Note 5**).

3. Set up the 30 μl SELEX binding reaction as follows: 2 μl of 3 μM SELEX library, 6 μl 5× binding buffer, 1 μl 1 μM Exd-HM, 2 μl of 1 μM Hox, 19 μl water (*see* **Note 5**).
4. Assemble control- and SELEX-binding reactions on ice (or in 4 °C cold room), and then incubate reactions for 20 min at room temperature. The EMSA gel from **step 1** should be pre-running while binding reactions are taking place. After the 20-min incubation stop pre-running gel, load samples into lanes, and run the gel for approximately 2 h at 150 V (4 °C).
5. After running the gel, remove it from glass plates, vacuum dry on Whatman paper, wrap the dried gel in Saran wrap, and tape the gel down in a phosphorimager cassette. Expose the gel at 4 °C. Short exposure times (1–2 h) should be sufficient for tracking the protein–DNA complex, but overnight exposure is also acceptable. After exposure, when scanning the phosphorscreen, adjust the phosphorimager to capture the entire gel (including extra space beyond all four gel borders).

3.3 Isolation and Elution of Bound DNA

1. To cut the appropriate SELEX "band" from the gel, print a 1:1 copy of the phosphorimager scan. Before printing, adjust the brightness and contrast of the scan to see all gel borders. Use scissors to cut exactly around gel borders of printed scan, and cut out a window from the print corresponding to the SELEX region to be isolated (window based on the actual band in tracking lanes). Align the gel cutout with the gel taped down in phosphorimager cassette; the free DNA probe from the "no-protein" lane should align approximately with bromophenol blue band for probes described here. Use marker to outline the SELEX band to be cut, remove the gel from cassette, and cut SELEX band out of gel using a new razor blade.
2. Remove Saran wrap from isolated band, and remove excess Whatman paper (though some carryover of Whatman paper into elution step is acceptable). Cut the isolated band into 3–4 pieces and place all pieces in a 1.5 ml tube. Add 1 ml of elution buffer to tube carrying the isolated gel pieces.
3. Elute at 37 °C overnight. Collect elution buffer from elution and add another 750 μl elution buffer to gel pieces. Elute for 2 h at 37 °C, then remove buffer, and combine with buffer from first elution.
4. Add an equal volume of phenol:chloroform to the eluted DNA solution in a 15 ml polypropylene Falcon tube. Vortex the mixture for 30 s, and then centrifuge for 3 min at 3,000 × *g*. Carefully remove the aqueous layer to a new tube (avoid interface).
5. Before ethanol precipitation of DNA, the aqueous layer should be split evenly among multiple 1.5 ml tubes because the current volume is too large for precipitation in a 1.5 ml tube

(which is optimal for viewing DNA pellet). To precipitate DNA, add 2 volumes of 100 % ethanol to each tube and store at −80 °C (or on dry ice) for at least 30 min. After 30 min, spin tubes at max speed (>15,000 × *g*) for 20 min (4 °C), and then wash pellet with 70 % ethanol. Remove all ethanol and dry pellets (do not over-dry).

6. Dissolve precipitated DNA in 200 μl TE pH 8. 200 μl is the total volume across all tubes, so if four tubes were used dissolve each pellet in 50 μl TE. Once pellets are dissolved in TE, pool all in one tube (200 μl final volume). Add 25 μl of 3 M sodium acetate pH 5.2, mix well, and then add 450 μl 100 % ethanol. Again, precipitate at −80 °C for at least 30 min, and then spin tubes at max speed (>15,000 × *g*) for 20 min (4 °C). Wash pellet with 70 % ethanol, dry pellet, and dissolve DNA in 20 μl TE pH 8.
7. Store DNA at −20 °C. Half of this eluted DNA will be amplified and used for the next round of selection and sequencing, and the other half will be stored as backup.

3.4 DNA Amplification and Preparation of Sequencing Library

1. A total of 10 μl eluted DNA will be amplified for an additional round of selection. This should be set up in five parallel 50 μl PCR reactions, each consisting of the following: 2 μl template DNA, 39.5 μl water, 5 μl of 10× *Taq* Reaction Buffer, 1 μl of 20 μM SELEX_SR0, 1 μl of 20 μM SELEX_SR1, 1 μl of 10 mM dNTPs, and 0.5 μl *Taq* polymerase. A master mix can be used to set up the five reactions. The PCR conditions are as follows: 2 min at 94 °C, then 15 cycles of 15 s at 94 °C, 15 s at 55 °C, and 30 s at 72 °C, followed by 1 min at 72 °C, and a hold at 4 °C (*see* **Note 6**).
2. After PCR, pool five reactions back to approximately 250 μl and purify using Qiagen's MinElute PCR Purification Kit. Follow standard Qiagen protocol, though note that 5 volumes of buffer PB in this case amounts to 1.25 ml, so the sample plus PB volume will be approximately 1.5 ml. Because the MinElute column capacity is ~700 μl, the entire sample cannot be put through the column at once; instead apply sample to the same column in three successive stages, approximately 500 μl each time, being sure to discard flow through after each spin. Once all samples have been added to column, proceed as per Qiagen's instructions. At the final step elute DNA in 13 μl buffer EB. Estimate DNA concentration using a NanoDrop spectrophotometer. Again, as described above with the initial library, adjust to a convenient concentration for SELEX (3 μM in this case), and check that the library is primarily a single band by acrylamide gel electrophoresis. This amplified library from round 1 (R1) of selection can be stored at −20 °C until round 2 of selection (R2, *see* **Note 7**) or generation of the sequencing library.

3. The amplified DNA from the previous step is then prepared for Illumina sequencing using limited cycle PCR to add the final adapter sequence (see Figs. 1 and 2a). The PCR is set up in five parallel 50 μl PCR reactions, each consisting of the following: 0.8 μl of 500 nM DNA template, 33.2 μl water, 10 μl of 5× Phusion HF Buffer, 2 μl of 20 μM SELEX_SR1, 2 μl of 20 μM SELEX_SR2, 1.5 μl of 10 mM dNTPs, and 0.5 μl of Phusion High-Fidelity DNA Polymerase. A master mix can be used to set up reactions. The PCR conditions are as follows: 30 s at 98 °C, then three cycles of 10 s at 98 °C, 30 s at 60 °C, and 15 s at 72 °C, followed by 10 min at 72 °C, and a hold at 4 °C.
4. Pool all five reactions and purify using the MinElute PCR Purification Kit (Qiagen) as described above in **step 2** of this section (elute in 10 μl buffer EB). Run the product on a 5 % TBE Acrylamide Gel (BioRad, *see* **Note 8**) with a 10 bp DNA ladder. Pre-run the gel for 10 min at 100 V, and then load sample and ladder lanes. When gel is complete (be careful not to run the product off of gel), stain the gel with ethidium bromide for 5–10 min. Visualize the PCR product on UV transilluminator. Two products will likely be visible: one higher mobility band corresponding to the selected DNA that has not acquired the full sequencing adapter (the DNA added by primer SELEX_SR2) and a lower mobility product containing the selected DNA plus all necessary Illumina adapter sequences. Cut out the band corresponding to the latter product.
5. Puncture the bottom of a 0.5 ml microfuge tube five times with a 21 gauge needle and stack the punctured tube in a 2 ml microfuge tube. Place the isolated gel slice from the previous step (**step 4**) in the 0.5 ml tube. Centrifuge the stacked tubes for 2 min at maximum speed on a benchtop centrifuge (room temperature); gel debris will be collected in the 2 ml microfuge tube. Discard the 0.5 ml tube, and add 200 μl of 1× NEB buffer 2 to the gel debris in the 2 ml microfuge tube. Elute DNA by rocking the 2 ml tube for 2 h at room temperature.
6. Transfer the eluate and gel debris to a Spin-X filter and centrifuge for 2 min at maximum speed on a benchtop centrifuge (room temperature). Discard the filter and gel debris, and transfer the eluate to a clean 1.5 ml microfuge tube. Add 1 μl of 20 mg/ml glycogen and 20 μl 3 M sodium acetate pH 5.2, mix well, then add 650 μl cold (−20 °C) 100 % ethanol, and vortex. Immediately centrifuge at 15,000 × *g* for 20 min at 4 °C.
7. After centrifugation, remove and discard supernatant, leaving pellet intact. Wash the pellet with 1 ml of room-temperature 70 % ethanol, remove and discard supernatant, and dry pellet. Measure DNA concentration using a NanoDrop

spectrophotometer. The sample is now ready for Illumina sequencing (*see* **Note 9**).

3.5 Modeling the Biases in the Initial Pool Using a Markov Model

In the remaining sections, we provide a conceptual overview of our computational analysis pipeline; detailed instructions on how to run it are provided in the online documentation that comes with the software.

Typically, there are significant sequence biases in the initial pool of random dsDNA oligonucleotides. These sequence biases are introduced during the synthesis of the ssDNA, double stranding, and/or PCR amplification steps. To account for these, we train Markov models using our custom software and assess their predictive accuracy.

1. First, we determine the largest K-mer length (K_{max}) for which the relative sample error (SE) in the observed count in the initial pool (R0) is smaller than 10 % for any K-mer. Since the relative sample error in a count N equals $1/\sqrt{N}$, this condition is satisfied when all the K-mers of a given length have counts ≥100. Given the significant biases in the R0 pool, this K-mer length has to be determined empirically through analysis of the R0 sequencing data, using our software. For the R0 data of [10], the longest K-mer length that satisfied this condition was $K_{max}=8$ (Table 1).
2. Next, we train Markov models (MM) of order zero through $K_{max}-1$. Each of these models can be used to predict an expected count for all K_{max}-mers (Table 1). For each order, we compute the fraction explained (R^2) of the variance in the observed counts of K_{max}-mers in one of the multiplexes of R0 by a MM built from the other R0 multiplex (cross-validation). This MM is subsequently used to accurately estimate the frequencies of all K-mers (for $K \geq K_{max}$) in the initial pool. In [10], we found that a fifth-order MM attained the highest level of cross-validation predictive accuracy ($R^2=0.992$; Fig. 3).

3.6 Determining the Effective Length of the DNA-Binding Site

For the particular TF or TF complex assayed, the number of base pair positions in the protein-DNA-binding site that contributes to the binding specificity is not known a priori, and their influence can extend beyond the protein-DNA interface over which direct contacts are made (e.g., through shape-mediated effects), so even a high-resolution co-crystal structure would not fully provide this information. Therefore, we use an information-theoretical approach that makes minimal assumptions to determine this effective length of the binding site. First, we apply our software to construct K-mer count tables for different lengths K up to the size of the variable region from the sequencing reads after one or more rounds of affinity-based selection (we will assume R2 in what follows). A

Table 1
Validation of the Markov model used to quantify the significant biases in the initial pool (R0)

8-mer	Observed count in R0	Expected count in R0
TTTTTTTT	6414	6366.4
ATTTTTTT	5081	5014.3
GTTTTTTT	5049	4978.2
TATTTTTT	4922	4859.7
TTTTTTTG	4885	4876.9
TTTTTTGT	4809	4821.2
TTATTTTT	4793	4760.1
TTTTTGTT	4788	4741.4
TTTGTTTT	4772	4695.6
⋮	⋮	⋮
CCCGCCCC	194	184.6
ACCCCCCC	190	189.7
CCCCCCCT	190	222.0
CCCGACCC	190	217.1
CCCCCCCG	189	191.1
CGCCCCCC	186	179.4
CCCCACCC	184	185.2
GCCCCCCC	179	188.1
CACCCCCC	174	189.9
CCCACCCC	165	186.0

Shown are the most and least frequent octamers (8-mers) in R0, in descending order. We chose 8-mers because they are the longest subsequences that all have an observed count greater than 100, so that the relative sample error (SE) in the observed count is never larger than 10 %. Also shown is the expected 8-mer count as predicted by a fifth-order Markov model trained on an independent replicate. The best agreement between the observed and expected counts is obtained using fifth-order Markov models

representative example for Exd-Lab and $K = 12$ is shown in Table 2. For each K, we convert the count table to a probability distribution by dividing by the sum over all K-mers (Table 2); in this process, all K-mers with a count <100 are lumped together into a single "background" category. Next, we compute the course-grained Kullback–Leibler divergence (D_{CGKL}) between the R2 distribution thus estimated and the Markov model constructed from R0:

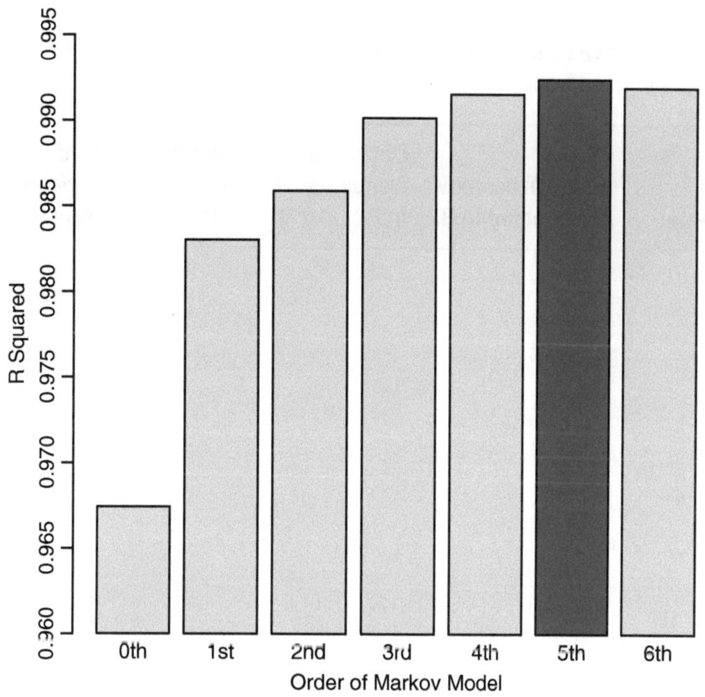

Fig. 3 Markov model optimization. To determine the optimal order for the Markov model of the initial pool, we quantified how a Markov model trained on one replicate of R0 predicted 8-mer counts in another replicate in terms of a coefficient of determination (R^2). Using the data from [10], we found that a fifth-order model has the best cross-validation performance. At lower order, the biases in R0 are not sufficiently captured; at higher order, the predictions degrade due to over-fitting

$$D_{CGKL}(K) = \sum_{i \in S_{100}(K)} \left(P_2(w_i) \log \frac{P_2(w_i)}{P_{MM}(w_i)} \right) + \left[1 - \sum_{i \in S_{100}(K)} P_2(w_i) \right] \log \left(\frac{1 - \Sigma_{i \in S_{100}(K)} P_2(w_i)}{1 - \Sigma_{i \in S_{100}(K)} P_{MM}(w_i)} \right)$$

Here, $S_{100}(K)$ denotes the set of K-mers or "words" w_i with counts ≥100 in R2, and $P_2(w_i)$ the frequency of w_i in R2. $P_{MM}(w_i)$ by contrast denotes the expected frequency of w_i in R0 as predicted using the Markov model; *see* Table 2 for an example. The intuitive interpretation of $D_{CGKL}(K)$ is that of the information gain associated with the transition from R0 to R2: How many more bits of information does it take to define the probability distribution in R2 after it has become more "structured" due to the affinity-based selection procedure? If K is chosen larger than optimal, the K-mer probabilities in R2 will not capture any additional sequence specificity; in fact, over-fitting to sampling errors will cause D_{CGKL} to decrease.

Table 2
A representative example illustrating our procedure for estimating relative affinities, using data for Exd-Lab binding from [10]

12-mer	Observed count in R2	Observed frequency in R2 (x10^{-3})	Estimated frequency in R0 (× 10^{-7})	Fold enrichment from R0 to R2	Estimated relative affinity	Standard error
GTAATCAATCAT	114136	1.6892	0.8735	19338.2	1.0000	0.0042
ATGATTGATTAC	135714	2.0086	1.1675	17203.7	0.9432	0.0036
AATGATTGATTA	145612	2.1551	1.3900	15504.2	0.8954	0.0033
TAATCAATCATT	99927	1.4789	1.0299	14359.6	0.8617	0.0039
GATGATTGATTA	116529	1.7247	1.3581	12699.3	0.8104	0.0034
GTCATCAATCAT	61953	0.9169	0.7348	12479.3	0.8033	0.0046
AATGATTGATGA	81746	1.2099	0.9830	12307.8	0.7978	0.0039
ATGATTGATGAC	67902	1.0050	0.8511	11808.3	0.7814	0.0042
ATGATTGATGAG	77316	1.1443	0.9984	11460.8	0.7698	0.0039
⋮	⋮	⋮	⋮	⋮	⋮	⋮
TTTTTGATATAT	102	0.0015	3.4059	4.4324	0.0151	0.0021
ATTGATATTTTT	103	0.0015	3.4431	4.4275	0.0151	0.0021
ATTATTTATTAT	103	0.0015	3.5303	4.3181	0.0149	0.0021
GTTTGTTTGAT	100	0.0015	3.4723	4.2623	0.0148	0.0021
TTTTTGATTGTT	117	0.0017	4.1203	4.2026	0.0147	0.0019
TTTTTTTGATTA	116	0.0017	4.2856	4.0060	0.0144	0.0019
ATTTATTATTTT	121	0.0018	4.5488	3.9369	0.0143	0.0018
TTTGTTTGATTT	113	0.0017	4.2748	3.9123	0.0142	0.0019
TTTTTTGATTTA	103	0.0015	4.2609	3.5777	0.0136	0.0019
TTTGTTTGTTTA	103	0.0015	4.5958	3.3170	0.0131	0.0018

Shown are the observed counts of all 12-mers in R2, the corresponding fraction in R2 (obtained by dividing by the total observed count), the expected frequency in the initial pool R0 (obtained using the fifth-order Markov model), the fold enrichment, the initial estimate of the relative affinity obtained by rescaling the fold enrichment and taking the square root, and the standard error for the relative affinity estimate

Thus, the optimal value of D_{CGKL} provides a good estimate of the effective binding site length (K_{opt}). For the Hox monomer data in [10], the largest information gain was achieved using 9-mers, while for Exd-Hox dimers a 12-mer table was found to be optimal (Fig. 4).

3.7 Calculating Relative Affinities Through Relative Enrichment of Motifs

1. Our ultimate goal will be to construct a table of relative affinities for all K-mers (of length K_{opt}) whose underlying counts are high enough to allow for their estimation (e.g., count at least 100 for a relative sample error <10 %). To this end, we use an equilibrium

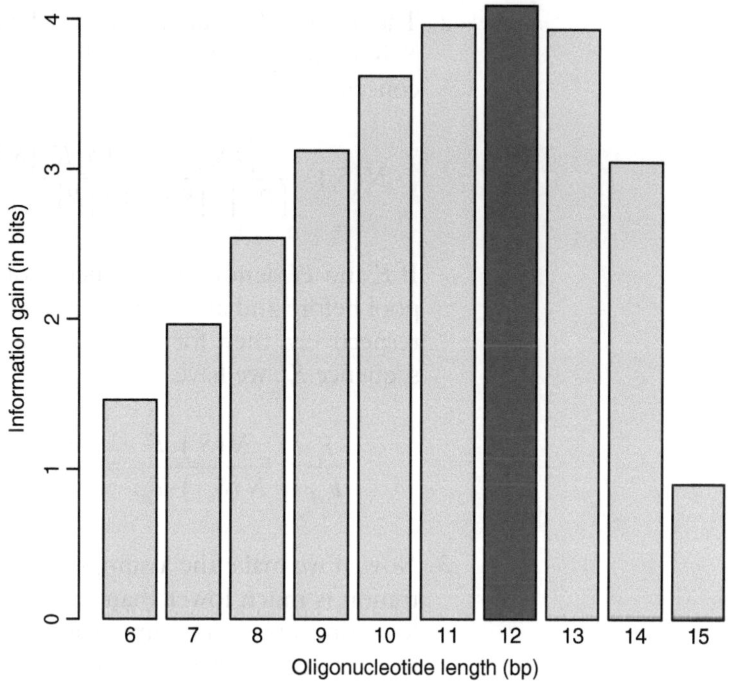

Fig. 4 Oligonucleotide length (K-mer) optimization. To determine the optimal number of base pairs over which to quantify relative affinity, we computed the information gain (Kullback–Leibler divergence) associated with two rounds of affinity-based selection (from R0 to R2). Using the Exd-Scr data from [10], we found that by counting 12-mers in the later round, we optimally capture the DNA-binding specificity of this complex

thermodynamics description of what happens during a single round of SELEX [15]. Our model considers simultaneous competing binding reactions between a protein or a protein complex P on the one hand, and a mixture of DNA molecules on the other. For each type of DNA molecule, with sequence S_i, the formation of the complex $P{:}S_i$ has forward and backward rates $K_{\mathrm{on}}(S_i)$ and $K_{\mathrm{off}}(S_i)$, respectively:

$$P + S_i \underset{k_{\mathrm{off}}(S_i)}{\overset{k_{\mathrm{on}}(S_i)}{\rightleftarrows}} PS_i$$

The strength of binding between the protein P and a DNA molecule of type S_i is quantified by the equilibrium association constant $K_a(S_i)$ (or, equivalently, the dissociation constant $K_{\mathrm{off}}(S_i)$):

$$K_a(S_i) = \frac{1}{K_d(S_i)} = \frac{[PS_i]}{[P][S_i]} = \frac{K_{\mathrm{on}}(S_i)}{K_{\mathrm{off}}(S_i)}$$

2. The fractional occupancy $N(S_i)$, defined as the probability that S_i is bound, is then given by ([P] denotes the *free* protein concentration)

$$N(S_i) = \frac{[PS_i]}{[PS_i]+[S_i]} = \frac{[p]K_a(S_i)}{1+[P]K_a(S_i)} = \frac{[P]}{K_d(S_i)+[P]}$$

If F_i and F_i' denote the frequency of oligomer S_i in the dsDNA pool before and after a given round of affinity-based selection, respectively, then for any pair of sequence S_i and a reference sequence S_{ref} we have

$$\frac{F_i'}{F_{\text{ref}}'} = \frac{N(S_i)\cdot F_i}{N(S_{\text{ref}})\cdot F_{\text{ref}}} = \left(\frac{K_d(S_{\text{ref}})+[P]}{K_d(S_i)+[P]}\right)\left(\frac{F_i}{F_{\text{ref}}}\right)$$

3. Now, if we make the assumption that the free protein concentration is much lower than the dissociation constant (i.e., the low-concentration regime where $[P] << K_d(S_i)$), then the relative affinity between the two sequences has a surprisingly simple relationship to the pre- and post-selection frequencies:

$$\frac{K_a(S_i)}{K_a(S_{\text{ref}})} = \frac{(F_i' / F_{\text{ref}}')}{(F_i / F_{\text{ref}})}$$

In other words, the relative affinity equals the *relative* enrichment in the pool. It is easy to see that after r rounds of SELEX starting from the initial pool (making the same no-saturation assumption) we have

$$\frac{K_a(S_i)}{K_a(S_{\text{ref}})} = \left(\frac{(F_i^{(r)} / F_{\text{ref}}^{(r)})}{(F_i^{(0)} / F_{\text{ref}}^{(0)})}\right)$$

Without loss of generality, we can choose S_{ref} to be highest-affinity sequence. All relative affinities will then have a value between zero and one. The frequency ratios in a later round r (where the medium-to-high-affinity DNA molecules are typically sequenced thousands of times) can be estimated from the unique-sequence counts as follows:

$$\frac{F_i^{(r)}}{F_{\text{ref}}^{(r)}} \approx \frac{N_i^{(r)}}{N_{\text{ref}}^{(r)}}$$

The same estimator cannot be used in round zero, as the counts for the same sequences would be too low to allow for reliable statistics. Fortunately, we can use the Markov model that was discussed above to predict the frequency ratios in round R0:

$$\frac{F_i^{(0)}}{F_{\text{ref}}^{(0)}} \approx \frac{P_{MM}(S_i)}{P_{MM}(S_{\text{ref}})}$$

4. Our goal is to infer relative affinities $\frac{K_a(S_i)}{K_a(S_{\text{opt}})}$ for all K-mers S_i of a length K that typically is smaller than the length of the DNA molecules in the protein-binding reaction. Strictly speaking, this requires a deconvolution, as the K-mer-binding site can occur at different offsets and directions within the DNA molecule. However, an approximate solution—which assumes that a single K-mer dominates the rate at which each DNA molecule is selected—is to use the above equation at the level of K-mer subsequences. In what follows, we will therefore consider F_i to denote the frequency of DNA molecules containing a specific K-mer S_i.
5. It is important to know the standard error of the relative affinity estimate. As above, $P_{MM}(S_i)$ denotes the expected frequency of S_i in R0 as computed using a Markov model. The error in the estimate of $\frac{K_a(S_i)}{K_a(S_{\text{opt}})}$ is dominated by the sample error in the count $N(S_i)^{(r)}$ in the later round. The error in the Markov model estimate is expected to be much smaller; however, to be conservative, we assume it to be of the same order. This yields the following equation for the standard error (SE) in the relative affinity estimate:

$$SE\left(\frac{K_a(S_i)}{K_a(S_{\text{opt}})}\right) = \frac{K_a(S_i)}{K_a(S_{\text{opt}})} \cdot \sqrt{\frac{2}{N(S_i)^{(r)}}}$$

6. Using the theory, statistics, and approximations described so far, we are ready to compute the relative enrichment from R0 to the later round for each K-mer, and then use it to generate an estimate of the relative affinity, and the error in this estimate. Table 2 shows the result based on the comparison of R0 and R2 for Exd-Lab, using the data from [10].

3.8 Refinement of Affinity Estimates

To validate consistent enrichment across the SELEX rounds, it is important to compare the fold enrichment from R0 to R1 with the rth-root of the fold enrichment from R0 to Rr for all K-mers. Typically, a consistent deviation from the straight line will be observed which is presumably due to a combination of binding

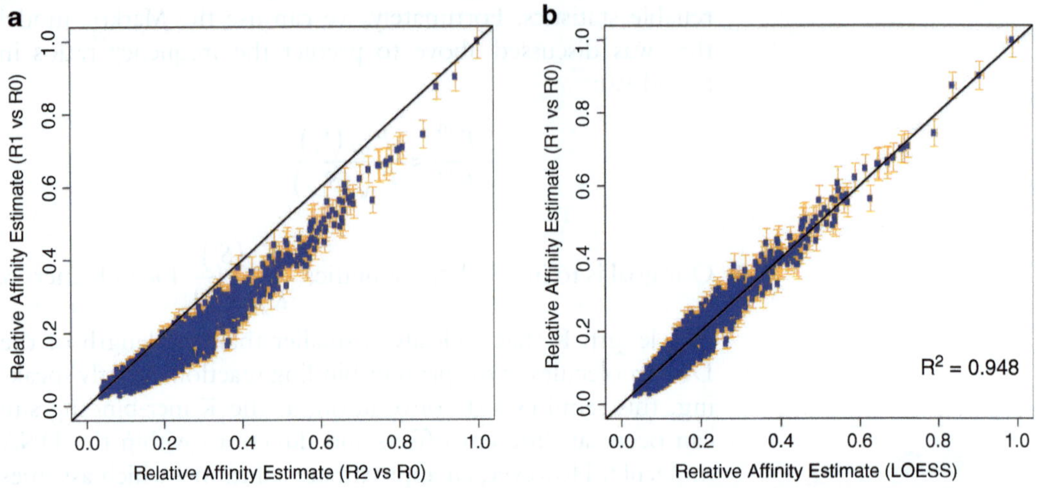

Fig. 5 Integration of data from multiple rounds of SELEX. To obtain relative affinity estimates that are both accurate and precise, we integrate information from multiple rounds of selection using LOESS regression. (**a**) Direct comparison between the normalized fold enrichment from R0 to R1 and the normalized square root of the fold enrichment from R0 to R2 of all 12-mers for Exd-Lab using the data from [10]. The deviation from the straight line is presumably due to a combination of binding saturation and buildup of PCR bias. These effects are expected to be less severe in the earlier round, and therefore R1/R0 is a more accurate predictor of relative affinity. However, since counts are lower in R1 than in R2, the value of R1/R0 is also less precise. The error bars denote the standard error in the estimate of the relative affinity, calculated as described. (**b**) Comparison between the R1-based affinity estimates and LOESS-based estimates resulting from integration of R1 and R2 data

saturation and PCR bias (Fig. 5a). We typically find that these effects are less severe in the earlier rounds, and therefore R1/R0 is the most accurate predictor of relative affinity (i.e., closest to the true value if there would be no random error). However, since counts are lower in R1 than in subsequent rounds, the value of R1/R0 is also less precise (i.e., has a larger error). To combine the respective advantages of the more accurate counts from R1 and the more precise counts from the later round, we integrate information from multiple rounds of selection using local (LOESS) regression. The idea is to use the ranking of K-mers in terms of their enrichment in the later round to which K-mers are expected to have similar affinity, and average over the noisy affinity estimates in the earlier round. The procedure results in estimates of relative affinity that are both accurate and precise (Fig. 5b).

1. LOESS regression requires the specification of certain parameters. To determine these, we compared the LOESS-corrected relative affinities to affinities determined using EMSA gel shifts for a small number of sequences in [10]. An optimal fit was obtained using a second-order polynomial with a smoothing span of 0.2 (see Fig. 2). We therefore used these parameter values in all our subsequent analyses.

Table 3
Refinement of relative affinities

	Relative affinity estimate			
12-mer	R1	R2	LOESS	Symmetrized
GTAATCAATCAT	1.0000	1.0000	1.0000	1.0000±0.0140
ATGATTGATTAC	0.9014	0.9432	0.9151	
AATGATTGATTA	0.8744	0.8954	0.8468	0.8601±0.0082
TAATCAATCATT	0.7439	0.8617	0.8003	
GTCATCAATCAT	0.7047	0.8033	0.7228	0.7402±0.0051
ATGATTGATGAC	0.6761	0.7814	0.6947	
GATGATTGATTA	0.7091	0.8104	0.7320	0.7369±0.0051
TAATCAATCATC	0.6627	0.7692	0.6793	
⋮	⋮	⋮	⋮	⋮
AATTAATTAATG	NA	0.0242	0.0240	0.0264±0.0133
CATTAATTAATT	NA	0.0225	0.0240	
TATAAGATTAAT	NA	0.0226	0.0239	0.0262±0.0134
ATTAATCTTATA	NA	0.0238	0.0239	
CATTAATTATAT	NA	0.0226	0.0238	0.0262±0.0134
ATATAATTAATG	NA	0.0236	0.0238	
AATATGATATAT	NA	0.0244	0.0236	0.0259±0.0136
ATATATCATATT	NA	0.0213	0.0236	
TTAATTAATAAT	NA	0.0208	0.0217	0.0242±0.0147
ATTATTAATTAA	NA	0.0209	0.0217	

Estimated affinities for 12-mers, before (A) and after (B) LOESS-based refinement and symmetrization over reverse complements. Entries with "NA" have counts that are too low (i.e., <100) for accurate affinity estimation

2. We found that our results are improved when we use the relative affinities themselves as weights in the polynomial fit. The reason for this improvement is that the density of the data points is higher for lower affinity sites (i.e., there are more ways to make a typical low-affinity site than a high-affinity one). Our weighting scheme compensates for this.
3. In theory, the binding affinity for any K-mer and its reverse complement should be identical, as they represent two equivalent ways of referring to the same double-stranded DNA molecule. However, in practice, the presence of remaining strand-specific biases and sample error will give rise to small deviations from this symmetry. To address this, we simply average the two values (*see* Table 3). In the rare cases where the non-symmetrized relative enrichment is negative after LOESS correction, we set the value to zero before averaging. Finally, we renormalize the table by dividing by the maximum symmetrized value.

4 Notes

1. The single-stranded oligonucleotide that serves as template DNA for the double-stranded SELEX library can be ordered from any number of commercial vendors. We have had success using oligonucleotides from Integrated DNA Technologies (IDT), with the randomized regions generated using the "hand mix" option.
2. The identity of the control oligonucleotides is dependent on the TF(s) being tested, and should contain the TF's consensus site if one is known. The control oligos should be the same size as the SELEX library to allow for accurate mobility tracking of the protein-DNA complex. We used permuted versions of the adapter sequences used in the SELEX library to keep GC content approximately equivalent between the control and SELEX probes while ensuring that the control DNA would not contaminate SELEX library during PCR amplification.
3. Radiolabeled DNA probes are used for tracking DNA mobility in EMSA. Nonradioactive alternative methods of DNA labeling are also available and suitable for EMSA, but not described here. Beyond EMSA, affinity purification (bead-based) of protein-DNA complexes can also be used for SELEX; these methods avoid radioactivity and do not require a known DNA-binding sequence for the TF being tested (no control probe), but do not allow for easy tracking of multimeric protein complexes.
4. The recipe here is for gel poured with a 14 cm-by-16 cm glass plates, with a 1.5 mm 10-well comb.
5. TF expression and purification must be optimized for each factor. In this case, His-tagged Hox proteins were expressed in BL21 bacteria and purified using Ni-chromatography [16]; Exd (also His-tagged) was co-purified with the HM domain of Hth [17]. DNA-binding proteins for SELEX can also be expressed using non-bacterial systems [14, 18]. All protein dilutions for EMSA reactions should be performed using the same buffer in which the protein is stored (often a dialysis or an elution buffer).
6. Cleanliness and good laboratory practices are important for the success of SELEX-seq, and especially important at the PCR and post-PCR stages; cross-contamination between samples is often possible because of common adapter sequences. Always use filter pipet tips, and start with a fresh box of microfuge tubes before each new experiment. Clean glass plates from EMSA gels first by washing with soap and water, and then with 50 % bleach (rinse thoroughly before use).

7. SELEX can be a reiterative protocol, with multiple rounds of selection, elution, and amplification, even though only the first round (R1) is described here. The number of rounds performed is influenced by trade-offs between sequence counts, the range of affinities represented, and selection bias. In general, data from the first or the second round of selection cover a wider range of relative affinities, while selected DNAs from later rounds are biased toward higher affinity motifs. However, individual sequence counts are lower in the first few rounds of SELEX, so up to five rounds or more might be necessary in some cases (for TFs targeting long DNA motifs, for example).
8. As described in **Note 6**, it is important to avoid cross-contamination during the SELEX procedure, especially if multiple proteins or complexes are being tested. For this reason, we use precast acrylamide gels (BioRad) for all DNA electrophoresis, as these gels leave less room for contamination. If pouring your own gels, be sure to clean plates thoroughly between experiments (*see* **Note 6**).
9. The details of Illumina sequencing are not described in this protocol because this is generally performed by a core facility. However, there are two points worth noting when discussing the sequencing of SELEX libraries with a core facility. First, the library described in this protocol is based on Illumina's small RNA sequencing library. This detail may impact the sequencing primer used, although Illumina is now using a "universal" sequencing primer mix that covers multiple library types, in the small RNA design. Second, Illumina sequencers do not perform well with libraries containing strong base bias immediately after the sequencing primer; calibration of the machine during the first few sequencing cycles depends upon the presence of signal for all four DNA bases. The library described here has extreme base bias immediately downstream of the sequencing primer (the TGG sequence immediately preceding the randomized region) and will cause the sequencing run to fail. To deal with this bias it is recommended that the sequencing reaction be spiked with Illumina's PhiX control library, which is not biased downstream of the sequencing primer and will allow the sequencing run to proceed without incident.

Acknowledgements

We thank the members of the Bussemaker and Mann labs for comments and feedback during the course of these studies. This work was supported by NIH grants R01GM054510, R01HG003008, U54CA121852, and P50GM071508; a John Simon Guggenheim Foundation Fellowship; and Columbia University's RISE program.

References

1. Hueber SD, Lohmann I (2008) Shaping segments: Hox gene function in the genomic age. Bioessays 30(10):965–979. doi:10.1002/bies.20823
2. Young T, Deschamps J (2009) Hox, Cdx, and anteroposterior patterning in the mouse embryo. Curr Top Dev Biol 88:235–255. doi:10.1016/S0070-2153(09)88008-3
3. Abramovich C, Humphries RK (2005) Hox regulation of normal and leukemic hematopoietic stem cells. Curr Opin Hematol 12(3):210–216
4. Mann RS, Lelli KM, Joshi R (2009) Hox specificity unique roles for cofactors and collaborators. Curr Top Dev Biol 88:63–101. doi:10.1016/S0070-2153(09)88003-4
5. Conlon FL, Fairclough L, Price BM, Casey ES, Smith JC (2001) Determinants of T box protein specificity. Development 128(19):3749–3758
6. Jones S (2004) An overview of the basic helix-loop-helix proteins. Genome Biol 5(6):226. doi:10.1186/gb-2004-5-6-226
7. Hollenhorst PC, McIntosh LP, Graves BJ (2011) Genomic and biochemical insights into the specificity of ETS transcription factors. Annu Rev Biochem 80:437–471. doi:10.1146/annurev.biochem.79.081507.103945
8. Joshi R, Passner JM, Rohs R, Jain R, Sosinsky A, Crickmore MA, Jacob V, Aggarwal AK, Honig B, Mann RS (2007) Functional specificity of a Hox protein mediated by the recognition of minor groove structure. Cell 131(3):530–543. doi:10.1016/j.cell.2007.09.024
9. Abu-Shaar M, Ryoo HD, Mann RS (1999) Control of the nuclear localization of Extradenticle by competing nuclear import and export signals. Genes Dev 13(8):935–945
10. Slattery M, Riley T, Liu P, Abe N, Gomez-Alcala P, Dror I, Zhou T, Rohs R, Honig B, Bussemaker HJ, Mann RS (2011) Cofactor binding evokes latent differences in DNA binding specificity between Hox proteins. Cell 147(6):1270–1282. doi:10.1016/j.cell.2011.10.053
11. Tuerk C, Gold L (1990) Systematic evolution of ligands by exponential enrichment: RNA ligands to bacteriophage T4 DNA polymerase. Science 249(4968):505–510
12. Ellington AD, Szostak JW (1990) In vitro selection of RNA molecules that bind specific ligands. Nature 346(6287):818–822. doi:10.1038/346818a0
13. Slattery M, Ma L, Negre N, White KP, Mann RS (2011) Genome-wide tissue-specific occupancy of the Hox protein Ultrabithorax and Hox cofactor Homothorax in Drosophila. PLoS One 6(4):e14686. doi:10.1371/journal.pone.0014686
14. Jolma A, Yan J, Whitington T, Toivonen J, Nitta KR, Rastas P, Morgunova E, Enge M, Taipale M, Wei G, Palin K, Vaquerizas JM, Vincentelli R, Luscombe NM, Hughes TR, Lemaire P, Ukkonen E, Kivioja T, Taipale J (2013) DNA-binding specificities of human transcription factors. Cell 152(1–2):327–339. doi:10.1016/j.cell.2012.12.009
15. Levine HA, Nilsen-Hamilton M (2007) A mathematical analysis of SELEX. Comput Biol Chem 31(1):11–35. doi:10.1016/j.compbiolchem.2006.10.002
16. Gebelein B, Culi J, Ryoo HD, Zhang W, Mann RS (2002) Specificity of Distalless repression and limb primordia development by abdominal Hox proteins. Dev Cell 3(4):487–498
17. Noro B, Lelli K, Sun L, Mann RS (2011) Competition for cofactor-dependent DNA binding underlies Hox phenotypic suppression. Genes Dev 25(22):2327–2332. doi:10.1101/gad.175539.111
18. Jolma A, Kivioja T, Toivonen J, Cheng L, Wei G, Enge M, Taipale M, Vaquerizas JM, Yan J, Sillanpaa MJ, Bonke M, Palin K, Talukder S, Hughes TR, Luscombe NM, Ukkonen E, Taipale J (2010) Multiplexed massively parallel SELEX for characterization of human transcription factor binding specificities. Genome Res 20(6):861–873. doi:10.1101/gr.100552.109

Chapter 17

DamID as an Approach to Studying Long-Distance Chromatin Interactions

Fabienne Cléard, François Karch, and Robert K. Maeda

Abstract

How transcription is controlled by distally located *cis*-regulatory elements is an active area of research in biology. As such, there have been many techniques developed to probe these long-distance chromatin interactions. Here, we focus on one such method, called DamID (van Steensel and Henikoff, Nat Biotechnol 18(4):424–428, 2000). While other methods like 3C (Dekker et al., Science 295(5558):1306–1311, 2002), 4C (Simonis et al., Nat Genet 38(11):1348–1354, 2006; Zhao et al., Nat Genet 38(11):1341–1347, 2006), and 5C (Dostie et al., Genome Res 16(10):1299–1309, 2006) are undoubtedly powerful, the DamID method can offer some advantages over these methods if the genetic locus can be easily modified. The lack of tissue fixation, the low amounts of starting material required to perform the experiment, and the relatively modest hardware requirements make DamID experiments an interesting alternative to consider when examining long-distance chromatin interactions.

Key words Dam methyltransferase, DamID, Drosophila, Chromatin, Long-distance interactions, Gene regulation

1 Introduction

DamID is a method where one targets the Dam methyltransferase (Dam) to a site in the genome (generally by fusing Dam to a chromatin-recruiting protein) and monitors DNA methylation (at GATC sites) near the site of targeting [1]. As DNA methylation is mostly nonexistent in flies, this method has proven to be quite powerful in localizing the binding sites of proteins in the *Drosophila* genome as an alternative to ChIP experiments. On a small scale, monitoring DNA methylation is done by simple restriction enzyme digestion using the methylation-sensitive *Dpn*II enzyme and Q-PCR with primers centered around methylation site being tested. For a wider, genomic scale experiment, the *Dpn*II enzyme can be replaced by the DpnI enyzme, which cuts at methylated GATC sites. Following DpnI digestion, the digested chromatin is size fractionated and DNA linkers are added to the DpnI-cleaved

Yacine Graba and René Rezsohazy (eds.), *Hox Genes: Methods and Protocols*, Methods in Molecular Biology, vol. 1196,
DOI 10.1007/978-1-4939-1242-1_17,

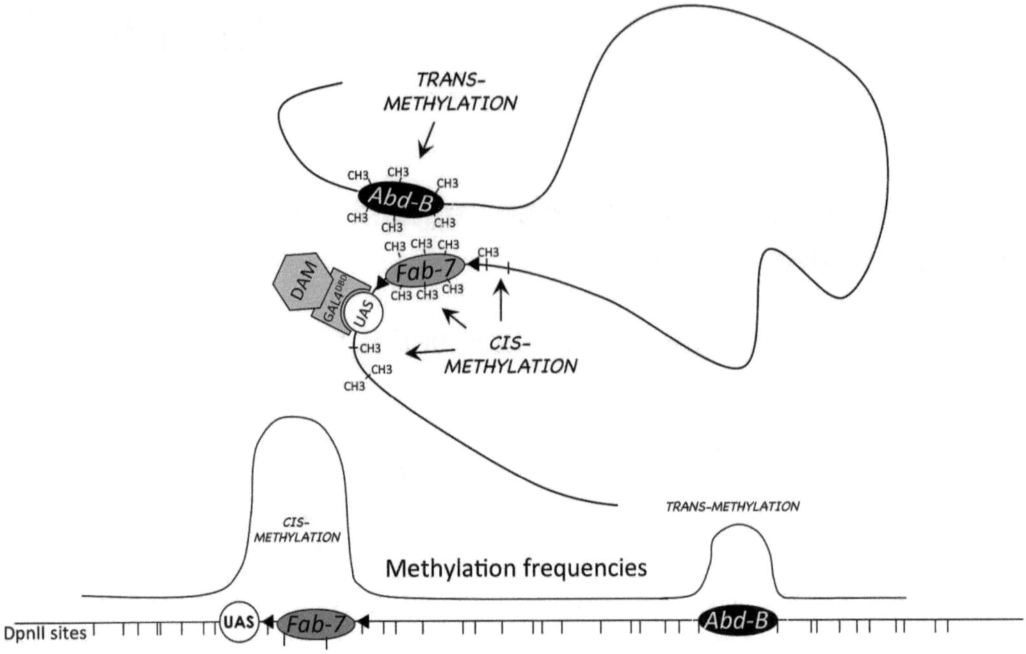

Fig. 1 DamID to study long-distance chromatin interactions. The *top panel* shows in a cartoon form the 37 kb long chromatin loop between the *Fab-7* boundary and the *Abd-B* promoter region as deduced from the DamID experiment described here. Dam was recruited at *Fab-7* by the mean of a fusion with the Gal4 DNA-binding domain (Gal4^{DBD}) and the introduction of UAS sites in the immediate vicinity of *Fab-7* by gene conversion (Cléard et al. [3]). The *bottom part* shows the methylation frequency (also in a cartoon form) of GATC sites as measured by resistance to *Dpn*II digestion across the region. In addition to a peak of methylation at the recruitment site (*cis-methylation*), a second of peak of methylation is detected in the vicinity of the *Abd-B* promoter (*trans*-methylation)

ends for subsequent analysis on microarrays or using a mass sequencing-based approach [2].

We adapted the DamID method for studying long-distance chromatin interactions involving the boundary elements of the *Drosophila* bithorax complex (BX-C; Fig. 1) [3]. Although boundary elements behave as insulators in transgenic assays (blocking enhancer-promoter interactions), it was clear from their location in the complex that they do not act as simple insulators within the BX-C. The *Fab-7* boundary element, for example, lies in between all of the enhancers of the *iab-5* and *iab-6* domains and their target *Abd-B* promoter [4]. A number of hypotheses have been proposed to account for the transgenic insulator activity of BX-C boundary elements. Many of these mechanisms, like enhancer trapping/promoter decoys and the creation of chromatin loops to prevent enhancer-promoter communication, hypothesize that boundary elements may specifically interact with other elements in the BX-C to form chromatin-chromatin interactions [5]. This idea of insulators mediating chromatin-chromatin interactions was supported by work on other elements with insulator activity, such as the *Su(Hw)* insulator from the gypsy retrotransposon [6, 7]. Thus, we were interested in understanding if the *Fab-7* boundary element contacted other elements in the BX-C to perform its function. To investigate this, we turned to the method of DamID. Previous studies had shown that a Dam methyltransferase fused to a Gal4 DNA-binding domain (Gal4^{DBD}) could methylate GATC sites within a range of a couple of kilobases from the initial site of targeting [1]. This "spreading" of methyltransferase activity suggested to us that DamID might be able to methylate interacting DNAs in trans to visualize DNA elements in close proximity to *Fab-7* within the nucleus. Of course the major difficulty of DamID is the requirement of a protein to target the methyltransferase to the specific region of the genome. At that time, outside of the GAGA factor, which binds to numerous sites in the genome, we knew nothing of other *Fab-7* proteins. Thus, we modified the sequence immediately adjacent to the *Fab-7* boundary to include 14 binding sites for the Gal4 protein and flanked the *Fab-7* boundary with loxP sites. Using this line, we were able to monitor long-distance chromatin interactions at various sites in the BX-C by monitoring DNA methylation levels away from the targeting site. As a control, we also performed experiments in the absence of *Fab-7* to determine which interactions were *Fab-7* dependent. Based on our results, we determined that the *Fab-7* boundary element interacted with an area near the *Abd-B* promoter (>35 kb away) in tissues where *Abd-B* was not expressed. In areas where *Abd-B* was expressed, this interaction was absent. Thus, using DamID, we were able to determine a role for *Fab-7* in long-distance chromatin interactions in vivo, and determine that this activity was regulated along the A-P axis of the fly [3].

One of the advantages of DamID as a method for probing long-distance interactions is its relative simplicity. If one follows the protocol described below, respecting the warnings mentioned in the text and footnotes, one can expect fairly reproducible results. The hardest part of DamID, and its most limiting element, is the creation of proper fly lines in which to do the experiments. While one can make Dam fusions to proteins that bind to a particular element of interest (for example to an insulator), we recommend, instead, spending the time to add heterologous binding sites near the element to be tested. We recommend this approach because of the controls it allows. First, one can remove the element while keeping the targeting sites to check for the specificity of the interactions. Second, one can use a chromosome without targeting sites, but expressing the same Dam fusion protein to characterize the nonspecific level of methylation with the fusion protein.

Obviously, making these modifications to the genome is time consuming and challenging. But, with the advent of numerous new gene targeting methods, these difficulties are not as hard to overcome as in the past.

Because of the experimental constraints of DamID, one must consider DamID in the context of other methods used to probe long-distance interactions. Even with the improvements and developments in other technologies such as 3C, 4C, 5C, and Chia-PET [8–12], DamID remains a powerful option to investigate long-distance chromatin interactions. Once again, its limitation is the fact that one needs to target Dam to a specific site in the genome. But, if this is possible, DamID has a number of positive advantages. First, although the quantitation is performed in vitro, it is monitoring marks that occurred in the living organism and does not suffer from artifacts stemming from fixation. Next, it requires very little materials to perform. Our non-genomic scale experiments were performed on DNA isolated from as little as three fly heads, which yielded enough DNA to be used for 40 experiments. Given that flies are small, and fly tissues even smaller, this can be a huge advantage. This advantage might even become more important if combined with tissue-specific expression of Dam fusion proteins. The creation of such tissue-specific forms of Dam has been problematic in the past because high-level expression of Dam (such as the levels one might expect from a UAS-controlled transgene) is lethal. This problem, however, has recently been overcome by the lab of Andrea Brand, who showed that low-level, tissue-specific expression of Dam can be achieved if Dam is made as the second coding sequence in a polycistronic transcript without an internal ribosome entry site under the control of a UAS promoter [13]. Thus, Dam can now be cell- or tissue-specifically expressed, precluding the requirement for tedious dissections. Given the low amount of tissue required to see signal, one might potentially be able to visualize interactions coming from only a small number of cells.

The technique described here has been described previously in Cleard et al. [3]. It is a method to determine long range interactions at selected GATC sites in the genome. As this method is relatively quick and easy to perform with standard molecular biology lab equipment, it is a protocol that should be usable by a wide variety of scientists. If a more genomic approach is required, please refer to Van Steensel et al. [2].

2 Materials

1. Standard cornmeal-agar fly growth media.
2. Fly stocks carrying a Dam fusion protein (with targeting sites inserted near the genomic elements to study if the fusion protein does not possess native targeting sites in the genome;

that is, for Gal-Dam fusions one must have UAS targeting sites in the genome near the element to be investigated). Control flies expressing the Dam fusion protein without targeting sites in the genome, or flies expressing Dam not fused to a DNA bind domain for background methylation controls. As DamID requires target sites in the genome for a Dam fusion protein, this review assumes the availability of a Dam fusion protein to a DNA-binding protein with binding sites near the element being studied. These can be Dam fusions to a DNA-binding protein that normally binds the element in question, or, as in our case, a Dam fusion to the Gal4 DNA-binding domain ($Gal4^{DBD}$) whose binding sites (UAS) have been added next to an element of interest [1].

3. A quantitative PCR apparatus like an ABI Prism SDS 7900 HT system.
4. 2× Power SYBR Green Master Mix (Applied Biosystems).
5. Applied Biosystems Primer Express Software package (Applied Biosystems).
6. Control primers designed to amplify a genomic region lacking Dam methylation sites (GATC) from the *D. melanogaster Gapdh1-F* gene (Dm Gapdh1-F) and from the *D. melanogaster* EF1g-F gene (DmEF1g-F) to be used to standardize the input DNA concentrations for each reaction.

   Dm Gapdh1-F: ATTTCGCTGAACGATAAGTTCGT (T_m=61.3 °C).

   Dm Gapdh1-R: CGATGACGCGGTTGGAGTA (T_m=63.1 °C).

   Dm EF1g-F: GTGTTCATGTCGTGCAATCTCA (T_m=62.1 °C).

   Dm EF1g-R: CGCCTTGCGCATCTTGT (T_m=62.1 °C).
7. Control primers designed to amplify a region of the genome containing a Dam methyltransferase site (GATC). These primers are used to determine the background level of methylation at a location not targeted by the Dam fusion protein. We use the following primers to amplify a region from the TBP gene:

   TBP-74 F: CAAAGGTGCGGCAGGAGAT (T_m=63.6 °C).

   TBP-155R: CCTATTTATGACTGCTTCTTGAACTTCTT (T_m=61.7 °C).
8. Fly homogenization buffer: 10 mM Tris-HCl pH 7.5, 10 mM EDTA, 100 mM NaCl, 0.5 % SDS, 100 μg/ml Proteinase K (diluted freshly from 20 mg/ml stock).
9. Restriction enzyme *Dpn*II.

3 Methods

3.1 Fly Growth Conditions

All flies were raised on a standard cornmeal-agar food mixture and grown at room temperature (22 °C). Higher growing temperatures are to be avoided due to the toxicity of Dam overexpression (*see* **Note 1**).

3.2 Genomic DNA Preparation from Whole Flies

This protocol is modeled after the protocol of Van Steensel and Henikoff [1].

1. Six adult flies per genotype to be tested (not forgetting the nontargeted control) (*see* **Note 2**) are placed individually in PCR tubes.
2. Standard micropipette tips containing 100 μl of fly homogenization buffer are used to crush the flies without expelling the held liquid. After the fly is sufficiently crushed, the held liquid should be expelled onto the fly remains.
3. Tubes are placed at 50 °C for 3 h.
4. Fly extracts are then phenol/chloroform and then chloroform extracted, retaining the aqueous phase (*see* **Note 3**).

3.3 DpnII Digestion of Genomic DNA

1. For each genomic DNA extraction, set up the following two reactions: Incubate reactions overnight at 37 °C.

*Dpn*II(-) *Control*
5 μl Genomic DNA
2 μl *Dpn*II buffer (NEB)
13 μl H_2O
20 μl Total volume
*Dpn*II *digest*
5 μl Genomic DNA
2 μl *Dpn*II buffer (NEB)
0.2 *Dpn*II (2 units)
12.8 μl H_2O
20 μl Total volume

2. Heat inactivate at 80 °C for 10 min.
3. DNA quality can be assessed using standard fly PCR reactions using the control primers, which are centered on regions containing no GATC (*Dpn*II) sites (the GAPDH and EF1G primers).

3.4 DNA Preparation and Digestion from Dissected Fly Parts

Genomic DNA was isolated from dissected fly parts using a similar protocol to that used for whole flies. The only modifications to this protocol were the following:

1. Three adult heads or three adult abdomens were pooled and used as sources for the genomic DNA.

2. *Dpn*II digestions were performed with or without 2 units of *Dpn*II enzyme in a reaction volume of 25 μl.

3.5 Selection of Oligonucleotides for Q-PCR

Oligonucleotides were selected using the default parameters of the Primer Express v. 2.0 software from Applied Biosystems, centered on single *Dpn*II cut sites in the area being examined. In general, primers are designed to have melting temperatures in the range of 61–64 °C and product lengths of <100 bp (generally ~50 bp). Expected amplicon sequences were then aligned against the *Drosophila* genome using BLAST to ensure that they were specific for the region being tested. Oligonucleotides were ordered from Invitrogen (Life Technologies). Each primer pair was then tested in standard PCR reactions using serial dilutions of genomic DNA (without *Dpn*II cutting) to verify strong and unique amplification products.

3.6 Calibration of DNA for Q-PCR

Each DNA sample is then calibrated to determine the optimum dilution of DNA to be used for amplification. Generally, 1/10, 1/40, and 1/100 dilutions are tested using the three sets of control primers. The aim of this step is to have all samples produce CT values (cycle threshold, *see* **Note 4**) between 10 and 35, where Q-PCR reaction is most precise [14]). These tests are carried out as normal Q-PCR reactions, whose conditions are described below (*see* Subheading 3.7 and **Note 5**).

3.7 Q-PCR Reactions

PCR reactions were performed in a 10 μl final volume and contained 1 μl diluted genomic DNA, 5 μl 2× Power SYBR Green Master mix, and 300 nM forward and reverse primers. Q-PCR reactions were performed on an SDS 7900 HT instrument (Applied Biosystems) with the following parameters:

1. 10 min at 95 °C.
2. 42 cycles of 15 s at 95 °C, and 1 min at 60 °C (no elongation time).

3.8 Preparation of Data

CT values and relative DNA amounts for each reaction were determined using the SDS v. 2.2.2 software package of ABI.

1. *Normalization of amplified products for varied input DNA amounts.* Input DNA amounts are determined by averaging the relative DNA amounts amplified from the two control reactions in the GAPDH and EF1G genes, respectively. Using these values, a normalization factor (NF) for the input DNA can be determined for each DNA sample. The amount of amplified DNA for each sample is determined by dividing the amount of DNA amplified in each reaction by the NF. This result is called the normalized DNA amounts (NDA).

2. *Determine relative methylation level.* The NDAs for each primer pair and each biological replicate are then grouped according to whether or not the DNA sample used was *Dpn*II digested. From within each DNA sample and primer pair group, a ratio of NDAs from digested and undigested DNA was made. This value is the relative methylation level (RML) for the position probed in the DNA sample.

3. *Determine background methylation levels.* Dam fusion proteins give a low level of background methylation without direct targeting to the DNA. This background methylation seems to more or less follow the DNA accessibility (*see* **Note 6**). In order to take this background methylation into account, experiments should be performed to determine the background methylation level over the sites to be examined. To do this, one can examine methylation levels in flies lacking the targeting sites (as in our case with the Gal-Dam fusions) or in flies expressing Dam without a DNA-binding domain. Note that as methylation levels are probably correlated to Dam expression levels, the latter experiment can be problematic if not using identical growth conditions, and constructs expressing Dam from the identical promoter integrated at the identical genomic locus.

   In order to control for background methylation levels, average methylation levels found in the untargeted control flies at each position probed are determined for each sample both with and without *Dpn*II treatment. The ratio of the background methylation level from *Dpn*II digested vs. *Dpn*II undigested samples is called the relative background methylation level (RBML).

4. *Calculation of relative methylation over background (RMoB):* RMoB is calculated for each primer pair location being analyzed. It is calculated using the simple formula of RMoB = RML − RBML.

5. *Calculation of the average relative methylation over background:* The average RMoB values for each point are determined by simply taking the mean average of the RMoB values of the biological replicates from each area probed. These values can then be plotted relative to genomic location.

3.9 Statistical Analysis

To determine significant changes in methylation levels two-factor ANOVA analysis [15] can be performed on the RMoB values using BX-C position as one variable and genotype as the second. To compare methylation levels in different tissues from the same genotypes, BX-C position can be used as one variable and tissue type as the second. If ANOVA analysis demonstrates a significant interaction between the two variables, ($p < 0.05$) a standard *t*-test can be performed on the values of each genotype at each position probed. A standard cutoff for significance is generally set at $p<0.05$. If methylation levels deviate from the average, a Scheffe *post hoc* analysis can be used [16].

4 Notes

1. For our experiments, we placed 14 Gal-4 UAS sequences adjacent to the *Fab-7* boundary and used a Gal4-DNA-binding domain-Dam fusion protein (Gal-Dam) to target the Dam methyltransferase to the *Fab-7* region. This Gal-Dam fusion protein was expressed from a transgene using the HSP70 promoter [1]. All Dam expression in our experiments depended solely on the leakiness of the HSP70 promoter at 22 °C, which provided sufficient Dam expression for visualizing methylation. Heat shocking these flies is lethal. In fact, it seems clear that even slightly increasing expression of the Gal-Dam fusion can have deleterious effects. Our attempts to make additional Gal-Dam expression lines were unsuccessful, with transformed flies dying as pupae. This suggests that the Gal-Dam line we used expresses particular low levels of the Gal-Dam fusion (Cléard, Galetti, and Maeda, unpublished).
2. We generally use a 1:1 male/female ratio of flies about 5 h post-eclosion.
3. Inactivation of proteinase K using heat can lead to denaturation of the DNA, reducing the efficiency of subsequent *Dpn*II digestion.
4. Cycle threshold is defined as the number of cycles required for the fluorescent signal to exceed a threshold level. The threshold level is determined by the background level of fluorescence and is generally set at ten standard deviations above the mean baseline fluorescence.
5. From our experience, a 1:80 dilution was best for DNA from whole flies while a 1:40 dilution was best for DNA isolated from dissected fly parts.
6. This can be seen, for example, by examining the methylation pattern in the BX-C without Gal4-UAS targeting. As can be seen in Fig. 2a (dark grey bar), methylation levels in flies expressing the Gal-Dam fusion protein but without UAS targeting sites at *Fab-7* show slightly higher levels of background methylation at *Fab-7*, *Fab-8*, and the *Abd-B* promoter, areas known to contain DNase I hypersensitive sites. This background methylation is probably the cause of Dam toxicity when overexpressed.

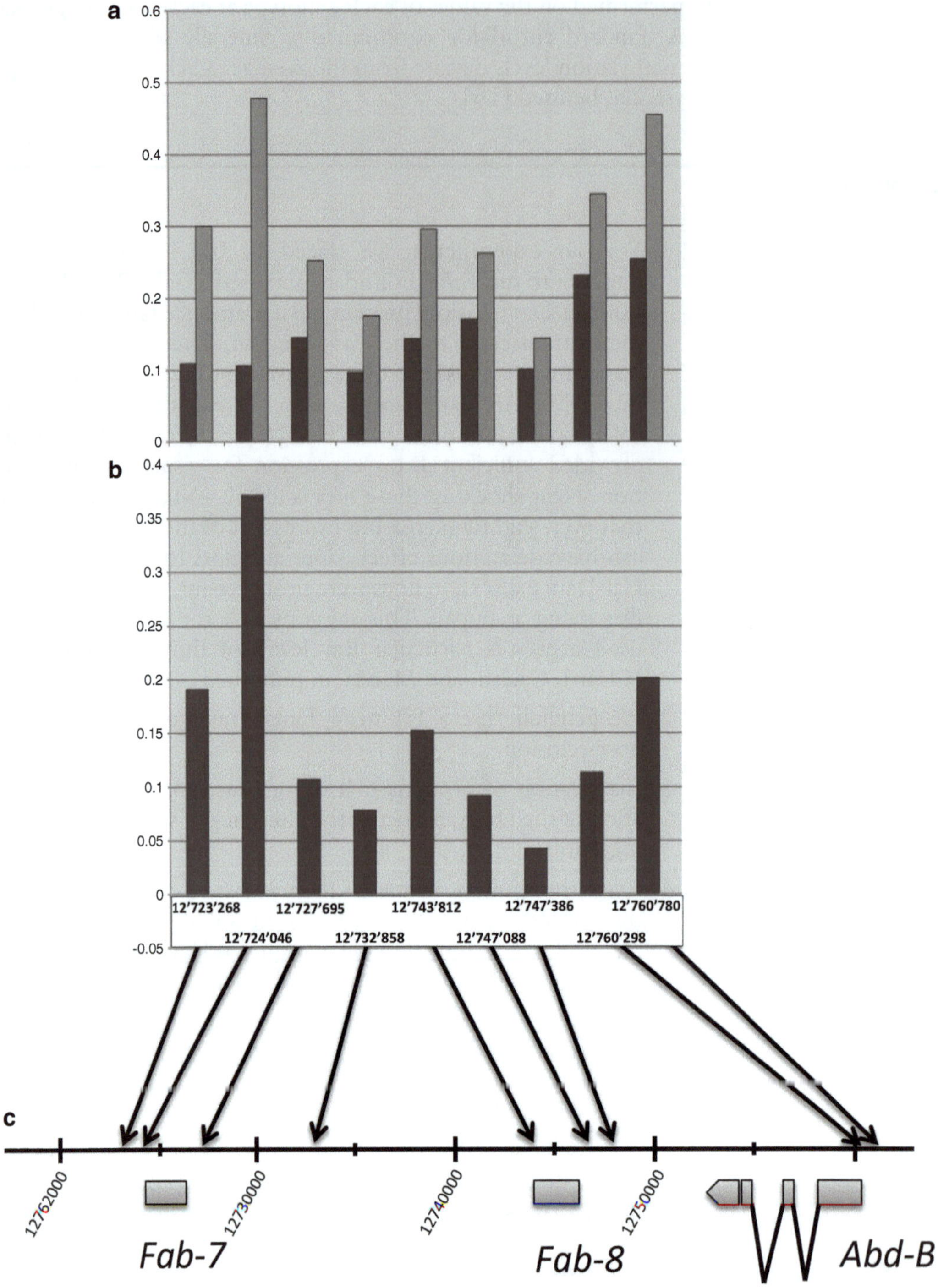

Fig. 2 Example results at different stages of preparation. (**a**, **b**) Show methylation levels at different stages of data treatment at the various points across the BX-C indicated in by the genome project coordinates located at the bottom of panel (**b**). These points are also indicated by the *arrows* shown in (**c**), relative to known landmarks in the BX-C. (**a**) Shows relative methylation levels in lines expressing the Gal-Dam fusion with targeting sites at *Fab-7* (*light grey bars*) and in lines expressing the Gal-Dam fusion but without targeting sites at *Fab-7* (*dark grey bars*). The nontargeted methylation levels indicate background levels of methylation. (**b**) Shows the final results after subtracting background methylation

References

1. van Steensel B, Henikoff S (2000) Identification of in vivo DNA targets of chromatin proteins using tethered dam methyltransferase. Nat Biotechnol 18(4):424–428. doi:10.1038/74487
2. van Steensel B, Delrow J, Henikoff S (2001) Chromatin profiling using targeted DNA adenine methyltransferase. Nat Genet 27(3):304–308. doi:10.1038/85871
3. Cleard F, Moshkin Y, Karch F, Maeda RK (2006) Probing long-distance regulatory interactions in the Drosophila melanogaster bithorax complex using Dam identification. Nat Genet 38(8):931–935
4. Maeda RK, Karch F (2006) The ABC of the BX-C: the bithorax complex explained. Development 133(8):1413–1422
5. Maeda RK, Karch F (2007) Making connections: boundaries and insulators in Drosophila. Curr Opin Genet Dev 17(5):394–399. doi:10.1016/j.gde.2007.08.002
6. Muravyova E, Golovnin A, Gracheva E, Parshikov A, Belenkaya T, Pirrotta VV, Georgiev P (2001) Loss of insulator activity by paired Su(Hw) chromatin insulators. Science 291(5503):495–498
7. Cai HN, Shen P (2001) Effects of cis arrangement of chromatin insulators on enhancer-blocking activity. Science 291(5503):493–495
8. Dekker J, Rippe K, Dekker M, Kleckner N (2002) Capturing chromosome conformation. Science 295(5558):1306–1311. doi:10.1126/science.1067799
9. Simonis M, Klous P, Splinter E, Moshkin Y, Willemsen R, de Wit E, van Steensel B, de Laat W (2006) Nuclear organization of active and inactive chromatin domains uncovered by chromosome conformation capture-on-chip (4C). Nat Genet 38(11):1348–1354. doi:10.1038/ng1896
10. Zhao Z, Tavoosidana G, Sjolinder M, Gondor A, Mariano P, Wang S, Kanduri C, Lezcano M, Sandhu KS, Singh U, Pant V, Tiwari V, Kurukuti S, Ohlsson R (2006) Circular chromosome conformation capture (4C) uncovers extensive networks of epigenetically regulated intra- and interchromosomal interactions. Nat Genet 38(11):1341–1347. doi:10.1038/ng1891
11. Dostie J, Richmond TA, Arnaout RA, Selzer RR, Lee WL, Honan TA, Rubio ED, Krumm A, Lamb J, Nusbaum C, Green RD, Dekker J (2006) Chromosome conformation capture carbon copy (5C): a massively parallel solution for mapping interactions between genomic elements. Genome Res 16(10):1299–1309. doi:10.1101/gr.5571506
12. Fullwood MJ, Ruan Y (2009) ChIP-based methods for the identification of long-range chromatin interactions. J Cell Biochem 107(1):30–39. doi:10.1002/jcb.22116
13. Southall TD, Gold KS, Egger B, Davidson CM, Caygill EE, Marshall OJ, Brand AH (2013) Cell-type-specific profiling of gene expression and chromatin binding without cell isolation: assaying RNA Pol II occupancy in neural stem cells. Dev Cell. doi:10.1016/j.devcel.2013.05.020
14. Applied_Biosystems (2004) Guide to performing relative quantitation of gene expression using real-time quantitative PCR.
15. Rohlf FJ, Sokal RR, Sokal RR (1981) Statistical tables, 2nd edn. Freeman, San Francisco, CA
16. Scheffé H (1961) The analysis of variance. Wiley, New York, NY

Chapter 18

cgChIP: A Cell Type- and Gene-Specific Method for Chromatin Analysis

Marios Agelopoulos*, Daniel J. McKay*, and Richard S. Mann

Abstract

Hox and other homeobox-containing genes encode critical transcriptional regulators of animal development. Although these genes are well known for their roles in the body axis and appendage development, little is known regarding the mechanisms by which these factors influence chromatin landscapes. Chromatin structure can have a profound influence on gene expression during animal body formation. However, when applied to developing embryos, conventional chromatin analysis of genes and *cis*-regulatory modules (CRMs) typically lacks the required cell type-specific resolution due to the heterogeneous nature of animal bodies. Here we present a strategy to analyze both the composition and conformation of in vivo-tagged CRM sequences in a cell type-specific manner, using as a system *Drosophila* embryos. We term this method cgChIP (*c*ell- and *g*ene-specific Chromatin Immunoprecipitation) by which we access and analyze regulatory chromatin in specific cell types. cgChIP is an in vivo method designed to analyze genetic elements derived from limited cell populations. cgChIP can be used for both the analysis of chromatin structure (e.g., long-distance interactions between DNA elements) and the composition of histones and histone modifications and the occupancy of transcription factors and chromatin modifiers. This method was applied to the Hox target gene *Distalless* (*Dll*), which encodes for a homeodomain-containing transcription factor critical for the formation of appendages in *Drosophila*. However, cgChIP can be applied in diverse animal models to better dissect CRM-dependent gene regulation and body formation in developing animals.

Key words *Hox*, Chromatin, Histone modifications, DNA looping, *Cis* regulatory modules (CRMs), Enhancer-promoter communication, Gene transcription, *LacI/lacO*

1 Introduction

Many developmentally regulated genes are active in a minority of the cells in a multicellular organism. Transcriptional regulation of these genes is typically governed by multiple CRMs that control only a subset of a gene's overall expression pattern. Each CRM can be differentially active with respect to cell type, developmental time, and cellular environment. Furthermore, the activity of each CRM often must be communicated over long distances to their cognate promoter

*Author contributed equally with all other contributors.

Yacine Graba and René Rezsohazy (eds.), *Hox Genes: Methods and Protocols*, Methods in Molecular Biology, vol. 1196,
DOI 10.1007/978-1-4939-1242-1_18, © Springer Science+Business Media New York 2014

in order to impact gene transcription. Current approaches to investigating the molecular events controlling gene regulation lack the necessary resolution to assess such dynamic and cell-type-specific events. For example, standard chromatin immunoprecipitation (ChIP) studies, when applied to developing embryos, provide an average of signals, corresponding to all cell types in which the factor of interest is expressed at the time of chromatin preparation. One major obstacle arises from technical difficulties associated with the ability to isolate specific cell types from a tissue that then can be used for chromatin and gene expression studies [1]. Consequently, research on animal models often lacks cell type resolution and thus cannot reveal biologically relevant differences if they exist [2]. Methods based on the ground-breaking approach capturing chromosome conformation (3C), including 4C, 5C, Hi-C, and ChIA-PET, have provided major insights into the analysis of the 3-dimensional interactions that exist between chromosomal elements [3–8]. However, when applied to intact organisms, such as a developing embryo, these methods cannot reveal cell-specific interactions. Furthermore, only ChIA-PET and the ChIP-loop assay [9] have the potential to combine transcription factor binding and/or histone modifications with chromatin conformation in the same experiment.

The above considerations prompted us to develop a method that combines cell specificity with the analysis of chromatin conformation, transcription factor occupancy, and histone modifications at specific loci. We call this method cgChIP (for *c*ell- and *gene-specific* ChIP). We reasoned that expression of a DNA-binding protein used in conjunction with its cognate-binding site would provide a molecular tag that could be used to immunoprecipitate target chromatin. We chose the bacterial DNA-binding protein LacI and its operator sequence *lacO* because LacI binds *lacO* with high specificity and affinity even in nucleosomal DNA [10]. In addition, the *lacI/lacO* system has previously been used in vivo in *Drosophila* and *C. elegans* for other purposes without adverse affects on development [11, 12]. In cgChIP, regulatory elements (CRMs and promoters) are tagged with *lacO*-binding sites and tissue-specific expression of an epitope-tagged LacI allows the immunoprecipitation of the lac*O*-tagged elements from specific cell types [2]. These elements can be analyzed directly (e.g., for looping) or re-precipitated (e.g., double immunoprecipitation) to analyze chromatin conformation and composition, respectively. cgChIP shares some features with other methods, such as iChIP [13], which depend on the tagging of DNA elements and partial purification of the desired chromatin fragment. However, an important aspect of cgChIP is the cell-type-specific expression of LacI, which allows the immunoprecipitation of the tagged DNA only from those cell types (Fig. 1a–d). Although the methods and examples outlined here have been optimized for *Drosophila*, the approach can readily be adapted to any system in which transgenes can be generated.

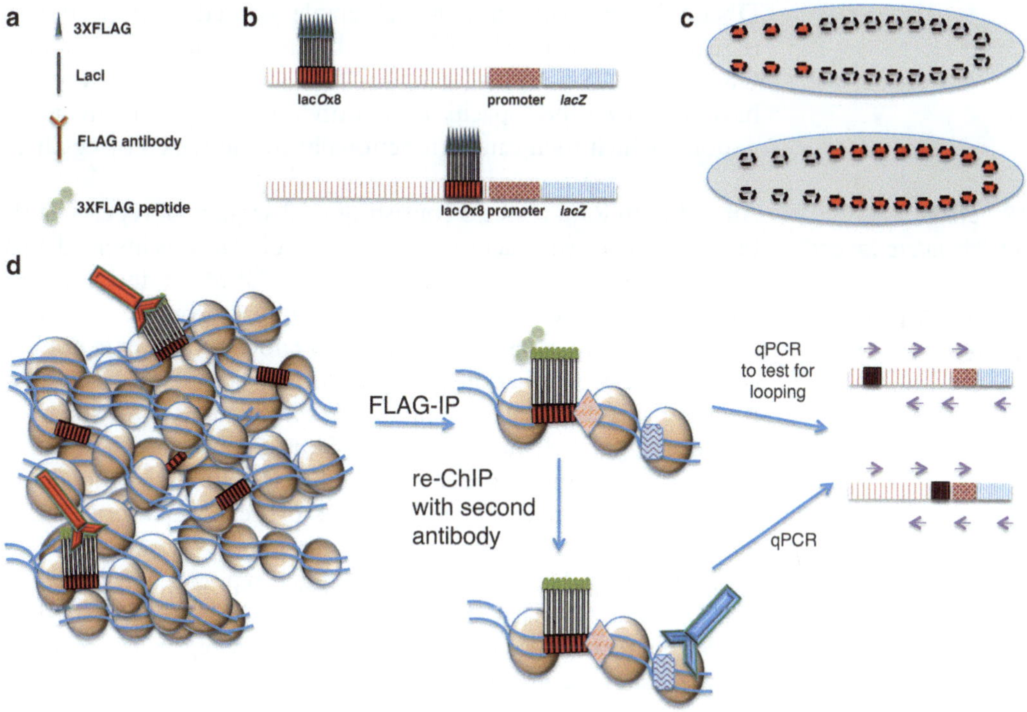

Fig. 1 A graphical synopsis of the cgChIP method. (**a**) Overview of the schematics. (**b**) Schematic representation of 3× FLAG::LacI chimeras bound on lac*O*-tagged CRM. The insertion of the lac*O* repeats can be either proximal (*bottom*) or distal (*upper*) from the transcription start site of the transgene. (**c**) Graphical representation of *Drosophila* embryos (viewed ventrally) expressing 3× FLAG::LacI (*red*) across thoracic (*upper*) or abdominal (*bottom*) tissues. (**d**) Embryonic lac*O*-tagged chromatin can be isolated by anti-FLAG precipitation and used as template either directly for cg looping experiments or re-precipitated and analyzed in double-cgChIP experiments. Both approaches are finally analyzed by semi- and quantitative PCR

2 Materials

2.1 CRMs of Interest

The identification and functional characterization of a regulatory element (CRM or promoter) is an essential prerequisite for carrying out cgChIP. CRMs integrate temporal and spatial information by binding groups of transcription factors [14]. Proximal or distal CRMs can regulate gene transcription in several manners, including controlling assembly of transcriptional pre-initiation complexes and regulating release of stalled RNA polymerase at a gene's promoter. For many developmentally regulated genes such as the *Hox* genes and their targets, in vivo *lacZ*- or GFP-based reporter gene studies typically provide the initial characterization of regulatory elements. In addition to classical analyses (http://redfly.ccr.buffalo.edu/), a large screen recently characterized more than 6,000 potential regulatory elements in *Drosophila* at several stages and in different tissues, providing a large number of novel CRMs [15–17]. In addition to these resources, histone modifications,

DNase I sensitivity, and formaldehyde-assisted identification of regulatory elements (FAIRE) can be used to identify candidate CRMs [18, 19]. Thus, technical and methodological advances have improved the capacity to identify regulatory elements in the genome which then can be functionally characterized by cgChIP.

2.2 CRMs and Promoters Tagged with lacO-Binding Sites and Fusion to lacZ

An ~300 bp long cassette consisting of 8 sequential repeats of the binding site for the bacterial repressor LacI (*lacO*) is inserted adjacent to the 5′ or the 3′ edge of the CRM of interest. The endogenous gene's promoter or a minimal promoter sequence (e.g., *hsp43*)—in cases where the endogenous promoter is not part of the CRM—and coding sequences for a nuclear version of β-galactosidase (β-Gal, encoded by *lacZ*) are fused downstream of the *lacO*-tagged element (Fig. 1b). Thus, β-Gal expression can be used as a reporter in order to monitor the ability of the cloned CRM to recapitulate the endogenous expression pattern, and to ensure that the *lacO*-binding sites do not interfere with CRM activity. Transgenic flies with these tagged reporter genes are generated using standard methods. Ideally, the tagged transgenes should generate the same expression pattern regardless of where the *lacO* sites are inserted, or where the transgene is integrated in the genome (Fig. 2a). In case of incorrect or leaky expression the *lacO*-tagged constructs should be redesigned, either by moving the location of the *lacO* or by decreasing the number of *lacO* repeats. Indeed, in pilot experiments, as few as 2 *lacO* repeats were sufficient for successful immunoprecipitation of tagged DNA (DJM and RSM, unpublished results).

cgChIP can also be used to monitor interactions between a distally located CRM and its promoter. In the specific example we have characterized, the ~12 kb region 5′ of the *Dll* promoter, which contains several CRMs, was cloned upstream of the β-Gal-coding sequence [2]. Constructs were generated in which either distal CRMs were tagged or the promoter was tagged (Fig. 1b). In both cases, expression of β-Gal was unaffected and closely matched the endogenous *Dll* expression pattern (Fig. 2a).

2.3 Fly Lines for Cell-Type-Specific Expression of Epitope-Tagged LacI

In addition to being objects for study, well-characterized CRMs serve as drivers for the cell-type-specific expression of LacI. This can be done directly (by fusing the CRM to LacI-coding sequences) or indirectly by, for example, the Gal4/UAS or the LexA/LexO system [11, 20–22]. The large availability of GAL4 "enhancer trap" fly lines serves as a standard source for such genetic tools. Further, the use of repressors (e.g., Gal80) and other intersectional methods (e.g., splitGal4) can be used to fine-tune expression patterns [23]. For cgChIP, these methods are used to drive the expression of an epitope-tagged LacI protein. In our experience, three tandem copies of the "FLAG" tag (3× FLAG) worked well, because of its small size and the availability of high-affinity commercially

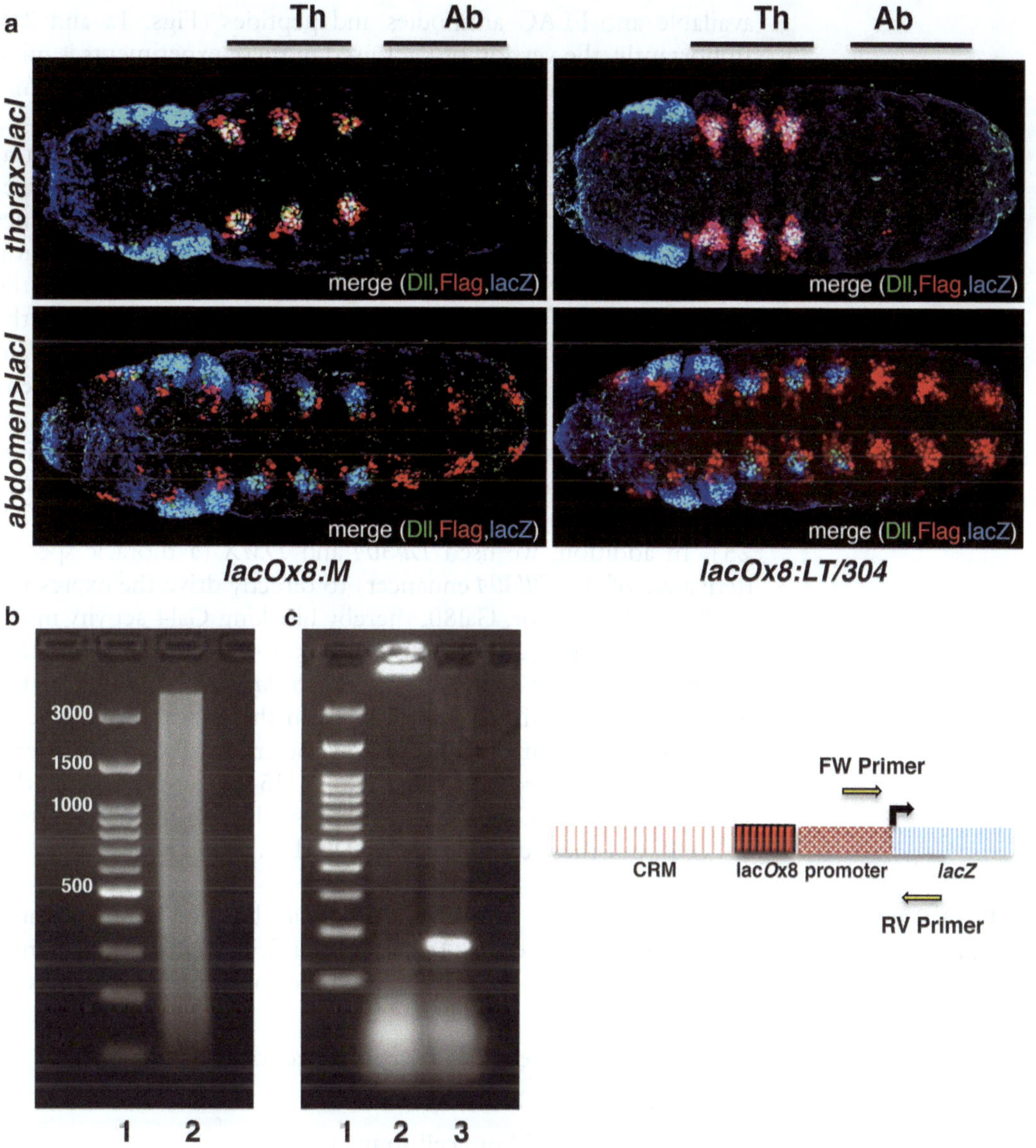

Fig. 2 Control steps required for an accurate chromatin preparation used for cgChIP assays. (**a**) Antibody-staining reactions depicting spatiotemporally accurate expression of 3× FLAG::LacI, endogenous gene, and the reporter gene of study. (**b**) Agarose gel electrophoresis of sheared isolated DNA derived from total embryonic chromatin shows a robust distribution ranged between 300 and 3,000 bp molecular weight (lane 1: 100 bp DNA marker; lane 2: sheared isolated DNA). (**c**) Control PCR for evaluation of chromatin cross-linking efficiency. Equal amounts of total chromatin and extracted DNA are amplified in PCR reactions with a set of primers hybridized at the promoter (forward) and the *lacZ* (reverse) sequences, respectively. The total chromatin-containing PCR reaction fails to amplify the above region (chromatin is detected inside the well). In sharp contrast, a robust/specific amplification signal is produced when extracted DNA from the same chromatin preparation is used as a template. The above quality control step ensures for the absence of non-cross-linked material inside the embryonic chromatin preparation (lane 1: 100 bp DNA marker; lane 2: PCR reaction with total chromatin; lane 3: PCR reaction with extracted DNA)

available anti-FLAG antibodies and peptides (Figs. 1a and 2a). Importantly, the version of LacI used in these experiments is missing its tetramerization domain, thus preventing LacI dimer-dimer interactions when expressed in vivo [24].

Ideally, 3× FLAG::LacI should be expressed in a pattern as precise as possible to analyze CRM structure and function in specific cell types. For *Dll*, we generated two 3× FLAG::LacI transgenes (referred as "*flag-lacI*" in the genotypes) to compare chromatin organization in cells where *Dll* is differentially regulated in the early embryo. To purify chromatin from *Dll*-expressing cells of the thoracic limb primordia, we used the early *Dll* enhancer, *Dll304*, to drive Gal4 and indirectly activate a *UAS-flag-lacI* transgene (Figs. 1c and 2a). To purify chromatin from the analogous population of *Dll* non-expressing cells in the abdomen, we used a derivative of *Dll304* that lacks the Hox repressor element (DME^{ACT}) to drive Gal4 and indirectly activate the *UAS-flag-lacI* transgene [2, 25]. In addition, we used *Dll304* and *DMX* (a thoracic specific derivative of the *Dll304* enhancer) to directly drive the expression of the Gal4 inhibitor, Gal80, thereby blocking Gal4 activity in the thoracic segments [2, 25]. This combination of transgenes (*DMX-Gal80, 304-Gal80*; DME^{ACT}*-Gal4,UAS-flag-lacI*) resulted in the expression of 3× FLAG::LacI in cells in the abdominal segments that have the potential to express *Dll* but normally do not due to Hox-mediated repression (Fig. 2a) [2, 25]. We note that in this genotype, 3× FLAG::LacI is also expressed in several nonabdominal cells, but these do not express *Dll* (Fig. 2a) [2].

2.4 Fly Culture Components

1. Commercial or homemade cages can be used to host a large population of adults. To provide sufficient circulation, homemade cages should have multiple openings covered with netting.
2. Agar-apple juice plates for egg collections (*see* **Note 1**).
3. Yeast paste (*see* **Note 1**).
4. Cell strainers: 70 μm cell strainer.
5. Paintbrush.
6. Wide-mesh metal strainer.

2.5 Chromatin Preparation Buffers and Solutions

1. Embryo wash buffer (EWB): 1× PBS, 0.01 % Triton X-100.
2. *n*-Heptane.
3. Fixing solution (FS): 1× PBS, 3.7 % formaldehyde.
4. Quenching solution (QS): 1× PBS, 0.125 M glycine.
5. Buffer A (BA): 0.25 % Triton X-100, 10 mM Na-EDTA, 0.5 mM Na-EGTA, 10 mM HEPES pH 7.9.
6. Buffer B (BB): 0.2 M NaCl, 1 mM Na-EDTA, 0.5 mM Na-EGTA, 10 mM HEPES pH 7.9.

7. Sonication buffer (SB): 1 mM Na-EDTA, 0.5 mM Na-EGTA, 10 mM HEPES pH 7.9.
8. Glycerol-containing storage buffer (GSB): 10 % Glycerol, 1 mM Na-EDTA, 0.5 mM Na-EGTA, 10 mM HEPES pH 7.9.

2.6 Immunoprecipitation Components, Solutions, and Apparatus

1. Magnetic beads: Protein G magnetic beads (Invitrogen).
2. Magnetic particle concentrator: MPC-S (Invitrogen) or an equivalent model like DynaMag™-2 (Invitrogen).
3. FLAG antibody: M2 (Sigma-Aldrich).
4. 3× FLAG peptide: Lyophilized powder (Sigma-Aldrich).
5. LacI antibody (Rockland antibodies and assays).
6. Reaction buffer (1× RIPA): 0.5 % Triton X-100, 140 mM NaCl, 0.07 % Sodium deoxycholate (deoxycholic acid), 10 mM HEPES pH 7.9.
7. Wash RIPA buffer (1× WRP): 0.5 % Triton X-100, 140 mM NaCl, 0.14 % Sodium deoxycholate (deoxycholic acid), 0.2 % SDS, 10 mM HEPES pH 7.9.
8. Elution buffer (EB): 200 μg/ml 3× FLAG peptide, 140 mM NaCl, 0.2 % SDS, 10 mM HEPES pH 7.9.
9. Phenol.
10. Chloroform.
11. Proteinase K solution: 0.25 mg/ml Proteinase K, 0.4 % SDS.
12. Precipitation mix: 0.2 mg/ml Glycogen, 0.3 M sodium acetate pH 5.2, 3 volumes ice-cold ethanol.

2.7 Semiquantitative and Real-Time PCR Components, Solutions, and Apparatus

1. Transgene-specific amplicon: In cases where *lacZ* is used as a readout, a reverse primer hybridizing within *lacZ* but close to the transcriptional start site can be combined with a forward primer complementary to the promoter sequence (Fig. 2c). This combination restricts the amplification to the transgenic and not the endogenous sequences. Thus the final amplification product derived absolutely from the tagged transgene.
2. Reference amplicon: A pair of primers hybridizing at an endogenous sequence that is not included into the tagged transgene is used as a negative reference.
3. Gene-specific amplicons: If the regulatory region under investigation is large (>10 kb), unique regions should be scanned by amplicons spaced every ~2 kb.
4. *Taq* DNA Polymerase.
5. Polyacrylamide gel electrophoresis system.
6. Alpha ^{32}P-dCTP.
7. Phosphorimager/screens.
8. qPCR application: 2× SYBR Green mix, 96-well plates.

3 Methods

The experimental procedure described below can be completed within 3 days assuming that transgenic fly lines are already available.

3.1 Fly Strain Development and Culture

Large populations must be established to enable the isolation of a sufficient amount of staged embryos (*see* **Note 2**). In our experience with *Dll*, at least six 500 ml population cages were required to collect a sufficient amount of staged embryos in a 2–3-day window. Importantly, in order to avoid any contamination each population should be initiated from a single vial and disposed of once they are no longer productive. In addition, once expanded the population should be retested prior to embryo collection by antibody staining (*see* **Note 3**). The adults should eclose approximately 2 days prior to the first collection in order to achieve efficient egg laying. Moreover, when multiple transgenic fly lines are analyzed in parallel, synchronization of egg laying helps to maximize the amount of embryos collected simultaneously.

3.2 Embryo Collections and Fixation

1. Collect embryos in ddH_2O with the help of a paintbrush from apple juice agar plates and pour them through a wide-mesh metal strainer followed by extensive wash with ddH_2O. Adults, body parts, and debris remain in the strainer and embryos are collected in a glass beaker. Transfer embryos to 70 μm cell strainers and wash with ddH_2O to remove agar and yeast.
2. Place the 70 μm cell strainer with embryos inside a Petri dish filled with bleach and shake for 3 min on a laboratory shaker operating at low speed (<50 rpm).
3. Wash embryos extensively with ddH_2O followed by EWB. This step is required in order to remove the bleach and chorion fragments.
4. Remove most of the moisture with laboratory wipes.
5. At this point a small subset of embryos should be used for immunostaining, to confirm the correct age of the embryos, the expression of the target and reporter genes, and the transcription factor(s) of interest. In addition, an anti-FLAG antibody staining should be performed to monitor the tissue-specific expression of the 3× FLAG::LacI protein (*see* **Note 3**).
6. Dechorionated embryos are placed in a Petri dish filled with heptane. Upon slight shaking, any debris, chorion fragments, and agar should be aspirated. Transfer "pure" embryonic mass with a glass pipette into a 50 ml glass bottle.
7. Balance the volume of heptane at 30 ml and add directly 10 ml of the fixing solution (FS). Close the cap tightly to avoid any leaks of the above harmful solutions.

8. Place the bottle at horizontal position on a shaker that operates at 200–250 rpm for 30 min at room temperature (*see* **Note 4**).
9. Transfer embryos into a Falcon tube and spin down for 5 min at 3,000 × *g*.
10. Aspirate fixing and heptane solutions.
11. Quench fixing reaction by washing with 50 ml quenching solution (QS).
12. Gently help embryos settle to the bottom of the tube with a glass pipette.
13. Aspirate quenching solution and transfer embryos into a new Falcon tube. Repeat **steps 11–13**.
14. Wash embryos three times with 50 ml of EWB for 3 min on ice. In between washes, spin down for 2 min at 3,000 × *g* and aspirate.
15. Dry embryos with the help of a glass pipette. Put the pipette tightly at the bottom of the Falcon tube and remove any remaining EWB without disrupting the pellet.
16. Place embryos at −80 °C for at least 4 h to pulverize. An overnight incubation at −80 °C is more effective for that aspect.

3.3 Chromatin Isolation for cgChIP

1. Incubate pulverized embryonic mass twice with 15 ml of buffer A (BA) for 10 min at room temperature. Place the Falcon tube at a semi-angled position on a laboratory rocker operating at medium speed. Spin down embryonic mass for 2 min at 1,700 × *g* and discard supernatant.
2. Incubate twice with 15 ml of buffer B (BB) for 15 min at 4 °C. Place the Falcon tube at a semi-angled position on a laboratory rocker operating at medium speed. Spin down embryonic mass for 2 min at 1,700 × *g* and discard supernatant.
3. Add 6 ml of sonication buffer (SB) and about 2 g of glass beads directly to the pellet. At this stage the structure of the embryonic mass is changed. The solid/compact structure of the pellet becomes loose and the embryos are no longer settled at the bottom of the tube. The above morphological change is a sign for the correct collection, dechorionation, fixing, and quenching procedures as described above.
4. Place the samples on ice and proceed with sonication. The sonicator's probe should be placed in the middle of the liquid at a semi-angle (~15° angle) in order to achieve efficient mixing of the sample during sonication. The sonication should be performed at maximum power for seven 40-s rounds with 2-min intermediate incubations on ice (*see* **Note 5**).
5. Isolate sheared chromatin by performing two rounds of centrifugation. Initially spin down at ~1500 × *g* in a microcentrifuge

for 10 min. The pellet includes the glass beads and the majority of the cellular debris, and should be discarded. Transfer the supernatant into new 1.5 ml tubes and spin for 20 min at ~13300 × *g* in a microcentrifuge at 4 °C (*see* **Note 6**). Transfer the chromatin-containing supernatant into new tubes and add an equal volume of GSB.

6. Store chromatin at −80 °C (*see* **Note 7**).

3.4 Titration and Quality Control of Chromatin Preparation

1. Calculate chromatin concentration (μg/μl) with a Nanodrop Spectrophotometer and digest 5 μg with proteinase K for 2 h at 55 °C. Reverse formaldehyde cross-linking by overnight incubation at 65 °C.
2. Isolate sonicated DNA by phenol/chloroform extraction and ethanol/glycogen-based precipitation. Resuspend precipitated DNA pellets in 100 μl of ddH_2O.
3. Analyze (~500 ng) sonicated DNA by agarose gel electrophoresis to determine the distribution of fragment sizes. The majority of the sonicated DNA should be more than 300 bp and less than 3,000 bp in order to ensure that shearing was effective but *lacO* sequences were not separated from the CRMs and promoter-reporter sequences (Fig. 2b).
4. The presence of non-cross-linked material in the chromatin sample can interfere with immunoprecipitation results. Therefore, it is important to evaluate the efficiency of chromatin cross-linking for each chromatin preparation by PCR (Fig. 2c). Amplify 10–50 ng of cross-linked chromatin and an equivalent amount of reverse-cross-linked/purified DNA side by side for 25 rounds of PCR. In our experiments with *Dll*, we used primers that specifically recognized the *lacO*-tagged transgene (*see* Subheading 2.7, **step 1**). Analyze PCR products by agarose gel electrophoresis. Chromatin samples that produce amplification products without prior proteinase K digestion and reversal of cross-linking should be discarded.

3.5 Cell-Type- and Gene-Specific Looping of Cis-Regulatory Modules

1. Pre-clear 500 μg of chromatin by incubation with 10 μl of magnetic beads in reaction buffer (1× RIPA) for 1 h at 4 °C. Set up pre-clearing reaction at a final volume of 800 μl.
2. Remove the magnetic beads with the help of an MPC-S and divide the reaction in two tubes.
3. Add 2.5 μg of anti-FLAG or IgG antibodies, respectively, and incubate for 4 h at 4 °C by continuous rocking.
4. Transfer the reactions at room temperature and couple antibody to the magnetic beads. Use 2 μl of beads per sample and incubate for 1 h by continuous rocking.
5. Immobilize bead/antibody/chromatin complexes with the help of an MPC-S and aspirate supernatant.

6. Wash beads twice with 1 ml of 1× WRP for 7 min at room temperature by continuous rocking and aspirate with the help of an MPC-S.
7. Dilute bead/antibody/chromatin complexes in 300 μl 1× WRP and transfer into a new tube with the help of a cut filter tip.
8. Wash beads once with 1 ml of 1× WRP at room temperature and mix gently for 5 min without rocking.
9. Aspirate and proceed to proteinase K digest, cross-linking reversal and DNA extraction/precipitation reactions as in Subheading 3.4, **steps 1** and **2** (*see* **Note 8**).
10. Design and perform ^{32}P-based radioactive PCR reactions and analyze the amplification products in 8 % acrylamide gels [2, 26] (*see* **Note 9**).
11. Design and perform SYBR Green-based qPCR in order to quantify the amplification signals (*see* **Note 9**).

3.6 Cell-Type- and Gene-Specific Double Immunoprecipitation

1. Collect and fix 8 g of staged embryos and isolate/titrate total chromatin as in Subheadings 3.1–3.3 in order to perform cell-type- and gene-specific chromatin analysis. An initial collection of 8 g of staged embryos is required in order to probe chromatin with five different antibodies at the final stage of the cgChIP protocol (*see* **Note 2**).
2. Pre-clear total chromatin as in Subheading 3.5, **steps 1** and **2**. Set up multiple pre-clearing reactions at a final volume of 800 μl by using 10 μl of magnetic beads for each 500 μg chromatin sample.
3. Isolate Flag:lacI/lac*O*-tagged chromatin by FLAG immunoprecipitation as in Subheading 3.5, **steps 3–5**. For an initial 8 g collection, a total of eight reactions can be set up.
4. Immobilize bead/antibody/chromatin complexes with the help of an MPC-S and aspirate supernatant.
5. Wash beads twice with 1 ml of 1× WRP for 7 min at room temperature by continuous rocking and aspirate with the help of an MPC-S.
6. Dilute bead/antibody/chromatin complexes in 300 μl 1× WRP and transfer into a new tube with the help of a cut filter tip. Aspirate with the help of an MPC-S.
7. Wash beads once with 1 ml of 1× WRP at room temperature and mix gently for 5 min without rocking. Aspirate with the help of an MPC-S.
8. Dilute Flag:lacI/lac*O*-tagged chromatin complexes in 600 μl of elution buffer (EB) and transfer into a new 0.6 ml tube.
9. Incubate for 4 h at 4 °C by continuous rocking.
10. Pre-clear eluted chromatin as in Subheading 3.5, **steps 1** and **2**.

11. Aliquot a small fraction for an additional control step (*see* Subheading 3.7, **step 2**).
12. Carry out the second round of immunoprecipitation for 12 h at 4 °C in a total volume of 300 μl. Set up the reaction in 0.6 ml tube in order to achieve efficient mixing during rocking. Incubate equal samples of chromatin with 2 μg of antibody against the transcription factor of interest side by side with equivalent amounts of IgG as negative or anti-LacI antibody as positive control, respectively.
13. Transfer the reactions at room temperature and couple antibody to magnetic beads as in Subheading 3.5, **step 4**.
14. Transfer the reactions into 1.5 ml tubes with the help of a cut filter tip.
15. Immobilize bead/antibody/chromatin complexes with the help of an MPC-S and aspirate supernatant.
16. Wash beads twice with 1 ml of 1× WRP for 7 min at room temperature by continuous rocking. Intermediately, aspirate with the help of an MPC-S.
17. Aspirate and proceed to proteinase K digest, cross-linking reversal and DNA extraction/precipitation reactions as in Subheading 3.4, **steps 1** and **2**. Analyze as in Subheading 3.5, **steps 10** and **11** (*see* **Note 9**).

3.7 Internal Controls for Evaluation of cgChIP Results

1. Ideally, cgChIP experiments should include several controls to assess possible contamination of LacI-tagged tissue-specific chromatin with chromatin derived from other cell types. In case of tissue-specific expression of transcription factors (e.g., *Hox* genes) additional controls may include anti-Ubx or anti-AbdA cgChIPs to test the purity in cases where thoracic derived chromatin is analyzed. Conversely, thoracic specific factors (e.g., *Dll*) may be used as a control for cgChIPs where abdominal derived chromatin is analyzed. If any contamination is observed (as evidenced by the presence of a factor that should not be there, for example, Dll in abdominal derived chromatin), the experiment should be aborted and new chromatin preparations should be carried out.
2. Additional controls are performed in order to ensure for the specificity of the immunoprecipitation and the efficiency of both rounds of cgChIP. For one, IgG samples are processed side by side with anti-FLAG immunoprecipitations, to detect any nonspecific enrichment during the first round of precipitation. Second, primers hybridizing across the sequence of the LacZ reporter gene are used in combination with gene-specific primers in order to ensure that transgenic and not endogenous sequences are precipitated and amplified (*see* Subheading 2.7, **step 1**). Third, upon 3× FLAG peptide-based elution of

LacI-tagged tissue-specific derived chromatin and prior to the second round of precipitation, a small fraction of the eluted material is processed as in Subheading 3.4, **steps 1** and **2**, and analyzed by PCR. The precipitated DNA is amplified with transgenic specific and off-target primer pairs in order to evaluate for the efficiency and specificity of FLAG immunoprecipitation. Fourth, at the second round of precipitation, IgG and anti-LacI are used to assess the quality and the efficiency of both rounds of cgChip. A weak LacI:IgG ratio suggests that the experiment was unsuccessful. In our experience, an eight-fold enrichment of LacI compared to IgG is satisfactory when LacI signal recovers at least 1 % of the input used at the second round of precipitation.

4 Notes

1. Apple juice containing agar plates and yeast paste must be freshly prepared and stored at 4 °C. The efficiency of laying is enhanced by pre-warm of the plates and yeast paste for 10–15 min at 25 °C. Avoid changes in apple plate batches and/or recipes during a single experiment. The fly-cage culture system is very sensitive and slight changes in conditions can dramatically reduce the rhythm of egg laying.
2. The exact amount of embryos required depends on the target gene, the embryonic tissue, and the developmental stage of the study. For example, leg primordial cells at stages 11–14 generate less than 200–300 cells from about 6,000 total embryonic cells. For *Dll*, we determined that 8 g of embryos were required to purify a sufficient quantity of chromatin derived from these cells. Preparation of chromatin in this manner provided enough template for five different immunoprecipitations in the second round of cgChIP.
3. Immunostaining of embryos can be performed by standard formaldehyde/heptane fixing and basic incubation steps as follows: (1) Blocking is carried out overnight in PBST with 5 % BSA at 4 °C. (2) Both the primary and the secondary antibody incubations are carried out for 12 h at 4 °C. For β-galactosidase and FLAG expression, commercially available antibodies are used.
4. It is important to place the bottle in horizontal position in order to maximize the interface between fixing solution and heptane where embryos equilibrate during mixing.
5. It is important to follow the electronic display of the sonicator in order to keep the output power at high levels by making slight changes of the Falcon/probe relative position. For example Fisher Sonic Dismembrator Model 100 sonicator has

an electronic display that provides information for the output power. In general the shearing procedure requires to be standardized specifically for each sonicator type.

6. It is important to perform sequential centrifugation. The first (low-speed) round results in separation of chromatin from other cellular debris and glass beads. Chromatin can be trapped inside these particles if a direct high-speed centrifugation step is performed.
7. GSB-diluted chromatin can be stored at −80 °C for at least 6 months.
8. The immunoprecipitated DNA can be stored at −20 °C until further use.
9. It is important to perform radioactive PCRs in order to evaluate the specificity and efficiency of both DNA looping and double-cgChIP readouts. This is especially important for developmental genes that are expressed in small populations of cells. Radioactive PCR products help with the design of amplicons and subsequent qPCRs. In addition, we suggest that interpretation of qPCR results should be based both on the percentage of recovery of input DNA for each antibody or IgG reaction and the fold enrichment of the antibody versus IgG ChIPs. By making both comparisons we reduce the possibility that background signals can be interpreted as true positives due to a high ratio of antibody to IgG ChIPs that may happen in cases where input DNA is poorly recovered.

Acknowledgments

This work was supported by GM058575 and GM054510 awarded to R.S.M. D.J.M. was supported by 5T32DK07328 and M.A. was supported by a long-term EMBO fellowship. We thank D. Arvanitis and S. Tsiftsoglou for critical proofreading.

References

1. Deal RB, Henikoff S (2011) The INTACT method for cell type-specific gene expression and chromatin profiling in Arabidopsis thaliana. Nat Protoc 6(1):56–68. doi:10.1038/nprot.2010.175. Epub 2010 Dec 16
2. Agelopoulos M, McKay DJ, Mann RS (2012) Developmental regulation of chromatin conformation by Hox proteins in Drosophila. Cell Rep 1(4):350–359. doi:10.1016/j.celrep.2012.03.003
3. Gavrilov A, Eivazova E, Priozhkova I, Lipinski M, Razin S, Vassetzky Y (2009) Chromosome conformation capture (from 3C to 5C) and its ChIP-based modification. In: Collas P (ed) Chromatin immunoprecipitation assays. Methods mol biology, vol 567. 2009/507/10 edn. Humana, New York, pp 171–188. doi:10.1007/978-1-60327-414-2_12
4. van Berkum NL, Dekker J (2009) Determining spatial chromatin organization of large genomic regions using 5C technology. In: Collas P (ed) Chromatin immunoprecipitation assays. Methods Mol Biol, vol 567. 2009/07/10 edn. Humana, New York,

pp 189–213. doi: 10.1007/978-1-60327-414-2_13

5. van de Werken HJ, de Vree PJ, Splinter E, Holwerda SJ, Klous P, de Wit E, de Laat W (2012) 4C technology: protocols and data analysis. In: Wu C, Allis D (eds) Nucleosomes, histones & chromatin Part B. Methods Enzymol vol 513. 2012/08/30 edn. Elsevier, Amsterdam, pp 89–112. doi: 10.1016/B978-0-12-391938-0.00004-5
6. Belton JM, McCord RP, Gibcus JH, Naumova N, Zhan Y, Dekker J (2012) Hi-C: a comprehensive technique to capture the conformation of genomes. Methods 58(3):268–276. doi:10.1016/j.ymeth.2012.05.001
7. Dekker J, Rippe K, Dekker M, Kleckner N (2002) Capturing chromosome conformation. Science 295(5558):1306–1311. doi:10.1126/science.1067799
8. Fullwood MJ, Liu MH, Pan YF, Liu J, Xu H, Mohamed YB, Orlov YL, Velkov S, Ho A, Mei PH, Chew EG, Huang PY, Welboren WJ, Han Y, Ooi HS, Ariyaratne PN, Vega VB, Luo Y, Tan PY, Choy PY, Wansa KD, Zhao B, Lim KS, Leow SC, Yow JS, Joseph R, Li H, Desai KV, Thomsen JS, Lee YK, Karuturi RK, Herve T, Bourque G, Stunnenberg HG, Ruan X, Cacheux-Rataboul V, Sung WK, Liu ET, Wei CL, Cheung E, Ruan Y (2009) An oestrogen-receptor-alpha-bound human chromatin interactome. Nature 462(7269):58–64. doi:10.1038/nature08497
9. Horike S, Cai S, Miyano M, Cheng JF, Kohwi-Shigematsu T (2005) Loss of silent-chromatin looping and impaired imprinting of DLX5 in Rett syndrome. Nat Genet 37(1):31–40. doi:10.1038/ng1491
10. Chao MV, Gralla JD, Martinson HG (1980) lac Operator nucleosomes. 1. Repressor binds specifically to operator within the nucleosome core. Biochemistry 19(14):3254–3260
11. Danzer JR, Wallrath LL (2004) Mechanisms of HP1-mediated gene silencing in Drosophila. Development 131(15):3571–3580. doi:10.1242/dev.01223
12. Gonzalez-Serricchio AS, Sternberg PW (2006) Visualization of C. elegans transgenic arrays by GFP. BMC Genet 7:36
13. Hoshino A, Fujii H (2009) Insertional chromatin immunoprecipitation: a method for isolating specific genomic regions. J Biosci Bioeng 108(5):446–449. doi:10.1016/j.jbiosc.2009.05.005
14. Istrail S, Davidson EH (2005) Logic functions of the genomic cis-regulatory code. Proc Natl Acad Sci U S A 102(14):4954–4959. doi:10.1073/pnas.0409624102
15. Jenett A, Rubin GM, Ngo TT, Shepherd D, Murphy C, Dionne H, Pfeiffer BD, Cavallaro A, Hall D, Jeter J, Iyer N, Fetter D, Hausenfluck JH, Peng H, Trautman ET, Svirskas RR, Myers EW, Iwinski ZR, Aso Y, DePasquale GM, Enos A, Hulamm P, Lam SC, Li HH, Laverty TR, Long F, Qu L, Murphy SD, Rokicki K, Safford T, Shaw K, Simpson JH, Sowell A, Tae S, Yu Y, Zugates CT (2012) A GAL4-driver line resource for Drosophila neurobiology. Cell Rep 2(4):991–1001. doi:10.1016/j.celrep.2012.09.011
16. Jory A, Estella C, Giorgianni MW, Slattery M, Laverty TR, Rubin GM, Mann RS (2012) A survey of 6,300 genomic fragments for cis-regulatory activity in the imaginal discs of Drosophila melanogaster. Cell Rep 2(4):1014–1024. doi:10.1016/j.celrep.2012.09.010
17. Manning L, Heckscher ES, Purice MD, Roberts J, Bennett AL, Kroll JR, Pollard JL, Strader ME, Lupton JR, Dyukareva AV, Doan PN, Bauer DM, Wilbur AN, Tanner S, Kelly JJ, Lai SL, Tran KD, Kohwi M, Laverty TR, Pearson JC, Crews ST, Rubin GM, Doe CQ (2012) A resource for manipulating gene expression and analyzing cis-regulatory modules in the Drosophila CNS. Cell Rep 2(4):1002–1013. doi:10.1016/j.celrep.2012.09.009
18. McKay DJ, Lieb JD (2013) A common set of DNA regulatory elements shapes Drosophila appendages. Dev Cell 27(3):306–318. doi:10.1016/j.devcel.2013.10.009
19. Thurman RE, Rynes E, Humbert R, Vierstra J, Maurano MT, Haugen E, Sheffield NC, Stergachis AB, Wang H, Vernot B, Garg K, John S, Sandstrom R, Bates D, Boatman L, Canfield TK, Diegel M, Dunn D, Ebersol AK, Frum T, Giste E, Johnson AK, Johnson EM, Kutyavin T, Lajoie B, Lee BK, Lee K, London D, Lotakis D, Neph S, Neri F, Nguyen ED, Qu H, Reynolds AP, Roach V, Safi A, Sanchez ME, Sanyal A, Shafer A, Simon JM, Song L, Vong S, Weaver M, Yan Y, Zhang Z, Lenhard B, Tewari M, Dorschner MO, Hansen RS, Navas PA, Stamatoyannopoulos G, Iyer VR, Lieb JD, Sunyaev SR, Akey JM, Sabo PJ, Kaul R, Furey TS, Dekker J, Crawford GE, Stamatoyannopoulos JA (2012) The accessible chromatin landscape of the human genome. Nature 489(7414):75–82. doi:10.1038/nature11232
20. Brand AH, Perrimon N (1993) Targeted gene expression as a means of altering cell fates and generating dominant phenotypes. Development 118(2):401–415
21. Brent R, Ptashne M (1985) A eukaryotic transcriptional activator bearing the DNA

specificity of a prokaryotic repressor. Cell 43 (3Pt 2):729–736, 0092-8674(85)90246-6
22. del Valle Rodriguez A, Didiano D, Desplan C (2011) Power tools for gene expression and clonal analysis in Drosophila. Nat Methods 9(1):47–55. doi:10.1038/nmeth.1800
23. Pfeiffer BD, Ngo TT, Hibbard KL, Murphy C, Jenett A, Truman JW, Rubin GM (2010) Refinement of tools for targeted gene expression in Drosophila. Genetics 186(2):735–755. doi:10.1534/genetics.110.119917
24. Robinett CC, Straight A, Li G, Willhelm C, Sudlow G, Murray A, Belmont AS (1996) In vivo localization of DNA sequences and visualization of large-scale chromatin organization using lac operator/repressor recognition. J Cell Biol 135(6 Pt 2):1685–1700
25. Gebelein B, McKay DJ, Mann RS (2004) Direct integration of Hox and segmentation gene inputs during Drosophila development. Nature 431(7009):653–659. doi:10.1038/nature02946
26. Agelopoulos M, Thanos D (2006) Epigenetic determination of a cell-specific gene expression program by ATF-2 and the histone variant macroH2A. EMBO J 25(20):4843–4853. doi:10.1038/sj.emboj.7601364

Chapter 19

Bimolecular Fluorescence Complementation (BiFC) in Live Drosophila Embryos

Marilyne Duffraisse, Bruno Hudry, and Samir Merabet

Abstract

Bimolecular fluorescence complementation (BiFC) is a powerful method for studying protein-protein interactions in different cell types and organisms. This method was recently developed in the fruit fly *Drosophila melanogaster*, allowing analyzing protein interaction properties in a physiologically relevant developing context. Here we present a detailed protocol for performing BiFC with the Venus fluorescent protein in live *Drosophila* embryos, taking the Hox-PBC partnership as an illustrative test case. This protocol applies to any transcription factor and split fluorescent protein in general.

Key words Hox, Transcription, BiFC, EMSA, Drosophila

1 Introduction

BiFC relies on the property of monomeric fluorescent proteins to be reconstituted from two separate non-fluorescent sub-fragments once these fragments are close enough in space [1]. This property has by now been used in diverse systems to validate interactions between different types of proteins [2]. Unlike other visualization methods such as fluorescence resonance energy transfer (FRET), BiFC does not require high levels of fusion protein expression, which allows analyzing protein protein interactions in physiological conditions. BiFC presents however the disadvantage of stabilizing the formation of protein complexes [3] and is therefore not appropriate for measuring protein interaction dynamics in vivo.

Virtually any monomeric fluorescent protein can be used for BiFC and different split sites are suitable for generating different complementary moieties. Although the BiFC toolbox is already quite large, proposing a palette of different fluorescent proteins [2], new variants are regularly generated in order to improve parameters such as the signal-to-noise ratio [4] or the time maturation rate [5, 6].

Yacine Graba and René Rezsohazy (eds.), *Hox Genes: Methods and Protocols*, Methods in Molecular Biology, vol. 1196,
DOI 10.1007/978-1-4939-1242-1_19, © Springer Science+Business Media New York 2014

Here we describe the experimental procedure for using BiFC in live *Drosophila* embryos with the Venus fluorescent protein. Venus is a variant of the green fluorescent protein (GFP) of the jellyfish *Aequorea victoria* and is one of the brightest monomeric fluorescent proteins for BiFC [7, 8]. Thus, Venus is well appropriate for revealing protein-protein interactions in the context of the high fluorescent background that characterizes the *Drosophila* embryo. In addition, Venus can be used in combination with the Cerulean fluorescent protein, allowing performing multicolor BiFC to visualize two different interactions simultaneously [7, 8].

2 Materials

2.1 Cloning Vectors with Venus Fragments

1. Fusion constructs are cloned into the pcDNA3 vector (http://www.addgene.org/vector-database/2092/) for in vitro transcription/translation and in pUAST (http://www.addgene.org/vector-database/4473/) or pUASTattB vectors for expression with the UAS/Gal4 system in fly (*see* **Note 1**).
2. All vectors allow for coding the N-terminal (VN: residues 1–173) or the C-terminal fragment of Venus (VC: residues 155–238) for fusions at the N- or the C-terminus of the proteins of interest. For fusions at the N- or the C-terminus, VN and VC fragments are cloned into the EcoRI-XhoI or the XhoI-XbaI restriction sites, respectively. For fusions at the N-terminus with the VC fragment add an additional start ATG codon at the 5′ end and remove the last stop codon for in-frame translation with the downstream gene. Along the same line, for fusions at the C-terminus with the VN fragment add a stop codon at the 3′ end for translation arrest. In this fusion topology, it is not necessary to remove the first ATG codon of VN since it will not be recognized as a start codon (*see* **Note 2**).
3. It is better to insert a short linker region in between the Venus fragment and the protein, especially for fusions with the long VN fragment. This linker has to be flexible and neutral, like the GGSGG sequence, and can be encoded by an oligonucleotide designed for PCR amplification and cloning of the cDNA (*see* **Note 3**).

2.2 Protein Production

TNT quick coupled in vitro transcription/translation system from rabbit reticulocyte lysate (Promega) (*see* **Note 4**).

2.3 Electromobility Shift Assays (EMSAs)

1. Glass plates for EMSA polyacrylamide gels are usually 20 cm × 20 cm (*see* **Note 5**).
2. PlusOne Repel-Silane (GE Healthcare), or equivalent solution of dimethyldichlorosilane.

3. Acrylamide/Bis-acrylamide, 30 % solution: Store at 4 °C.
4. Ammonium persulfate 10 % solution in water (APS): Store aliquots at −20 °C.
5. *N,N,N,N'* tetramethyl-ethylenediamine (TEMED): Store at room temperature.
6. To inhibit sticking of acrylamide gels and glass plates, use a paper soaked in the Repel solution to cover the glass plate face that will be in contact with the gel. Let it dry for 5 min before assembling plates.
7. Assemble the plates and device for polyacrylamide gel electrophoresis according to instructions provided by the manufacturer (for example, Hoefer vertical electrophoresis system SE400 can be used).
8. Prepare acrylamide gel solution (final volume 80 ml): Mix 4 ml of TBE 10× (106 g Tris base, 55 g boric acid, 40 ml of 0,5 M EDTA pH 8.0), 8 ml of acrylamide/Bis-acrylamide, 67 ml of water in an Erlenmeyer, and 670 μl of APS. Add 56 μl of TEMED for polymerization. Mix and immediately pour between the glass plates. Remove air bubbles by gently mixing. Insert a gel comb. Wait for 20 min before complete polymerization of the gel (*see* **Note 6**).
9. Binding buffer 5×: 50 mM Tris–HCl (pH 7.5), 5 mM $MgCl_2$, 250 mM NaCl, 2.5 mM EDTA, 2.5 mM DTT, 0.25 mg/ml poly (dI-dC), 20 % glycerol (*see* **Note 7**).
10. Annealing DNA probe: Mix 40 μl (1 μg/μl) of each oligonucleotide in 1:1 ratio, 9 μl of TE 10× (Tris 100 mM, EDTA 10 mM) pH 8, and 1 μl water and place for 10 min at 70 °C in a thermo-block or a water bath. Then, switch off the temperature set and let it slowly cool down till reaching room temperature (can take up 3–4 h). Store at −20 °C. Probes can be kept at −20 °C for months.
11. ^{32}P labeled probe: Each probe is labeled with 40 μCi ^{32}P αATP. Mix 1.5 μl of annealed DNA probe (100 μM), 20 μl of sterile water, 3 μl of DNA polymerase 10× buffer, 1.5 μl of DNA polymerase I Klenow, and 4 μl of fresh ^{32}P α dATP (*see* **Note 8**). Incubate for ½ h at 37 °C. Purify the probe using Illustra MicroSpin G50-columns (GE Healthcare) (*see* **Note 9**). Elute in a final volume of 30 μl. Use a scintillation counter to estimate the efficiency of labelling. Store the probe at −20 °C.
12. Autoradiography film (Hyperfilm MP, GE Healthcare).

2.4 Fly Lines

1. Transgenesis with pUAST or pUASTattB vectors are performed in *yw* flies through *P*-element or ΦC31 integrase-mediated system [9], respectively.

2. *Gal-4* lines correspond to artificial constructs or *P* insertions (*see* **Note 10**).
3. Adult flies can be established with both *UAS-VN* and *UAS-VC* fusion transgenes and the *Gal4* driver by using a ***tubulin*** (***tub***)-*Gal80* balancer chromosome (*see* **Note 11**). Alternatively, adult flies bearing the *UAS-VN* and *UAS-VC* transgenes can be crossed with *Gal4* lines for direct BiFC analysis.

2.5 Embryo Collection

1. Petri dishes–diameter 5.6 cm.
2. Prepare Petri dishes for egg laying. Mix 9 g agar, 0.5 g Nipagin (antifungal and antibacterial agent), and 300 ml of water in a 500 ml Erlenmeyer. Boil in a microwave oven. Leave the solution to cool for few minutes. Add 100 ml of fruit juice and 10 g of sugar and boil again for few minutes. Pour in Petri dishes. Let it cool down and solidify. Invert the plates and store them at 4 °C. Plates can be kept at 4 °C for 1 month.
3. Baker's yeast: Reconstitute yeast with water. Store at 4 °C.
4. Cage: Polypropylene Tricorn Beaker (100 ml).
5. Bleach.
6. Voltalef oil 10S (VWR): Poly(chlorotrifluoroethylene) 800.

2.6 Immunostaining

1. Fixation solution. Mix 800 µl of water, 100 µl of PBS 10× (see below), and 100 µl of formaldehyde 37 %.
2. N-heptane.
3. Absolute ethanol.
4. Methanol.
5. Phosphate buffer saline (PBS 10×): 137 mM NaCl, 2.7 mM KCl, 10 mM Na_2HPO_4, 1.76 mM KH_2PO_4. Adjust pH to 7.4.
6. Tween 20.
7. PBS containing 0.1 % tween 20 (PBT): Prepare PBT mixing 100 ml of PBS 10×, 899 ml of water, and 1 ml of Tween 20.
8. Bovine serum albumin 10 % (BSA).
9. PBT containing 1 % BSA (PBT/BSA).
10. Chicken anti-GFP (Promega; 1/500) (*see* **Note 12**).
11. Alexa Fluor® 488 Goat Anti-Chicken IgG (H+L) (Molecular Probes, Life Technologies; 1/500).
12. Vectashield Mounting Medium (Vector Laboratories).
13. Confocal microscope and Image Acquisition Software (Image J for example; http://rsb.info.nih.gov/ij/).

For Venus fluorescence, argon laser (488 or 514 nm) is used for excitation.

3 Methods

3.1 Determining the Best Suitable Fusion Topologies for BiFC

In the case of transcription factors, the influence of fusion topologies on protein-protein and protein-DNA interactions can quickly be measured in vitro by electromobility shift assays (EMSAs) (*see* **Notes 13** and **14** and Fig. 1). This approach allows determining the best appropriate fusion topologies for subsequent BiFC assay in vivo.

1. Binding reaction: 2–4 μl (depending on the amount of produced proteins previously quantified with ^{35}S-methionine) of programmed lysates are incubated with 5 μl of binding buffer 5×, 1 μl of radioactive probe, and x μl of sterile water in a total volume of 20 μl for ½ hour at 4 °C (*see* **Note 15**).
2. Pre-run the polyacrylamide gel at 190 V during the binding reaction time in TBE 0.5×.

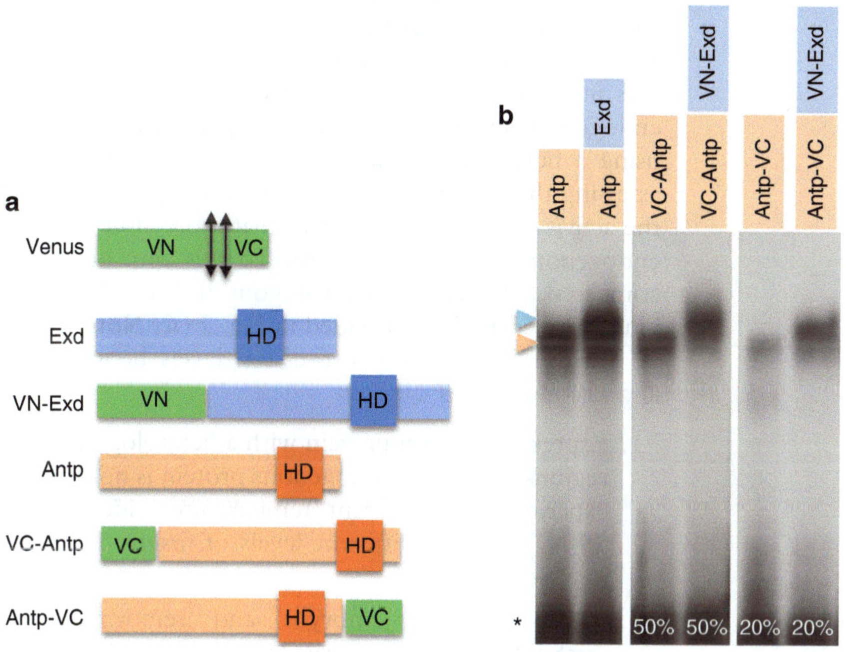

Fig. 1 Influence of fusion topologies on interaction properties of the Hox protein Antennapedia (Antp) in vitro. (**a**) Scheme of fusion constructs with N- (VN) or C- (VC) terminal fragment of Venus and Extradenticle (Exd) or Antp, respectively. Position of the homeodomain (HD) is indicated. (**b**) Band shift assay with non fused or fused proteins, as depicted above the gel. Monomer or dimer binding is indicated by *light orange* or *light blue arrowhead*, respectively. *Asterisk* shows free radioactive probe (corresponding here to the *Dllcon* probe [13]). The level of DNA binding with fusion proteins is indicated as a percentage of DNA-binding activity observed with non-fused proteins. Fusion of the VC fragment at the C-terminus of Antp affects more strongly DNA binding and complex formation with Exd than fusion at the N-terminus

3. Load the total volume of reaction into the gel. Keep a lane for a migration marker.
4. Run at 200 V for 5 min.
5. Run for approximately 1.5 h at 160 V.
6. Remove glass plates.
7. Put the gel onto a paper.
8. Dry the gel under a vacuum air system at 80 °C for ¾ h.
9. Place the gel in a cassette with autoradiography film at −80 °C overnight.
10. Reveal the film and place a new film for shorter or longer exposition time if necessary.

3.2 Assessing the Expression Level of Fusion Constructs In Vivo

Even in the case of a *P-Gal4* insertion reproducing the endogenous expression pattern of interest, expression levels can vary according to the genomic insertion site of the *UAS* fusion construct (*see* **Note 16**). It is therefore necessary to measure the expression level induced by the *UAS/Gal4* system in order to adjust it as closely as possible to the expression level of the endogenous gene. This can be achieved in two steps. The first step consists in comparing expression levels between the fusion and the endogenous proteins. To this aim, quantifications are performed using a heterologous *Gal-4* driver. The second step consists in assessing the expression level of the fusion proteins resulting from the activity of the *P-Gal4* driver that more closely reproduces the endogenous expression pattern. To this aim, quantifications are compared between the heterologous and the *P-Gal4* drivers. An illustrative example is provided in Fig. 2 (*see* **Note 17**).

Step 1: Determining expression levels between endogenous and fusion proteins:

1. Express the fusion protein with a heterologous *Gal4* driver in regions where the endogenous protein is not present. Do this experiment at different temperatures (ideally 22, 25, and 29 °C) to have different levels of fusion protein expression (*see* **Note 18**).
2. After collecting embryos and removing the chorion (*see* Subheading 3.3) proceed to the fixation by placing embryos in a 1.5 ml tube containing 500 μl of the fixation solution and 500 μl of N-heptane. Place on soft agitation (around 20 rpm) for 25 min. Remove the fixation solution (lower phase) and add 500 μl of methanol. Place on strong agitation (vortex) for 1 min. Let embryos fall down and remove the solution completely. Rinse three times with absolute ethanol and keep embryos in ethanol at −20 °C.
3. Immunostaining against your protein (*see* **Note 19**): Rinse embryos three times with absolute ethanol, twice with methanol,

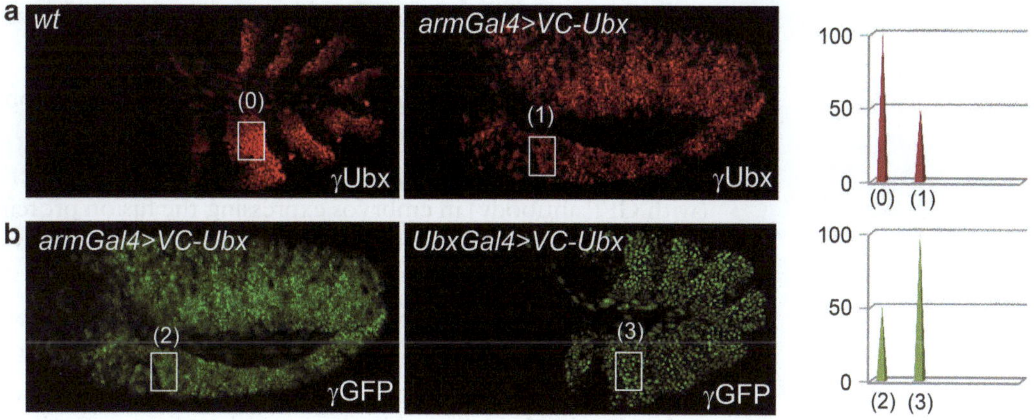

Fig. 2 Comparing the expression level between the endogenous Ultrabithorax (Ubx) and the induced VC-Ubx fusion protein in the *Drosophila* embryo. (**a**) Endogenous Ubx expression is quantified in wild-type (*wt*) embryos (*box 0*) with an anti-Ubx antibody. Expression of VC-Ubx under the control of the *armadillo* (*arm*)-*Gal4* driver is quantified in segments where endogenous Ubx is not present, with both the anti-Ubx (*box 1*) and anti-GFP (*box 2*) antibodies. (**b**) Quantification in *box 2* is compared to the expression level of VC-Ubx induced with the *Ubx-Gal4* driver and measured with the same anti-GFP antibody (*box 3*). Comparing *box 3* with *box 2* allows deducing the expression level of VC-Ubx induced by *Ubx-Gal4* with regard to endogenous Ubx in wild-type embryos. Experimental parameters can thus be modified accordingly (the UAS/Gal4 system is more active according to the temperature). Here crosses are realized at 25 °C. Under these conditions, ectopic expression of VC-Ubx with *arm-Gal4* corresponds to 50 % of endogenous levels (*upper graph*). Under the same conditions, *Ubx-Gal4* induces two times more expression levels of VC-Ubx than *arm-Gal4* (*lower graph*). Thus, using *Ubx-Gal4* at 25 °C is appropriate for inducing expression levels close to endogenous Ubx levels

and five times with PBT and place them for soft agitation in PBT+BSA 1 % for 1 h. Rinse twice with PBT and add your antibody at the requested dilution in PBT/BSA 1 % overnight at 4 °C.

4. Reveal with a fluorescent secondary antibody. Rinse embryos five times with PBT, and add the secondary anti-species antibody coupled to a fluorochrome at the requested dilution in PBT for 1 h at room temperature. Rinse five times with PBT and mount embryos in Vectashield in between slide and cover slide.

5. Compare the fluorescent staining between ectopic and endogenous places in the embryo. A minimum of ten different embryos at the desired stage should be analyzed in each experimental condition, using a confocal microscope under fixed parameters of acquisition.

6. Analyze pictures using ImageJ (or another) software (*see* **Note 20**) and compare expression levels between endogenous and experimentally induced fusion proteins (boxes 0 and 1 in Fig. 2, respectively). Values corresponding to endogenous expression levels should be determined at this step.

Step 2: Determining expression levels induced by the *P-Gal4* driver:

1. Collect embryos expressing the fusion protein under the control of the *P-Gal4* driver at various temperatures.
2. Do the immunostaining against the VN or the VC fragment (with GFP antibody) in embryos expressing the fusion protein under the control of either the heterologous *Gal-4* (as previously) or *P-Gal4* driver.
3. Quantify and compare the expression levels between the two *Gal-4* drivers.
4. Choose the experimental condition allowing *P-Gal4*-induced expression levels comparable to the endogenous ones, indirectly through comparison with the heterologous *Gal-4* driver conditions known to correspond to endogenous-like expression levels (boxes 2 and 3 in Fig. 2).

3.3 BiFC in Live *Drosophila* Embryos

1. Once experimental parameters have been established with the *UAS/Gal4* system, flies can be crossed *en masse* for collecting embryos.
2. Do not use old flies for having good-quality embryos. Classically collect flies a week before using them for egg collection. Crosses between males and females are set up 2 days before starting egg collection. The population is composed of approximately one male for four females. Egg collection should not continue over more than 8 days (*see* **Note 21**).
3. Depending on the embryonic stage of interest, flies are placed in a cage various times. 40 flies (32 females/8 males) typically produce more than 100 embryos during one overnight egg laying (*see* **Note 22**).
4. Petri dishes with eggs are then placed at 4 °C for 28 h (*see* **Note 23**).
5. Remove the protective chorion of embryos by recovering them with bleach, directly in the Petri dish for 1 min.
6. Collect embryos by placing them onto a filter and rinse with water.
7. Put 2/3 drops of halocarbon oil 10S on a glass slide.
8. Take embryos with forceps and dispatch them within the halocarbon oil 10S. Place a cover slide (*see* **Note 24**).
9. Process to BIFC analysis under a confocal microscope (*see* **Note 25**). Images can be acquired over several hours in the live embryo, until the end of embryogenesis. Illustrative pictures are provided in Fig. 3.
10. *See* **Note 26** about controls for BiFC signal specificity.

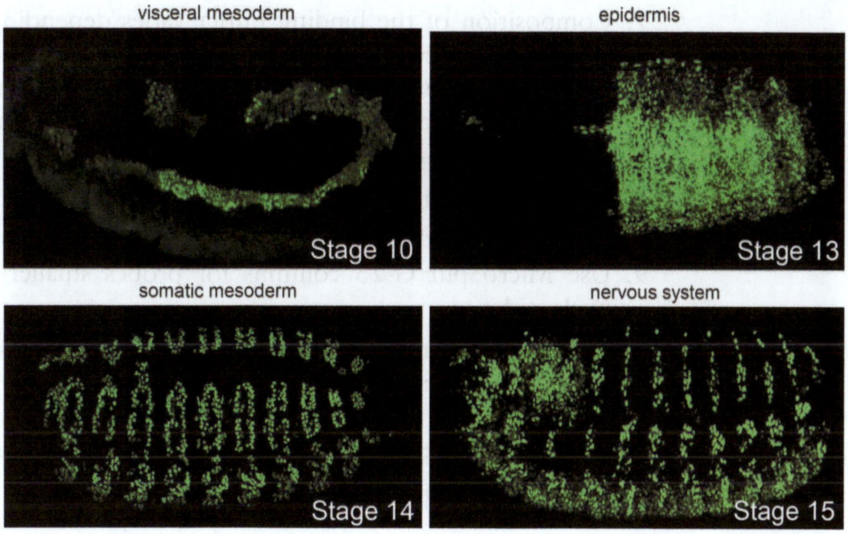

Fig. 3 Assessing BiFC in various tissues between the Hox protein AbdominalA (AbdA) and its cofactor Extradenticle (Exd). AbdA and Exd are, respectively, fused to the VC or the VN fragment at their N-terminus. This fusion topology was shown to be best appropriate for BiFC in vivo [8]. Interaction between AbdA and Exd fusion proteins is visualized in the visceral mesoderm, epidermis, somatic mesoderm, or nervous system with the *bagpipe-Gal4*, *abda-Gal4*, *rp238-Gal4*, and *elav-Gal4* drivers, respectively. Note that BiFC is observed in the presence of endogenous AbdA in all except the epidermis tissue. Images were acquired for various developmental stages, as indicated

4 Notes

1. A recent work described a collection of more than 1,000 ORF fly lines that can be used for BiFC in *Drosophila* upon epitope tag swapping [10].
2. pUAST vectors are also compatible with the Gateway system for fusions with the N- or the C-terminal fragment of the yellow fluorescent protein [11].
3. Alternatively the linker region can correspond to an epitope tag such as the HA or the Flag sequence.
4. The TNT system can be used with either T3 or SP6 promoter. Other kits are based on the TNT T7 insect cell extract or SP6 wheat germ expression system. Proteins synthesized with the rabbit reticulocyte lysate are not kept longer than 4 weeks at −20 °C before experimentation in vitro.
5. A minimum size of 18 cm × 18 cm is recommended for migration of multimeric protein complexes on DNA.
6. Gel polymerization induces a retraction: Keep aside some acrylamide solution in ice to complete and recover the comb if necessary. The gel can be kept at 4 °C overnight before using it for EMSA the day after.

7. Composition of the binding buffer varies depending on the transcription factor family and especially on the type of DNA-binding domain. Here the binding buffer is suitable for homeodomain-containing transcription factors but maybe not for other types of DNA-binding domains.
8. DNA probes can also be labeled by ^{32}P γATP with a T4 polynucleotide kinase.
9. Use MicroSpin G-25 columns for probes smaller than 20 nucleotides.
10. The *Gal4* driver should reproduce as closely as possible the expression pattern of at least one of the two candidate interacting partners. To this aim, a collection of *P-Gal4* lines is available in fly stock centers. Alternatively, *P-Gal4* lines can be generated by swapping experiments with other types of *P-element* [12]. Heterologous *Gal4* drivers are also required to indirectly deduce the expression level resulting from the activity of *P-Gal4* drivers with regard to endogenous proteins (Fig. 2).
11. Balancing the *P-Gal4* driver over a *tub-Gal80* chromosome allows visualizing BiFC only in embryos homozygous for the *P-Gal4* insertion. In case the *P-Gal4* insertion corresponds to a null allele, BiFC is therefore observed in the absence of endogenous gene products.
12. Chicken anti-GFP recognizes full-length GFP and its derivatives, as well as the VN and VC fragments. These recognition properties are not systematic, especially for monoclonal GFP antibodies, which usually recognize the VC and not the VN fragment.
13. If possible, different DNA-binding sites should be tested in EMSAs to select most neutral fusion topologies.
14. For non-DNA-binding proteins preliminary tests for assessing the influence of fusion topologies can rely on co-immunoprecipitation (co-ip) experiments from transfected cells. In this assay, UAS-VC and UAS-VN plasmids are co-transfected with a Gal-4-encoding plasmid. Co-ip experiments can be performed with a specific antibody against the protein or with antibodies recognizing an epitope tag (ideally in between the Venus fragment and the protein, working as a linker region). Along the same line, an alternative preliminary test can consist on BiFC analysis in cell culture. In any case, care should be taken regarding the expression levels, and weak Gal-4-expressing plasmids are recommended to avoid artificial and saturated protein-protein interactions in cells.
15. Before performing EMSA, the synthesis rate of each protein construct is estimated by doing the TNT reaction with ^{35}S-methionine. The amount of protein produced generally

depends on the expression vector but can also differ a bit between constructs based on a same expression vector.

16. We observed that random insertions by classical P-element transgenesis usually led to stronger expression levels than several landing sites we used with the ϕC31 integrase-mediated system.
17. When possible, functional assays are also required to verify that the fusion topology does not affect protein activity. This could be performed by rescue experiments of mutant phenotypes or by assessing known regulatory functions [8].
18. Embryo collection, fixation, and staining should be performed in parallel for all conditions.
19. Alternatively, if no antibody is available for your protein of interest, qRT-PCR could be envisaged for comparing mRNA expression levels in the absence or the presence of the fusion transcript.
20. The number and intensity of all pixels can be measured using the ImageJ software.
21. We observed that 15-day-old flies start to lay a non-negligible proportion of non-fertilized eggs. This proportion increases with the age. Along the same line, keeping virgin females more than 2 weeks before starting crosses with males will reduce the proportion of good-quality embryos for BiFC observations.
22. Discard larvae from the filter before mounting embryos in between slide and cover slide.
23. A minimum time of 6 h (for chromophore maturation) is required to start seeing fluorescent signals in the *Drosophila* embryo [8]. Depending on the strength of the interaction between the two fusion proteins, this time delay can be sufficient or not for quantifying BiFC signals. We noticed that a time maturation delay of 28 h at 4 °C has little incidence on embryo viability [8]. In any case, comparison between different BiFC signals should be performed with identical maturation times.
24. Slides can be kept for few hours at 4 °C before observation under the microscope.
25. In the case of multicolor BiFC, excitation and emission wavelengths of Venus and Cerulean are 515 nm/528 nm and 433 nm/475 nm, respectively.
26. Specificity of BiFC signals should be verified by using proteins mutated in motifs known to alter the interaction with the candidate partner. Alternatively, specificity can also be validated by co-expressing a non-fused (cold) partner: this protein will titrate the BiFC complex resulting from the association between the two fusion partners, leading to a diminution of fluorescent signals [8].

Acknowledgement

This work was supported by the ARC (Association pour la Recherche sur le Cancer), and FRM (Fondation pour la Recherche Médicale).

References

1. Kerppola TK (2008) Bimolecular fluorescence complementation (BiFC) analysis as a probe of protein interactions in living cells. Annu Rev Biophys 37:465–487
2. Kodama Y, Hu CD (2012) Bimolecular fluorescence complementation (BiFC): a 5-year update and future perspectives. Biotechniques 53:285–298
3. Hu CD, Chinenov Y, Kerppola TK (2002) Visualization of interactions among bZIP and Rel family proteins in living cells using bimolecular fluorescence complementation. Mol Cell 9:789–798
4. Shaner NC, Lambert GG, Chammas A, Ni Y, Cranfill PJ et al (2013) A bright monomeric green fluorescent protein derived from Branchiostoma lanceolatum. Nat Methods 10:407–409
5. Zhou J, Lin J, Zhou C, Deng X, Xia B (2011) An improved bimolecular fluorescence complementation tool based on superfolder green fluorescent protein. Acta Biochim Biophys Sin (Shanghai) 43:239–244
6. Kim J, Zhao T, Petralia RS, Yu Y, Peng H et al (2012) mGRASP enables mapping mammalian synaptic connectivity with light microscopy. Nat Methods 9:96–102
7. Shyu YJ, Liu H, Deng X, Hu CD (2006) Identification of new fluorescent protein fragments for bimolecular fluorescence complementation analysis under physiological conditions. Biotechniques 40:61–66
8. Hudry B, Viala S, Graba Y, Merabet S (2011) Visualization of protein interactions in living Drosophila embryos by the bimolecular fluorescence complementation assay. BMC Biol 9:5
9. Bischof J, Maeda RK, Hediger M, Karch F, Basler K (2007) An optimized transgenesis system for Drosophila using germ-line-specific phiC31 integrases. Proc Natl Acad Sci U S A 104:3312–3317
10. Bischof J, Bjorklund M, Furger E, Schertel C, Taipale J et al (2013) A versatile platform for creating a comprehensive UAS-ORFeome library in Drosophila. Development 140: 2434–2442
11. Gohl C, Banovic D, Grevelhorster A, Bogdan S (2010) WAVE forms hetero- and homo-oligomeric complexes at integrin junctions in Drosophila visualized by bimolecular fluorescence complementation. J Biol Chem 285: 40171–40179
12. de Navas L, Foronda D, Suzanne M, Sanchez-Herrero E (2006) A simple and efficient method to identify replacements of P-lacZ by P-Gal4 lines allows obtaining Gal4 insertions in the bithorax complex of Drosophila. Mech Dev 123:860–867
13. Gebelein B, Culi J, Ryoo HD, Zhang W, Mann RS (2002) Specificity of Distalless repression and limb primordia development by abdominal Hox proteins. Dev Cell 3: 487–498

Chapter 20

Hox Protein Interactions: Screening and Network Building

Isabelle Bergiers, Barbara Lambert, Sarah Daakour, Jean-Claude Twizere, and René Rezsohazy

Abstract

Understanding the mode of action of Hox proteins requires the identification of molecular and cellular pathways they take part in. This includes to characterize the networks of protein-protein interactions involving Hox proteins. In this chapter we propose a strategy and methods to map Hox interaction networks, from yeast two-hybrid and high-throughput yeast two-hybrid interaction screening to bioinformatic analyses based on the software platform Cytoscape.

Key words Hox proteins, Yeast two-hybrid screening, High-throughput yeast two-hybrid, Affinity co-purification, Interaction networks, Cytoscape

1 Introduction

Hox proteins are well-known transcription factors involved in the control of key developmental processes in bilaterian embryos. Hox genes are active while the main body axis elongates and is patterned concomitantly with and further to gastrulation [1–3]. Hox genes also contribute to pattern the proximo-distal and then anteroposterior axes of developing limbs [4]. They also participate to the organogenesis of the mammary gland [5, 6], kidney [7, 8], digestive and respiratory tracts [9–12], or cardiovascular system [13]. In postembryonic development, up to adulthood, some Hox genes are also implicated in the establishment of neural circuits [14–17], some control cell stemness or differentiation [18], like in hematopoiesis [19, 20], and consequently, Hox genes have also been highlighted as oncogenes or oncosuppressors [21]. A handful of reports finally also established Hox genes as being active very early during embryogenesis, Hox mRNAs and proteins having been detected in mammalian oocytes, zygotes, prior to the main zygotic genome activation [22].

Hox proteins therefore appear to be involved in a vast repertoire of biological processes and are supposed to be integrated in a

Yacine Graba and René Rezsohazy (eds.), *Hox Genes: Methods and Protocols*, Methods in Molecular Biology, vol. 1196, DOI 10.1007/978-1-4939-1242-1_20, © Springer Science+Business Media New York 2014

multitude of molecular pathways in distinct cell types and lineages. Characterizing these pathways is crucial to understand the mode of action of Hox proteins. Since a decade, several screens to identify extensive gene sets under the control of Hox proteins have been performed, mainly in the Drosophila and mouse model organisms, or in human in the context of cell differentiation and oncogenic processes [23–29]. Understanding the function and modes of action of Hox proteins also necessitates to identify their protein interaction partners [30, 31]. It has now become clear that transcription factors never act alone, but in the context of multimolecular complexes governing the chromatin conformation, promoter and enhancer accessibility, assembly of transcription initiation complexes, polymerase recruitment, poising, and activation. Accumulating data increasingly support that Hox proteins are also involved in non-transcriptional molecular processes [32]. It is therefore of primary interest to gain a large if not exhaustive view of the interactions Hox proteins establish.

Two major approaches have been widely used to identify protein-protein interactions. The first approach relies on affinity co-purification assays that allow identifying multimolecular complexes. The second approach aims at identifying binary interactions and relies on two-hybrid assays. In such assays candidate interactors are fused to proteins or protein domains which once put together or reassembled through the partner interaction result in a detectable readout, like cell growth, fluorescence emission, or reporter gene activation. Several variants of such assays have been developed since the late 1980s, including the probably most popular yeast two-hybrid (Y2H) assay involving the DNA-binding domain (DB) and activation domain (AD) of the Gal4 transcription factor which are separately fused to the interactors. Upon their binding, the interactors bring Gal4AD and Gal4DB together. The "reconstituted" Gal4 transcription factor enables target genes to be activated which for example results in yeast auxotrophy relief and growth [33, 34]. Similar two-hybrid assays have been developed in bacteria or animal cells as well, involving distinct split and fusion proteins, like the split-ubiquitin assay or MAPPIT, a cytokine receptor-based two-hybrid method [35, 36].

While candidate interactors have been identified, additional and distinct interaction tests are required to independently validate the screenings outcome. In fact, while screening methods are prone to miss large sets of interactions, in particular when starting from assays in heterologous systems (like screening for animal protein interactions by Y2H), an important issue is to identify and discard from the datasets possible false-positive interactions. So, typically, further to Y2H screens, proteins identified as possible interactors will be assayed in an autologous context (mouse cells, insect cells, embryos, ...) by affinity co-purification for example, or the more recently developed bimolecular fluorescence complementation assay [37, 38].

Once interactions have been confirmed by one or several techniques, interaction networks can be built in order to highlight possible molecular pathways and cellular processes the proteins under study are involved in [39, 40].

We recently performed interaction screens for two mouse Hox proteins [41, 42]. In the context of this chapter we describe the methods referring to Y2H screens, validation by affinity co-purification, and network building we used.

2 Materials

2.1 Interaction Screening by Yeast Two-Hybrid (Y2H) Assay and High-Throughput Yeast Two-Hybrid Assay (HT-Y2H)

2.1.1 Yeast Strains

1. MaV103(*MAT*a) and MaV203 (*MATα*) genotype: *leu2-3, 112*; *trp1-901*; *his3-200*; *ura3-52*; *ade2-101*; *gal4Δ*; *gal80Δ*; *LYS2::GAL1-HIS3*; *URA3::SPAL10-URA3*; *GAL1-lacZ*; *cyh2*^{R}; *can*^{R}.

   The *Saccharomyces cerevisiae* strains MaV103 (*MAT*a) and MaV203 (*MATα*) are deleted for the *GAL4* and *GAL80* genes encoding Gal4p and its repressor Gal80p. These strains contain three Gal4p-inducible reporter genes: *HIS3*, which confers growth on media lacking histidine; *LacZ*, for colorimetric detection of Gal4p activity; and *URA3* that allows growth on media lacking uracil. These strains are resistant to canavanine and cycloheximide (CYH), and are sensitive to the drug 5-fluoro-orotic acid (5-FOA) [34, 43, 44].

2. Y8930 (*MAT*α) and Y8800 (*MAT*a) genotype: *leu2-3, 112*; *trp1-901*; *his3-200*; *ura3-52*; *ade2-101*; *gal4Δ*; *gal80Δ*; *GAL2-ADE2*; *LYS2::GAL1-HIS3*; *met2::GAL7-lacZ*; *cyh2*^{R})

   Strains Y8800 (MATa) and Y8930 (MATα) also contain deletions of the *GAL4* and *GAL80* genes and three Gal4p-inducible reporter genes, *HIS3*, *LacZ*, and *ADE2*. These strains were generated by adding the cycloheximide resistance marker to the PJ69-4 strains [45].

   MaV and Y strains therefore differ in one over three reporters: MaV strains use *URA3* while Y strains use *ADE2* reporters. In addition, Y strains need less 3-amino-1,2,4 triazole (3AT) (1 mM) than MaV strains (20 mM). 3AT is a competitive inhibitor of the *HIS3* reporter, allowing background reduction on selection media plates (*see* **Note 1**).

2.1.2 Expression Vectors

1. pGBT9 plasmid, GenBank Accession #: U07646; https://www.addgene.org/vector database/2828/, Clontech.
2. pGAD424 plasmid, GenBank Accession #:U07646; http://www.addgene.org/vector-database/2819/; Clontech.
3. ORFeome v8.1 library of vectors; http://horfdb.dfci.harvard.edu/.
4. Gateway® pDONR™221 or pDONR™223 vectors (Life Technologies).

Table 1
HT-Y2H scoring controls (adapted from [34])

	pDEST-AD	pDEST-DB	Score
Control 1	Empty	Empty	Background
Control 2	Human E2F1 aa 342–437	CYH2-human pRB aa 302–928	Weak (negative on CYH control plates)
Control 3	Mouse Jun aa 250–325	Rat Fos aa 132–211	Strong
Control 4	Empty	Yeast Gal 4	Very strong
Control 5	Drosophila E2F aa 225–433	Drosophila DP aa 1–377	Strong
Control 6	CYH2-Drosophila E2F aa 225–433	Drosophila DP aa 1–377	Strong (negative on CYH control plates)

5. Expression plasmids for HT-Y2H are Gateway® (or any other high-throughput compatible) vectors (*see* **Note 2**). For HT-Y2H, we use pDEST-DB and pDEST-AD to fuse the DNA-binding (DB) or activating (AD) domains of the yeast transcriptional factor Gal4 N-terminally to the open reading frames (ORFs) of interest, respectively [34].
   Entering inserts into Gateway® pDONR™ vectors requires BP clonase® and transferring inserts into pDEST™ vectors requires LR clonase® enzyme (Life Technologies).
6. pDEST-DB and pDEST-AD vectors for ORFs used as controls for Y2H (*see* Table 1).
7. Quant-iT™ Picogreen® dsDNA quantitation assay kit (Life Technologies).

2.1.3 Media and Reagents

1. YPD medium: 10 g Yeast extract, 20 g peptone, 20 g glucose, 1 l H_2O (for plates of solid medium: 20 g agar).
2. Yeast transformation buffer: 0.2 M lithium acetate, PEG 40 %, DTT 100 mM, pH 5.0.
3. Salmon sperm DNA, 10 mg/ml.
4. Synthetic defined (SD) minimal medium plates: 7 g yeast nitrogen base without amino acids (Sigma-Aldrich), 20 g glucose, 20 g agar, 1 L H_2O.
5. Synthetic complete medium (Sc): According to the yeast strain auxotrophies, supplement SD medium by adding 40 mg adenine (Ade), 20 mg uracil (Ura), 20 mg histidine (His), 60 mg leucine (Leu), 30 mg lysine (Lys), 20 mg methionine (Met), 40 mg tryptophan (Trp).
   For pGAD424 vector selection, use Sc medium without Leu (Sc-Leu).

For pGBT9 vector selection use Sc medium without Trp (Sc-Trp).

For pDEST-DB vector selection, use Sc medium without Leu (Sc-Leu).

For pDEST-AD vector selection, use Sc medium without Trp (Sc-Trp).

For Y2H reporters selection prepare Sc without His (Sc-His) to select for His synthesis, Sc without Ade (Sc-Ade) to select for Ade synthesis, and Sc without His and Ade (Sc-His-Ade) for both His and Ade synthesis.

6. Cycloheximide 1 mg/ml.
7. 3-amino-1,2,4 triazole (3AT) powder.

   For high-throughput experiments plates need to be compatible with robotic handling and the 96-head (*see* **Note 3**).
8. Zymolase 20 T: Working concentration (made fresh each time): 2.5 mg/ml and dissolved in 0.1 M sodium phosphate buffer (pH 7.4).
9. E-Gel 96 agarose, 1 % (Life Technologies) for high-throughput DNA electrophoresis. This system allows electrophoresis of 96 samples at a time.

2.1.4 Primers

1. Oligonucleotide primers to amplify the ORF coding for your proteins of interest.
2. Oligonucleotides to amplify ORFs of interest and designed to enter the Gateway® pDONR™ plasmids:

   5′-GGGGACAACTTTGTACAAAAAAGTTGGCATGnnnnnnn-3′

   5′-GGGGACAACTTTGTACAAGAAAGTTGAnnnnnnn-3′

where 5′-GGGGACAACTTTGTACAAAAAAGTTGGCATG-3′ corresponds to the AttB1 recombination site; 5′-GGGGACAACTTTGTACAAGAAAGTTGA-3′ corresponds to the AttB2 recombination site; and 5′-nnnnnnn-3′ corresponds to sequences specific to the ORF to amplify.

3. Oligonucleotides to verify pDEST-AD-ORF and pDEST-DB-ORF vectors:

   AD: 5′-CGCGTTTGGAATCACTACAGGG-3′

   DB: 5′-GGCTTCAGTGGAGACTGATATGCCTC-3′

   Term: 5′-GGAGACTTGACCAAACCTCTGGCG-3′

2.1.5 β-Gal Filter Assay

1. Z-buffer: To make a liter, add 16.1 g $Na_2HPO_4{\cdot}7H_2O$, 5.5 g $NaH_2PO_4{\cdot}H_2O$, 0.75 g KCl, and 0.246 g $MgSO_4{\cdot}7H_2O$; the Z buffer is autoclaved to sterilize.
2. 4 % X-gal: Dissolve 40 mg in 1 ml of *N,N*-dimethylformamide. Store at –20 °C wrapped with foil.

3. ß-Gal dyeing solution (made fresh each time): For each plate, prepare 5 ml of Z buffer with 120 μl of 4 % X-Gal, 13 μl of 2-mercaptoethanol.
4. Liquid nitrogen.
5. Nitrocellulose membranes.
6. 3MM Whatman filter paper.

2.1.6 Robotic Instruments

For high-throughput liquid handling, the following equipment can be used.

1. A 384/96-head + 2–8 channels hit picking arm (for example a Tecan EVO 100, 150, or 200, *see* **Note 4**).
2. A multidrop for reagent dispensing (glycerol, liquid media, …) (for example a Thermo Multidrop combi).
3. Large-capacity incubators and shakers for yeast cell culture.
4. A bar code reader for data transfer to computers.
5. Archival storage capacity at −80 °C, −20 °C, and +4 °C.

2.2 Interaction Validation by Affinity Co-purification

1. Expression vectors for Flag-tagged and glutathione-S-transferase (GST) fusion proteins. pDEST™ vectors are available to express Flag or GST fusion proteins and can be generated from pDONR™ plasmids by the Gateway® system (Life Technologies).
2. COS7 (ATCC® CRL-1651™) or HEK293T (ATCC® CRL-11268™) cells.
3. Dulbecco's modified Eagle's medium (DMEM), low glucose or high glucose, supplemented with L-Glutamine (Gibco/Invitrogen).
4. Fetal bovine serum (Gibco/Invitrogen).
5. 25 or 75 cm² cell culture flasks.
6. Six-well cell culture plates.
7. Penicillin 10,000 IU/ml and 10,000 μg/ml streptomycin mix (Gibco/Invitrogen).
8. TransFectin™ (BioRad).
9. Cell lysis buffer: Tris–HCl pH 7.5 20 mM, NaCl 120 mM, EDTA 0.5 mM, NP40 0.5 %, glycerol 10 %.
10. Phosphate buffer saline (PBS) 10×.
11. Complete™ protease inhibitor (Roche).
12. Cell scrapers.
13. Laemmli loading buffer for SDS-PAGE: 10 % SDS, 30 % glycerol, 350 mM Tris-Cl pH 6.8, 600 mM DTT, 0.1 % bromophenol blue.

14. Nitrocellulose membranes Hybond™C (Amersham).
15. Glutathione-agarose beads (Sigma), preconditioned in H_2O or PBS according to the manufacturer's instructions (*see* **Note 5**).
16. SDS-PAGE and electrophoresis device and Western blot transfer device, Mini Protean Tetra Cell (BioRad).
17. Anti-Flag and anti-GST antibodies.
18. Cell culture CO_2 incubator.

3 Methods

3.1 Interaction Screening by Yeast Two-Hybrid Assay (Y2H)

Y2H is a regular and well-established interaction screening method allowing binary interaction detection and thereby interactome characterization. In regular molecular biology laboratory settings, it is possible to either screen interactions one by one or screen libraries of Gal4DB and Gal4AD fusion proteins representing entire proteomes. Alternatively, systematic and large-scale pairwise interaction screenings are also feasible but require specific equipment and robotics to perform most of the screening steps in an automated way (*see* Subheading 3.2). The method described here below corresponds to a standard Y2H interaction assay.

3.1.1 Plasmid Design and Construction

The proteins you want to assay for possible interactions should be expressed as fusion proteins with the Gal4AD and Gal4DB. For regular binary interaction tests we used the pGBT9 (Trp+) plasmid for Gal4DB fusions and pGAD424 (Leu+) for Gal4AD fusions. PCR amplify the open reading frame (ORF) corresponding to the Hox protein and to the candidate interactors to assay with primers containing restriction sites to clone the PCR products in both pGBT9 and pGAD424 plasmids (*see* **Note 6**).

3.1.2 Two-Hybrid Yeast Auxotrophy Complementation

1. *Yeast transformation by the one-step lithium acetate method.* Spin down 1 ml of fresh cells from a stationary-phase culture into liquid YPD medium by a quickspin in a table microcentrifuge (*see* **Note 7**). Discard the supernatant and add 1 ml of H_2O. Mix well and spin down (quickspin in a table microcentrifuge). Dry well. To the pellet, add 100 μl of lithium acetate transformation buffer, 5 μl of salmon sperm carrier DNA, and 5 μl (typically about 0.1–1 μg) expression plasmid. Incubate for 30 min at 45 °C, and then chill for 5 min on ice. Plate the suspension onto Sc-Leu plates for pGAD424 vector transformation or Sc-Trp plates for pGBT9 vector transformation. Incubate overnight at 28 °C.
2. *Yeast amplification.* Transformed yeast must grow for a couple of days at 28 °C onto Sc-Trp for yeast transformed to express DB fusion proteins (pGBT9) and onto Sc-Leu for yeast transformed to express AD fusions (pGAD424).

3. *Yeast mating*. Resuspend a yeast-DB or -AD "clot" into liquid Sc medium, –Trp or –Leu, respectively. Mix 5 μl of each yeast strain and drop off onto Sc-Trp-Leu plate making streaks. Incubate at 28 °C for 24 h.
4. Pick a colony and replicate onto Sc-His-Trp-Leu, Sc-Ade-Trp-Leu, and Sc-His-Ade-Trp-Leu plates to select for two-hybrid complementation. Incubate at 28 °C (*see* **Note 8**).
5. *Reading results*. Positive controls make you sure that your media support yeast growth once auxotrophy complementation is sustained by the interaction between AD and DB fusion proteins. The negative controls ensure that yeast growth is not possible unless both Gal4DB and Gal4AD fusions are put together. In particular, the control with the DB fusion only (AD empty vector) is critical. This allows verifying that the ORF fused to Gal4DB has no intrinsic ability to stimulate transcription (what is referred to as an autoactivator). Yeast growing only onto Sc-His-Trp-Leu medium is suggestive of a weak interaction between the hybrid proteins, while yeast growth onto both Sc-His-Trp-Leu and Sc-Ade-Trp-Leu media is indicative of a strong interaction.

3.2 Interaction Screening by High-Throughput Yeast Two-Hybrid Assay (HT-Y2H)

The sequencing of model organisms and the human genomes and the emergence of novel high-throughput cloning strategies have allowed large-scale mapping of protein-protein interactions for systematic analysis of biological processes. One of the novel resources emerging from the human genome sequence is the human ORFeome (ORFeome v8.1, *see* http://horfdb.dfci.harvard.edu/), i.e., the complete set of protein-encoding genes, cloned as full-length ORFs [46–49]. Traditional cloning strategies using restriction endonucleases and ligase enzymes do not allow automation and transfer of several genes in different expression vectors. The human ORFeome collection is thus cloned in a "Gateway®" format allowing site-specific recombinational robotic transfer of, virtually, any ORF in any expressing vector. For HT-Y2H, the human ORFeome collection was N-terminally transferred by Gateway® into pDEST-AD and pDEST-DB encoding the Gal4 activating and DNA-binding domains, respectively. The resulting individual clones were transferred in yeast strains by large-scale transformation assay as previously described [34, 50] (*see* below for a detailed protocol).

3.2.1 Preparation of Baits and Prey Clones

1. The first step consists of PCR amplification of the ORFs of interest using specific primers flanked with the following AttB1 and AttB2 Gateway® sites: 5′-GGGGACAACTTTGTACAAAAAAGTTGGCATG-3′ and 5′-GGGGACAACTTTGTACAAGAAAGTTGA-3′.
2. PCR fragments are verified by agarose electrophoresis.

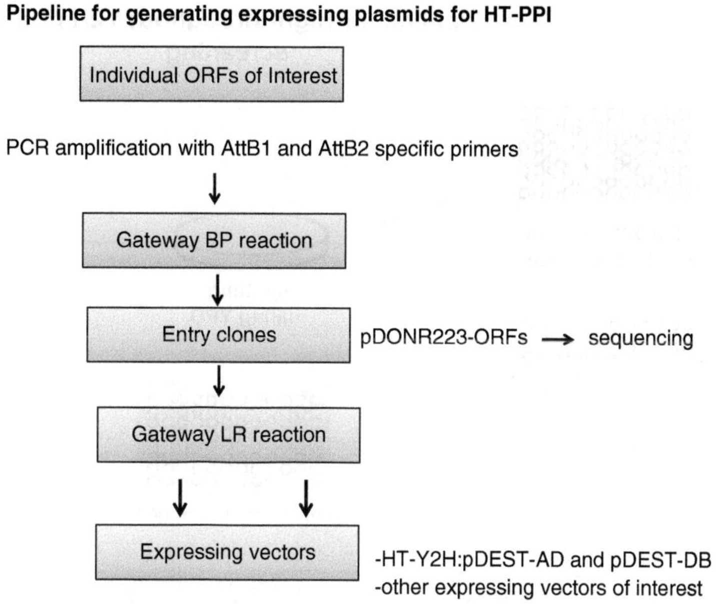

Fig. 1 Pipeline for generating expressing plasmids for high-throughput protein-protein interactions: ORFs of interest are amplified by PCR using AttB1 and AttB2 primers, transferred into pDONR223 to generate entry clones, which are sequence-verified before subcloning into an expression vector of interest

3. The amplified ORFs are then transferred by Gateway® BP clonase® recombination into a pDONR vector. We routinely use pDONR223 or pDONR221, which confer spectinomycin or kanamycin resistance, respectively (*see* **Note 9**).
4. Expressing vectors are then generated by Gateway® LR clonase® recombination into pDEST-AD and pDEST-DB vectors for HT-Y2H or any other Gateway®-compatible destination vector of interest (Fig. 1).
5. Y2H-expressing plasmid DNA are quantified using the fluorescent marker Picogreen® and quality controlled by PCR using the following primers: AD (forward primer specific to the AD vector), DB (forward primer specific to the DB vector), and Term (reverse primer, for both AD and DB vectors).
6. Y2H AD- and DB-expressing vectors are introduced by high-throughput yeast transformation in different strains: Y8800 or MaV103 (*MAT*a) and Y8930 or MaV203 (*MAT*α), respectively.
7. AD-ORFs and DB-ORF yeast-expressing colonies are selected on solid media lacking tryptophan (Sc-Trp for AD-ORFs) or leucine (Sc-Leu for DB-ORFs).

3.2.2 Removal of Autoactivating DB-ORFs

To remove autoactivator DB-ORFs responsible for the activation of a reporter gene in the absence of an interaction with an AD-ORF partner from the DB-ORF array sets, each yeast clone Y8930

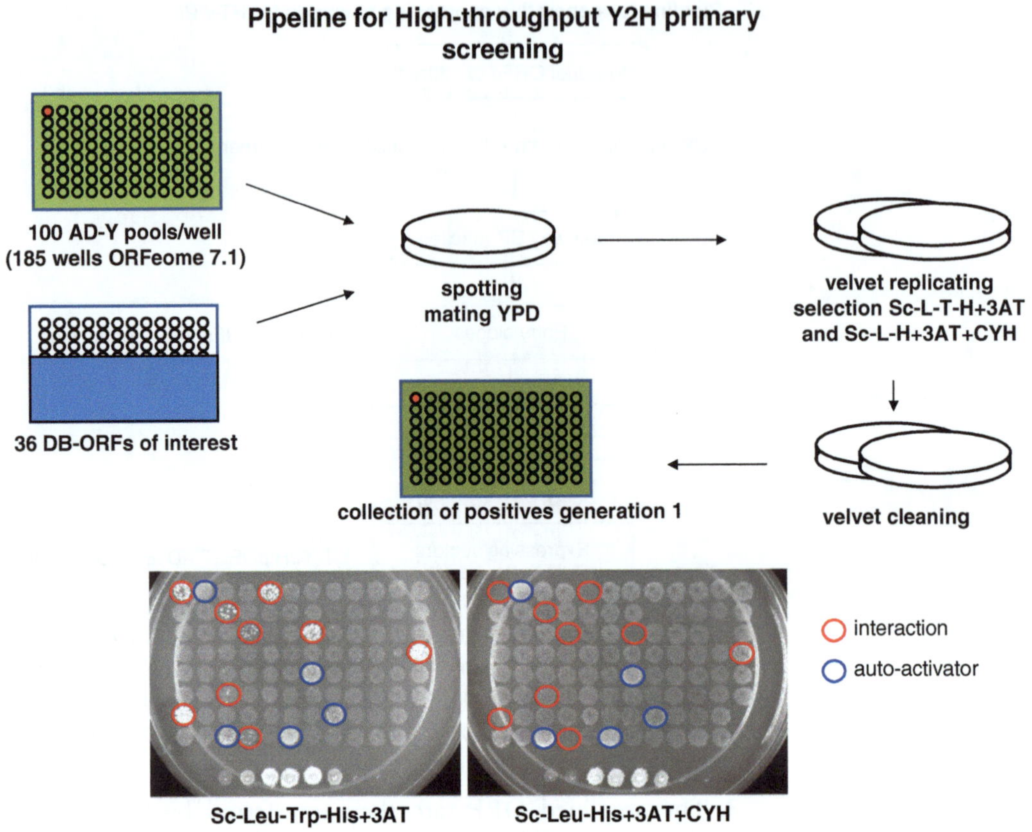

Fig. 2 Primary screening pipeline: AD-Y and DB-X containing haploid yeasts are mated onto YPD media; diploid cells are then replicated on selection and control media plates. Positives are present on the selection plate and absent in the control plate containing cycloheximide (CYH)

(or MaV203) transformed with a DB-ORF is mated with the Y8800 (or MaV103) yeast strain transformed with an empty AD vector. Diploid cells showing activation of the *HIS3* reporter (growth on the selection plate Sc-Leu-Trp-His + 1 mM of 3AT for the Y strains or +20 mM of 3AT for the MaV strains) are considered autoactivators and removed before the screening steps by re-arraying DB-ORFs expressing yeasts from glycerol stocks using a robotic multichannel arm.

3.2.3 HT-Y2H Screening

To ensure high quality and reliable protein-protein interactions data sets, the high-throughput Y2H screening consists of three important steps outlined in Figs. 2, 3, and 4.

1. *Primary screening* (Fig. 2). In this step, distinct yeast clones (*MATα*), containing individual DB-ORFs of interest arrayed in a 96-well plate, are mated to minipools of yeast containing 100 AD-human ORFs (*MAT*a) from the ORFeome collection on YPD plate media. Alternatively, each DB-human ORF from the ORFeome collection is mated to a pool of AD-ORFs

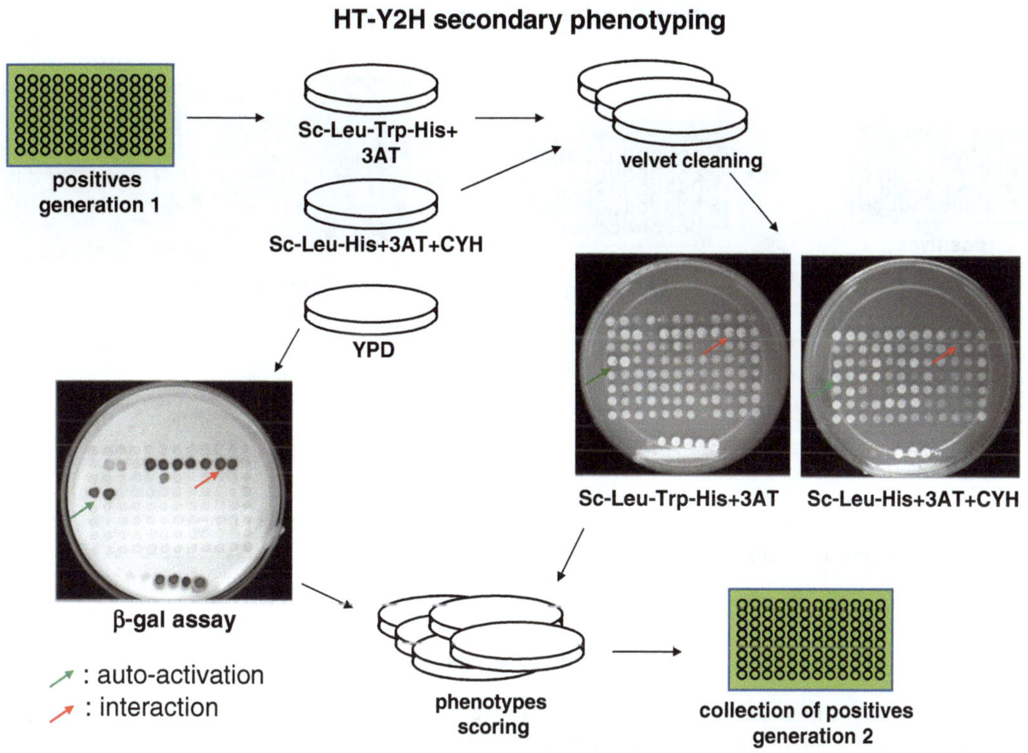

Fig. 3 HT-Y2H secondary phenotyping: Primary screening positives are grown in Sc-His liquid media and spotted on selection, control, and YPD-rich media plates. Selection and control plates are then "cleaned" by placing them on a piece of velvet stretched over a replica-plating block, to remove excess yeast. Plates are then left at 30 °C for at least 3 days and positive colonies are collected. YPD plates serve for β-galactosidase activity assay by scoring blue colonies on the nitrocellulose filter (see text for details)

of interest. After mating yeast are replicated to Sc-Leu Trp His selection plates and Sc-Leu-His + 3AT + 1 mg/ml of cycloheximide (CYH, *see* **Note 10**). Three to five days post-selection, primary positive diploid cells are picked and used for secondary phenotyping.

2. *Secondary phenotyping* (Fig. 3). Primary positives are grown in liquid selecting media (Sc-Leu-Trp-His) to saturation (at least 24 h) and spotted on selection and control plates: Sc-Leu-Trp-His + 1 mM 3AT; Sc-Leu-His + 1 mM 3AT + 1 mg/ml CYH; Sc-Leu-Trp-Ade + 1 mM 3AT; and Sc-Leu-Ade + 1 mM 3AT + 1 mg/ml CYH for the Y strains. For MaV strains, we use 20 mM 3AT and *GAL1::lacZ* as a second phenotyping marker, to determine the transcriptional activity using β-galactosidase filter assay. For this assay, a nitrocellulose filter is placed onto a YPD yeast plate. After 12 h of growth, attached yeast cells are lysed by placing the filter in liquid nitrogen for 30 s. The filter is then placed on a ß-gal dyeing solution-soaked 3MM Whatman paper at 37 °C. Positive colonies turn blue, suggesting a β-galactosidase activity [48].

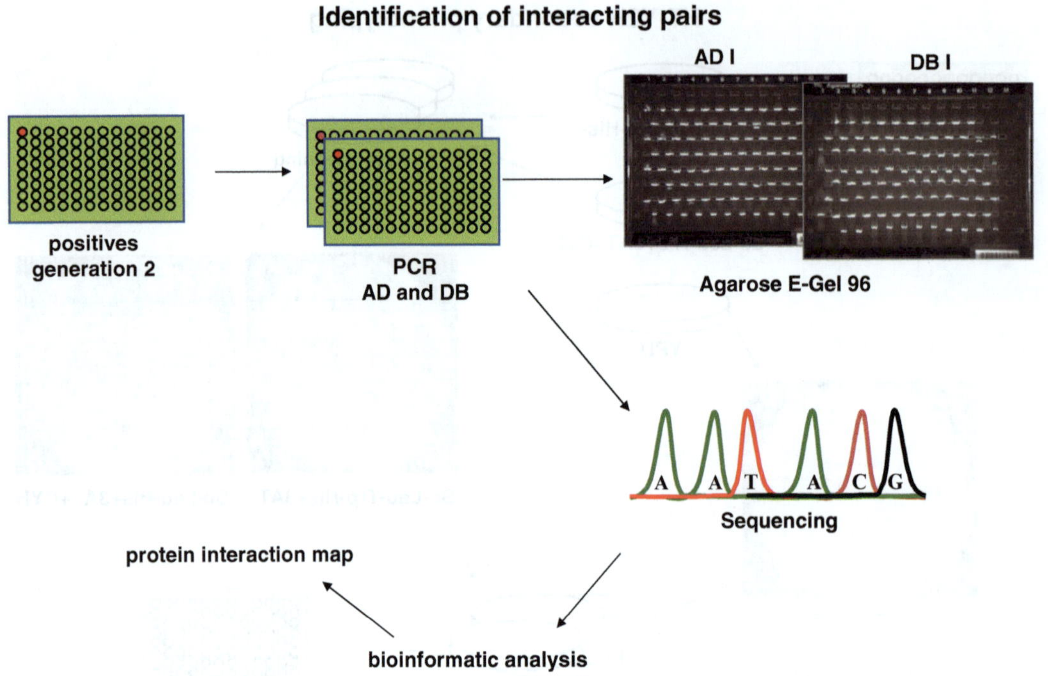

Fig. 4 Positive clones from secondary phenotyping are grown on YPD media plates, lysed, and used for PCR reactions using AD and Term or DB and Term primers. PCR fragments are then sequenced and ORFs identified by bioinformatic sequence analysis to build a protein–protein interaction map

3. *Scoring Y2H readouts.* To determine candidate positive clones we systematically add to selection and control plates a collection of six diploid control strains described in Table 1 (adapted from [34]). We usually score positive colonies showing a reporter transcriptional activity level equal or higher to that of control 2 (pDEST-AD-E2F1 + pDEST-DB-CYH2-pRB).
4. *Identification of interacting pairs* (Fig. 4): Positive yeast clones are grown overnight on YPD plates. A small amount of yeast cells is lysed using zymolase, an enzyme preparation from *Arthrobacter luteus*, which effectively lyses live yeast walls [34]. For each positive diploid, two PCRs are then performed to individually amplify DB-ORF (primers DB and Term) and AD-ORF (primers AD and Term). PCR products are then verified on E-Gel 96 agarose 1 %, and used as template in sequencing reactions to obtain identities of interacting pairs.
5. *Retesting interacting pairs.* Candidate Y2H pairs should be verified by fourfold independent experiments. To this end, the location of each clone is identified in the ORFeome database and corresponding fresh archival yeast stocks cultured in Sc-Leu (DB-ORFs) or Sc-Trp (AD-ORFs) liquid media, mated on YPD media plates and transferred fourfold, on two phenotyping selection (His3 and Ade2; or His3 and Gal1::lacZ) and

control media (+CYH). High-quality interactions usually score positive on at least three out of four independent plates for each phenotyping.

3.3 Interaction Validation by Affinity Co-purification

To validate interaction we use an affinity co-purification of a Flag-tagged and glutathione S-transferase (GST)-tagged proteins co-expressed in transfected mammalian cells.

3.3.1 Plasmid Design and Construction

PCR amplify the ORFs corresponding to the Hox protein and to the candidate interactors to assay with primers containing restriction sites to clone the PCR products in an expression vector with the Flag-coding sequence or the GST-coding sequence (*see* **Note 11**). Empty vectors will also be used as controls; in particular the vector expressing the unfused GST protein is of primary importance to control the specificity of the affinity co-purification (*see* Subheading 3.3.3, **steps 3** and **4**).

3.3.2 Cell Transfection

1. *Culture cells*: COS7 and HEK293T cells are maintained in DMEM low glucose or high glucose, respectively, supplemented with glutamine, 10 % fetal bovine serum, 100 IU/ml penicillin, and 100 μg/ml streptomycin. Cell lines are maintained in 25 or 75 cm^2 cell culture flasks at 37 °C in a humidified, 5 % CO_2 atmosphere.
2. *Transfection*: For transient transfection, 1.4×10^5 (COS7) or 4×10^5 (HEK293T) cells are plated into six-well plates. Twenty-four hours after plating, cells are transfected with TransFectin™ reagent. 1.5 μg of Flag-tagged protein expression vector and 3 μg of GST fusion protein expression vector were mixed with 250 μl of serum-free medium and added to a mix of 1 μl of TransFectin™ and 250 μl of serum-free medium. A control condition for the affinity co-purification consists in co-expressing the Flag-tagged protein with unfused GST protein.

3.3.3 Affinity Co-purification

1. *Cells lysis*: Forty-eight hours after transfection, cells are washed with PBS 1× and lysed for 30 min on ice by gentle shaking in cell lysis buffer containing Complete™ protease inhibitor. Scrap and pipet up and down five to ten times. Centrifuge for 5 min at 13,000 × *g* (in a table microcentrifuge) to clear the lysates. Remove supernatant carefully and keep on ice or store at −80 °C.
2. *Controlling protein expression*: Prepare SDS-PAGE gels, run 10 μl of cell lysate supernatant, and proceed with Western blotting to determine the expression level of Flag and GST fusion proteins (*see* **Note 12**).
3. *Affinity co-purification of proteins on glutathione-agarose beads*: Take 40 μl of glutathione-agarose beads per interaction to be tested and transfer them to a 1.5 ml Eppendorf

tube (*see* **Notes 13** and **14**). Beads must be collected by 30-s centrifugation at 600×*g* (table microcentrifuge). Gently resuspend beads with 500 μl of ice-cold lysis buffer with protease inhibitors, leave on ice for 1–2 min to let the beads settle down at the bottom of the tubes, and carefully remove the supernatant without touching the beads. Resuspend and wash the beads again twice with 500 μl of ice-cold lysis buffer with protease inhibitors. Cleared lysates are then incubated overnight at 4 °C with glutathione-agarose beads. Centrifuge at 600×*g* for 30 s at 4 °C (table microcentrifuge). Beads are washed three times with 500 μl of ice-cold lysis buffer containing protease inhibitors, letting beads sediment on ice between washes. Beads are then supplemented with Laemmli loading buffer for SDS-PAGE and boiled for 5 min at 95 °C. Samples are centrifuged and loaded on denaturing SDS-PAGE gel for analysis by Western blotting. In addition, the supernatants resulting from bead washes can be analyzed by Western blot as well. This allows visualizing the proteins that have not been affinity purified or that have been lost by the washes.

4. *Reading results*: The samples that have been run on SDS-PAGE must be analyzed by Western blotting. First, detection of the GST epitope is required to check that the GST fusion proteins were properly bound to the glutathione beads and remained intact (not degraded). Second, detection of the Flag epitope reveals the amount of Flag-tagged protein that has co-purified thanks to its affinity binding to the GST-fused interacting protein. As a corollary, no or very little Flag-tagged protein should be retrieved from the beads that have been incubated with the lysates obtained with the unfused GST protein (control condition, *see* **Note 15**).

3.4 Network Building

Once the interaction screens have been performed and candidate interactors identified, the challenge is to find its way among the huge amount of data generated, and the even huger amount of publicly available data, in order to figure out what could be their relevance. Data from protein-protein interaction screens as well as protein-DNA or other interaction screens may be combined and integrated in networks. These networks are afterwards analyzed to extract relevant information.

To build up an interactomic network around the protein(s) of interest several tools will be needed. Among them, the visualization software will be the first interface between the scientist and the interactomic tables. In that respect, the Java application Cytoscape [39] is one of the most easy-to-use and complete application. With the help of various additional plugins, it allows the visualization and analysis of interactomic networks.

In this section, we present how to import known interactions from publicly available databases [51, 52] and how to create a network involving the experimentally identified interactions (*see* Subheadings 3.1–3.3) with Cytoscape [39]. We then analyze this network using specific plugins to extract the relevant information. The procedure described here corresponds to a network built up from human and mouse protein interactions. It can be adapted to other species, implying to import proteome databases corresponding to the species under consideration.

3.4.1 Cytoscape Download and Installation

1. Access the open-source software platform Cytoscape download web page by clicking on the following link: http://www.cytoscape.org/download.html.
2. Provide requested information, accept terms of use, and click on "Proceed to Download".
3. Download the Latest 2.x Series Cytoscape v2.8.3 suitable for your platform.
4. Execute the installation application and follow the instructions to install Cytoscape on your computer.
5. Click on the application icon to open a new Cytoscape session (*see* **Note 16**).
6. In the File Menu, click on "Save as…" to save the Cytoscape session into a .cys file. This file will contain all Cytoscape information of a given session (network data, graphic information, …).

3.4.2 Plugin Installation

1. In your Cytoscape session, click on the Plugins tab and select Manage Plugins (*see* **Note 17**).
2. In the Manage Plugin window, using the Search tool box, find "PSICQUICUniversalClient v.0.31" and click on "Install".
3. In the Manage Plugin window, using the Search tool box, find "CyThesaurus-ID-Mapping v.1.31" and click on "Install".
4. In the Manage Plugin window, using the Search tool box, find CentiScaPe1.21 and click on "Install".
5. In the Manage Plugin window, using the Search tool box, find MCODE v1.32 and click on "Install".

*3.4.3 Complete Genome and Proteome Database Import (See **Note 18**)*

1. Download the "UniProt secondary ID mapping" table from the following location: ftp://ftp.uniprot.org/pub/databases/uniprot/knowledgebase/docs/sec_ac.txt.
2. Save the text file from your web browser onto your computer using the File Menu and the "Save as…" function.
3. In the File Menu of the Microsoft Excel application, click on "Open…" and select the saved text file.

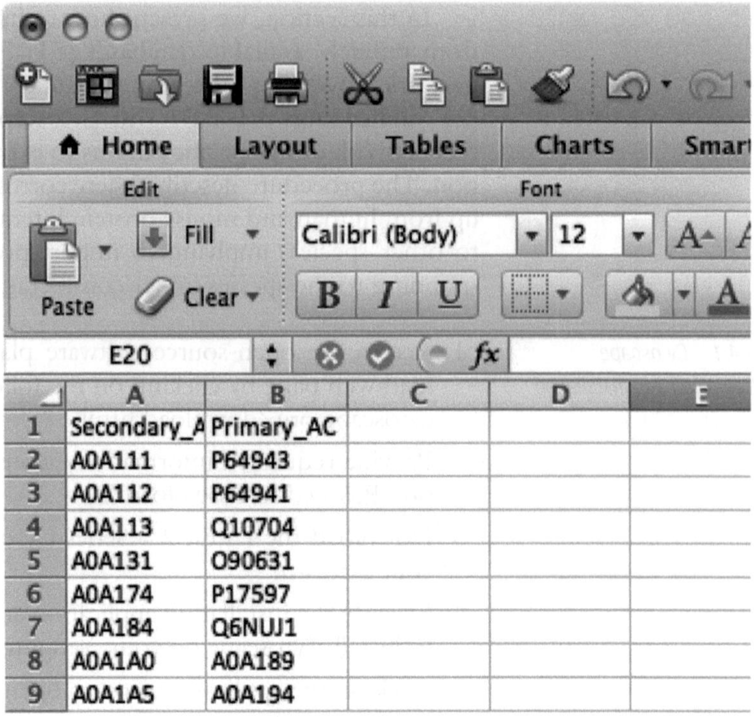

Fig. 5 Screen capture corresponding to the Microsoft Excel table generated while importing genome and proteome databases (Subheading 3.4.3), at **step 5**

4. Opening the text file in Excel, choose "Delimited data", then "Tab" and "Space" delimiter, and then "General" or "Standard" column data format.
5. Once the text file has been opened with Microsoft Excel, delete the first lines to keep only one line with the names of the columns (Primary_AC or Secondary_AC), that you will have to reformat slightly, and all the following accession numbers (Fig. 5).
6. Then, save the new file as usual in a ".txt" format (by default it will be named sec_ac.txt).
7. Go to the "UniProt ID mapping" table from the following location: ftp://ftp.uniprot.org/pub/databases/uniprot/current_release/knowledgebase/idmapping/by_organism/.
8. Select "ORGANISM_idmapping_selected.tab.gz" where "ORGANISM" will depend on your species of interest. For example, download "MOUSE_10090_idmapping_selected.tab.gz" or "HUMAN_9606_idmapping_selected.tab.gz" and save these files on your computer (*see* **Note 19**).
9. Extract the ".tab" files from the ".gz" archive files and save them on your computer.
10. In the File Menu of the Microsoft Excel application, click on "Open…" and select the saved ".tab" files.

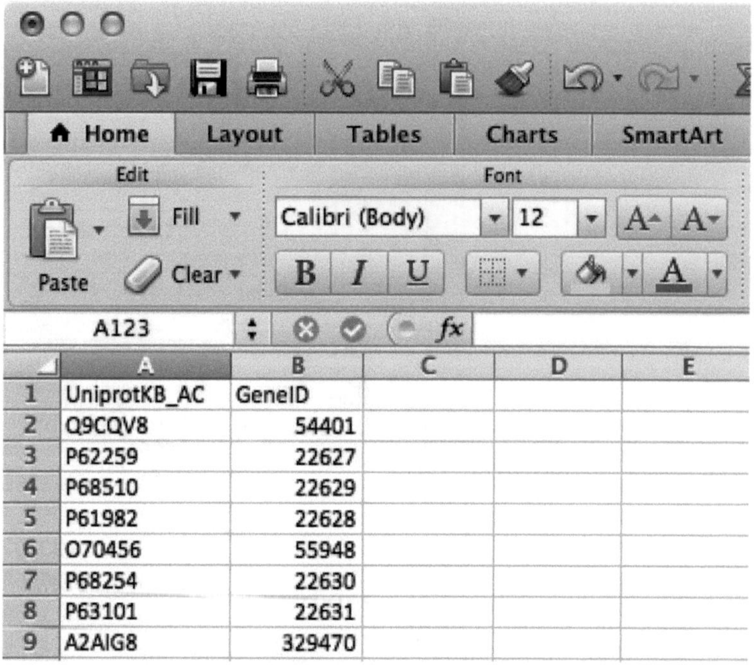

	A	B
1	UniprotKB_AC	GeneID
2	Q9CQV8	54401
3	P62259	22627
4	P68510	22629
5	P61982	22628
6	O70456	55948
7	P68254	22630
8	P63101	22631
9	A2AIG8	329470

Fig. 6 Screen capture corresponding to the Microsoft Excel table generated while importing genome and proteome databases (Subheading 3.4.3), at **step 13**

11. Opening the text files in Excel, choose "Delimited data", then "Tab" and "Space" delimiter, and then "General" or "Standard" column data format.
12. Keep only the mapping columns (column A for UniprotKB_AC and column C for GeneID) and delete all the others.
13. Create a new line with the names of the columns (UniprotKB_AC and GeneID) (Fig. 6, *see* **Note 20**).
14. Save as a text file.
15. Download the "HomoloGene" table of NCBI from the following location: ftp://ftp.ncbi.nih.gov/pub/HomoloGene/current/homologene.data (*see* **Note 21**).
16. Save the text file on your computer using the File Menu and the "Save as…" function.
17. In the File Menu of the Microsoft Excel application, click on "Open…" and select the saved text file.
18. Opening the text file in Excel, choose "Delimited data", then "Tab" and "Space" delimiter, and then "General" or "Standard" column data format.
19. Keep only the mapping columns (column A for HomoloGene group ID or HID, column B for Taxonomy_ID, and column C for Gene_ID) and delete all the others.

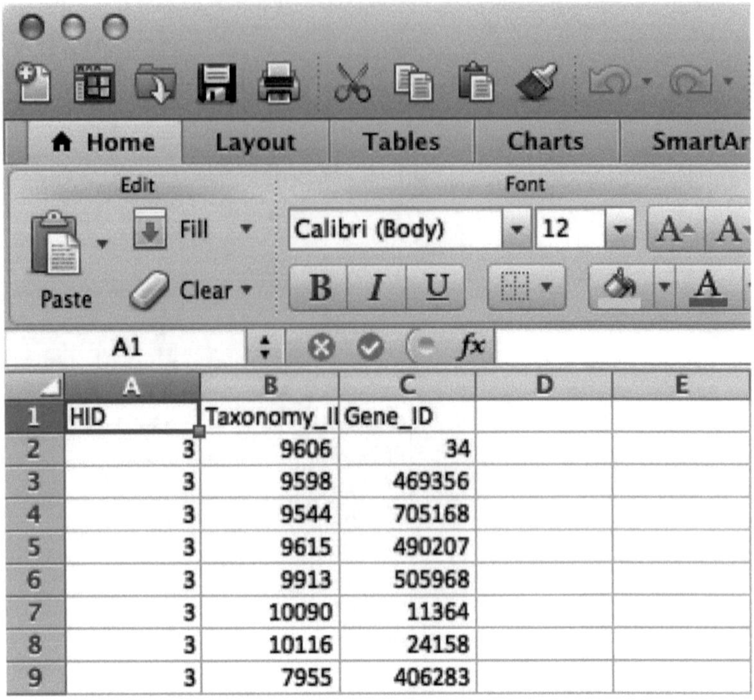

Fig. 7 Screen capture corresponding to the Microsoft Excel table generated while importing genome and proteome databases (Subheading 3.4.3), at **step 20**

20. Create a new line with the names of the columns (HID, Taxonomy_ID, and Gene_ID) (Fig. 7, *see* **Note 22**).
21. Save as a text file.
22. In Microsoft Excel, sort the HomoloGene table according to Taxonomy_ID.
23. Select the lines corresponding to your species of interest (for instance, for human, all lines with Taxonomy_ID corresponding to 9606) and create new Excel files containing only those lines.
24. Save species-specific HomoloGene tables as text files.
25. Download HUGO Gene Nomenclature Committee (HGNC) data from the following website: http://www.genenames.org/cgi-bin/hgnc_stats.
26. Click on "Custom" download of the "Total by Locus Group" section of "protein-coding gene".
27. Deselect all column data except "Approved Symbol" and "Entrez Gene ID" from the "Curated by the HGNC" section.
28. Leave all other default settings and click on SUBMIT.
29. Save the text file on your computer using the File Menu and the "Save as…" function.

3.4.4 Import Data from Databases Using PSICQUIC Universal Web Service Client

1. In your Cytoscape session, click on the File Menu, and select "Import" and "Network from Web Services".
2. Select "PSICQUIC Universal Web Service Client" as a Data Source.
3. In the Query tab, paste this:

   ((((((**PROTEIN SYMBOL** or **GeneID**) AND (species:**TAXID**)) AND NOT detmethod: "predictive text mining") AND NOT detmethod: "inferred by curator") AND NOT detmethod: "interologs mapping") AND NOT species:632) AND NOT detmethod:"unspecified method"

   where "**PROTEIN SYMBOL** or **GeneID**" corresponds to the list of names of your proteins of interest separated by "OR", and the **TAXID** should be your species of interest (e.g., 10090 for *Mus musculus*, 9606 for *Homo sapiens*, *see* **Notes 23** and **24**).
4. In the Search Property tab, use the following properties:

   Block size: 2000

   Timeout (sec.): 6000

   Query Mode: GET_BY_QUERY
5. Click on Search.
6. Agree to create a new network from the search results and select the databases of interest. Do not merge the networks at this stage since PSICQUIC import databases use variable protein identifiers to map the interactions; just click on "OK" (*see* **Notes 25** and **26**).

*3.4.5 ID Conversion and Unification Using CyThesaurus-ID-Mapping (See **Notes 27** and **28**)*

1. In your Cytoscape session, click on the Plugins tab and select "CyThesaurus".
2. Click on "ID Mapping Resources Configuration".
3. Select "Local/Remote Files".
4. In Data Source, select the "UniProt secondary ID mapping text file" (*see* Subheading 3.4.3, file saved at **step 6**).
5. Leave the default advanced options and click on OK.
6. Then close the "ID Mapping Resources Configuration" window.
7. On the Cythesaurus plugin window, select all networks.
8. As a Source ID type, select "Primary_AC" and "Secondary_AC".
9. As a Source Key Attribute, select "PSIMI25.uniprotkb.top".
10. As a Target ID type, select "Primary_AC".
11. As Target New Attribute, type the name of the mapping attribute (Primary_AC).

12. Select “Keep the first target ID only”.
13. Click on OK and close (*see* **Note 29**).
14. In your Cytoscape session, click on the Plugins tab and select “CyThesaurus”.
15. Click on “ID Mapping Resources Configuration”.
16. Select “Local/Remote Files”.
17. In Data Source, select the “ORGANISM_idmapping_selected.tab” text file (*see* Subheading 3.4.3, file saved at **step 14**).
18. Leave the default advanced options and click on OK.
19. Repeat **steps 16–18** for all organisms.
20. Select all the organism mapping tables and click on CLOSE.
21. Select all networks.
22. As a Source ID type, select all “UniProtKB_AC”.
23. As a Source Key Attribute, select “Primary_AC” (created at **step 11**).
24. As a Target ID type, select “GeneID”.
25. As Target New Attribute, type the name of the mapping attribute (GeneID).
26. Select “Keep the first target ID only”.
27. Click on OK and close (*see* **Note 30**).
28. In your Cytoscape session, click on the Plugins tab and select “CyThesaurus”.
29. Click on “ID Mapping Resources Configuration”.
30. Select “Local/Remote Files”.
31. In Data Source, select the general HomoloGene table text file (*see* Subheading 3.4.3, file saved at **step 21**).
32. Leave the default advanced options and click on OK.
33. Select the mapping table and click on CLOSE.
34. Select all networks.
35. As a Source ID type, select “Gene_ID”.
36. As a Source Key Attribute, select “GeneID” (created at **step 25**) and any other attributes corresponding to gene identifiers. This can be done by adding lines (click on “insert”) to link the Source ID type (“Gene_ID”) to additional Source Key Attributes (*see* **Note 31**).
37. As a Target ID type, select “HID”.
38. As Target New Attribute, type the name of the mapping attribute (HID).
39. Select “Keep the first target ID only”.
40. Click on OK (*see* **Note 32**).

41. In your Cytoscape session, click on the Plugins tab and select "CyThesaurus".
42. Click on "ID Mapping Resources Configuration".
43. Select "Local/Remote Files".
44. In Data Source, select a species-specific HomoloGene table text file (*see* Subheading 3.4.3, files saved at **step 24**, *see* **Note 33**).
45. Leave the default advanced options and click on OK.
46. Repeat **steps 43–45** if more than one species-specific HomoloGene file has been generated.
47. Select the mapping table and click on CLOSE.
48. Select all networks.
49. As a Source ID type, select "HID".
50. As a Source Key Attribute, select "HID" (created at **step 38**).
51. As a Target ID type, select "Gene_ID".
52. As Target New Attribute, type the name of the mapping attribute (e.g., Human_Gene_ID).
53. Select "Keep the first target ID only".
54. Click on OK and close (*see* **Note 34**).

3.4.6 Network Merging

1. In your Cytoscape session, click on the Plugins tab and select "Advanced Network Merge".
2. Choose union operation.
3. Select all networks.
4. Click on "Advanced Network Merge".
5. For all networks, select Human_Gene_ID as Matching attributes and click on MERGE.
6. Close the window (*see* **Note 35**).
7. In your Cytoscape session, click on the Plugins tab and select "CyThesaurus".
8. Click on "ID Mapping Resources Configuration".
9. Select "Local/Remote Files".
10. In Data Source, select the HGNC file created at **step 29**, Subheading 3.4.3.
11. Leave the default advanced options and click on OK.
12. Then close the "ID Mapping Resources Configuration" window.
13. On the Cythesaurus plugin window, select all networks.
14. As a Source ID type, select "Entrez Gene ID.
15. As a Source Key Attribute, select "Human_Gene_ID" created at **step 52** of Subheading 3.4.5.

16. As a Target ID type, select "Approved_Symbol".
17. As Target New Attribute, type the name of the mapping attribute "Approved Symbol".
18. Select "Keep the first target ID only".
19. Click on OK and close.

3.4.7 Network Processing

1. In your Cytoscape session, click on the Plugins tab, and select "Network Modifications" and "Remove Duplicated Edges".
2. Select the merged network.
3. Select "Ignore edge direction" and "Create an edge attribute with number of duplicated edges".
4. Click on OK.
5. In your Cytoscape session, click on the Plugins tab, and select "Network Modifications" and "Remove Self-Loops".
6. Select the merged network.
7. Click on OK.
8. In your Cytoscape session, click on the Plugins tab, and select "CentiScaPe1.21".
9. In the CentiScaPe1.21 Menu of the Control Panel, select "Degree" and start the calculation.
10. In the CentiScaPe1.21 Node Degree section of the Results Panel, put the threshold at 2, select "more/equals threshold", and click on "update".
11. In your Cytoscape session, click on the Select tab, "Nodes", and "Invert Node Selection".
12. Delete all those nodes using "Delete" key (*see* **Note 36**).
13. Repeat **steps 9–12** until there is no more node to delete.

3.4.8 Network Layout

1. In your Cytoscape session, click on the Select tab, select "Nodes" and "Deselect all nodes", then click on the Layout tab, and select "Cytoscape Layouts" and "Spring Embedded".
2. In the Control Panel, select "VizMapper" tab.
3. Modify the network layout according to your preferences.
4. In Node Visual Mapping, select "Node Label" and select the "Approved Symbol" column created at Subheading 3.4.6, **step 17**. All nodes are now labeled with their official approved symbol.
5. Visualization parameters can be modified by selecting the Plugin tab, "NetworkAnalysis", and "Visualization parameter". If this is done after network analysis, network analysis parameters can be used. For example, a gradient of node color can be attributed to the nodes according to their degree calculated at Subheading 3.4.7, **step 9**. For this, select "CentiScape Node

Degree" in the "NetworkAnalyzer—Visualize Parameters" window and click on APPLY. One color will be attributed to each node according to its degree. This can also be done for the node size and for all parameters generated under Subheading 3.4.9.

6. The network (or zoomed-in areas) can be exported in PDF files. Selecting the File tab, choose "Export" and "Export network view as graphics", and tick the box "Export".

*3.4.9 Network Analysis (See **Note 37**)*

1. In your Cytoscape session, click on the Plugins tab, and select "CentiScapePe1.21".
2. In the CentiScaPe1.21 Menu of the Control Panel, select all and start the calculation.
3. In the CentiScaPe1.21 tab of the Results Panel, you can modify parameters of the centrality values to identify characteristic nodes.
4. In your Cytoscape session, click on the Plugins tab, and select "MCODE" and "Start MCODE".
5. In the MCODE Menu of the Control Panel, under the "Find cluster(s)" section, select "in Whole Network", leave the advanced options selected by default, and click on "Analyze".
6. Identified clusters will be shown in the MCODE section of the Results Panel. Parameters and thresholds can be changed according to the network and the objectives of the study.

4 Notes

1. The *LacZ* reporter is not useful for the Y strains, as negative controls turn blue very quickly, making it difficult to identify true-positive colonies.
2. Classical restriction enzyme cloning is not adapted for automation of large-scale experiments because of restriction option variation between ORFs of interest.
3. We routinely use Greiner Bio-One OneWell plates and our robotic station plate holders have been customized accordingly.
4. The robot can be used for all high-throughput liquid-handling operations including large-scale generation of plasmid minipreps, yeast plating on selective media (96 samples at a time), cherry picking, and rearraying and combining reagents, creating glycerol stocks and high-throughput bacterial and yeast transformation.
5. Glutathione-sepharose beads (GE Healthcare) can alternatively be used.

6. Clone your ORF into the yeast vectors ensuring that they inserted in frame, N- or C-terminally, with the sequence of Gal4DB or Gal4AD ORFs.
7. Starting with colonies taken off from a solid medium plate, yeast growth in liquid YPD medium (at 28 °C; with 200–250 rpm shaking) takes at least 16 h to reach stationary phase (OD 600 > 1.5).
8. Positive controls: Mate yeasts transformed with vectors expressing AD and DB fusions known to interact. Typically, Hoxa1 is known to dimerize; AD-Hoxa1 and DB-Hoxa1 fusions therefore provide good controls. Background growth can however result from the fact that the DB-Hoxa1 can complement auxotrophy markers due to its intrinsic ability to activate transcription and provide GAL2-ADE2 and GAL1-HIS3 activation. Mutant Hoxa1 proteins having lost the ability to transactivate could then be used as controls [53]. Negative controls: Mate yeasts transformed to express DB fusions (pGBT9) with yeast transformed with empty AD vector (pGAD424). Also mate yeasts transformed with an empty DB vector with yeasts expressing AD fusions.
9. All cloned entry clones should be verified by sequencing.
10. These control plates are important to identify autoactivators that can arise during Y2H selection. In fact, the pDEST-AD vector harbors the *CYH2*-negative selection marker. In the presence of CYH, protein expression from the pDEST-AD vector is suppressed. Thus, de novo autoactivator exhibits a CYH+ phenotype, while true positives are CYH−.
11. Clone your ORF into the expression vectors ensuring that they inserted in frame, N- or C-terminally, with the sequence of Flag- or GST-coding sequences. Alternatively, if you entered your ORF in the Gateway® system, just recombine the entry clones you generated to destination plasmids available in Gateway® to express Flag or GST fusion proteins.
12. For regular SDS-PAGE and Western blot procedures, we use the SDS-PAGE and Western blotting device Mini Protean Tetra Cell and nitrocellulose membranes Hybond™C, and we follow the instructions and recipes provided by the manufacturers. For Flag and GST epitope detection, we use an anti-Flag primary antibody (M2) (#F1804, Sigma) at 1:500 dilution, and an anti-GST primary antibody (GST-2) (#G1160, Sigma) at 1:300 dilution. A goat anti-mouse secondary antibody, HRP-conjugated, is used at a 1:3,000 dilution (#sc-2005, Santa Cruz). Finally, to control protein load in samples, we also reveal Western blots with HRP-conjugated anti-β-actin antibody, used at a 1:3,000 dilution (#A3854, Sigma).
13. Always keep samples cold, on ice.

14. To pipet the beads cut the extremity of a 1 ml pipet tip. This allows gentle pipetting, avoiding shearing and damaging beads.
15. If Flag-tagged protein is detected in the samples obtained with the unfused GST protein, this means that the Flag-tagged protein aspecifically binds either to the beads or to the GST. To avoid the aspecific binding of proteins to the beads, beads can be saturated by preincubation and washing with the lysis buffer containing 200 μg/ml of bovine serum albumin (BSA).
16. A Cytoscape session appears as several panels displaying Cytoscape tools and network data. As a default setup, two panels will be opened in a new session: the "Control Panel" and "Data Panel". The remaining window constitutes the working area in which you will be able to open new panels and to visualize your networks.
17. Using Cytoscape, distinct plugins are required to build an interaction network involving both original experimental data and publicly available data. Interaction data are compiled in various accessible databases which do not use a common referencing for proteins, although PSI-MI interchange format for molecular interaction data, developed by the Proteomics Standards Initiative of the Human Proteome Organization (HUPO), is now widely used to facilitate molecular interaction data exchange in a standardized manner [54]. PSICQUIC (Proteomics Standards Initiative Common QUeri InterfaCe) is an interface connecting numerous databases and allows searching them for protein-protein interactions. With CyThesaurus, the protein identifiers and attributes peculiar to each database will be mapped to new attributes to unequivocally and exhaustively identify interactions in building up a network with Cytoscape. CentiScape is a Cytoscape plugin that calculates network centralities (i.e., node-to-node properties: some proteins—nodes in the network—being highly connected while not others, some proteins having a central position in the network, ...), which then define the topological parameters of the network. MCODE is a Cytoscape plugin allowing the identification of clusters, or highly interconnected regions, within a network. In the newly opened "Manage Plugins" window, you can visualize already installed Plugins as well as available Plugins. By clicking on a specific Plugin, you can access all related information. It is important to check that the Plugin you want to use has been verified for the Cytoscape version used. For further information regarding the plugins: PSICQUIC, *see* [55, 52]; CyThesaurus, *see* [56]; CentiScape, *see* [57]; MCODE, *see* [58].
18. As most of the databases use different protein identifiers, protein name, or gene identifier, converting tables will be needed. They can be downloaded directly from the Internet.

As a complete import of every database will require considerable space memory, it is recommended to use short updated versions of each database. Download directory can be found directly on the websites of interest.

19. ".gz" files are compressed files you will need to unzip.
20. For more information on the file format and contents, please refer to the "README" text file available here: ftp://ftp.uniprot.org/pub/databases/uniprot/current_release/knowledgebase/idmapping/README.
21. It might be useful to combine interaction data from closely related species considering that interactions between homologous proteins will be conserved. This especially holds for human and mouse interactomes. The HomoloGene table will allow to identify homologues between species and to provide them with a common identifier so as to define only one node in the network. Although such "interspecies" networks are obviously not biologically relevant, they can hold an important heuristic value.
22. For more information on the file format and contents, please refer to the "README" text file available here: ftp://ftp.ncbi.nih.gov/pub/HomoloGene/README.
23. Both protein symbols and GeneID can be used at the same time to ensure sufficient coverage of the databases. If working on mouse or human, one can find it interesting to search in both species databases. As for protein identifiers, "TAXID" are then separated by "OR".
24. Building up networks could be performed following the present procedure in an iterative way. Starting with a handful of interactors of your protein of interest which you enter in the process as query, after completion of a first round of network building, you can reenter as a query your protein and all the interactors fished out from databases; you will then search for interactors of interactors.
25. Some of the databases can be inactive due to server connection issues. Others can be useless depending on the research area. Choose the ones you are interesting in and take notes of inactive servers.
26. Remember that protein or gene IDs can differ according to the database. Thus, ID conversion has to be adapted depending on the chosen databases.
27. PSICQUIC import databases use variable protein identifiers. The ID conversion will unify protein ID to allow proper merging of the networks. The final protein ID has to be chosen depending on the purpose of the study. For mouse/human interspecies network, GeneID is probably the best choice as the homologous conversion between species can be easily performed using the HomoloGene database of NCBI.

Note that doing so, the protein isoform information will be lost as all isoforms will be related to the same gene.

28. Some interactions are still identified using secondary, obsolete, UniProtKB. As a first step, one should change these secondary UniProtKB into primary UniProtKB.
29. A new attribute column will be created and filled with the Primary UniProt ID of the proteins.
30. A new attribute column will be created and filled with the GeneID of the gene encoding the proteins.
31. Depending on the databases from which interaction data are extracted, distinct attributes will have to be added as "Source Key Attribute" at **step 36**. BioGrid for instance does not use UniProtKB identifiers. Therefore, calling "UniProtKB identifiers" as a source ID at **step 22** will not call any protein from BioGrid. Therefore, in this case, no GeneID attribute could be generated at **step 25**. In BioGrid tables, GeneID identifiers can be found in the "PSIMI25.entrezgene/locuslink.top" attribute column and has to be selected. So at this step, check that you selected all possible source key attributes referring to Gene IDs pertaining to the distinct interaction databases so that all nodes will be mapped by the conversion procedure.
32. A new attribute column will be created and filled with the HID of the genes.
33. For further analyses on the built-up network (Gene Ontology analyses, pathway analyses, etc.), if working on a mouse/human interspecies network (*see* **Note 21**), it is usually easier to work with human identifiers, since there are larger amount of tools available, so select the Human HomoloGene table text file.
34. A new attribute column will be created and filled with the human GeneID of the genes for the different species selected during the mapping.
35. A new network will be created in the Cytoscape session. This network corresponds to the fusion between all the data coming from the databases selected while importing interatomic data with PSICQUIC.
36. **Steps 9–12** aim at removing nodes which are not connected to at least two other nodes.
37. A CentiScape and MCODE combination constitutes a good starting point for the network analysis. CentiScape allows the identification of the most relevant nodes in the network. High-centroid-value nodes are known to control and connect the different clusters of the network. These clusters can be directly identified through the MCODE analysis. Taken together, those results will give you an idea of the main proteins of your network and as a consequence cues to help you in the choice

of the favorite interacting partners you will experimentally study in the future. CentiScape and MCODE are two examples of useful plugin for network analysis. Plenty of other plugins can be found in the Cytoscape Plugin Manager. More information about Cytoscape plugins can be found directly from the Cytoscape app store in the following link: http://apps.cytoscape.org/. GO enrichment analyses, network comparisons, and pathway identification analyses are only some examples of the tools available. Pay attention to use only plugins verified for the Cytoscape version you are using.

Acknowledgements

We are grateful to the Belgian Foundation against Cancer and to the Télévie program coordinated by the Belgian National Fund for Scientific Research (FNRS) for their support. I.B. and B.L. held a FRIA fellowship from the FNRS. S.D. is a "Leon Fredericq" fellow. J.C.T. is Research Investigator of the FNRS.

References

1. Forlani S, Lawson KA, Deschamps J (2003) Acquisition of Hox codes during gastrulation and axial elongation in the mouse embryo. Development 130:3807–3819
2. Mallo M, Wellik DM, Deschamps J (2010) Hox genes and regional patterning of the vertebrate body plan. Dev Biol 344:7–15
3. Wellik DM (2009) Hox genes and vertebrate axial pattern. Curr Top Dev Biol 88:257–278
4. Zakany J, Duboule D (2007) The role of Hox genes during vertebrate limb development. Curr Opin Genet Dev 17:359–366
5. Chen F, Capecchi MR (1999) Paralogous mouse Hox genes, Hoxa9, Hoxb9, and Hoxd9, function together to control development of the mammary gland in response to pregnancy. Proc Natl Acad Sci U S A 96:541–546
6. Garin É, Lemieux M, Coulombe Y et al (2006) Stromal Hoxa5 function controls the growth and differentiation of mammary alveolar epithelium. Dev Dyn 235:1858–1871
7. Wellik DM (2011) Hox genes and kidney development. Pediatr Nephrol 26:1559–1565
8. Di-Poi N, Zakany J, Duboule D (2007) Distinct roles and regulations for HoxD genes in metanephric kidney development. PLoS Genet 3:e232
9. Aubin J, Dery U, Lemieux M et al (2002) Stomach regional specification requires Hoxa5-driven mesenchymal-epithelial signaling. Development 129:4075–4087
10. Zacchetti G, Duboule D, Zakany J (2007) Hox gene function in vertebrate gut morphogenesis: the case of the caecum. Development 134:3967–3973
11. Boucherat O, Montaron S, Berube-Simard FA, Aubin J, Philippidou P, Wellik DM, Dasen JS, Jeannotte L (2013) Partial functional redundancy between Hoxa5 and Hoxb5 paralog genes during lung morphogenesis. Am J Physiol Lung Cell Mol Physiol 304:L817–830
12. Mandeville I, Aubin J, LeBlanc M et al (2006) Impact of the loss of Hoxa5 function on lung alveogenesis. Am J Pathol 169:1312–1327
13. Zaffran S, Kelly RG (2012) New developments in the second heart field. Differentiation 84:17–24
14. Oury F, Murakami Y, Renaud JS et al (2006) Hoxa2- and rhombomere-dependent development of the mouse facial somatosensory map. Science 313:1408–1413
15. Geisen MJ, Di Meglio T, Pasqualetti M et al (2008) Hox Paralog Group 2 Genes Control the Migration of Mouse Pontine Neurons through Slit-Robo Signaling. PLoS Biol 6:e142
16. Di Bonito M, Narita Y, Avallone B, Sequino L, Mancuso M, Andolfi G, Franze AM, Puelles L, Rijli FM, Studer M (2013) Assembly of the auditory circuitry by a Hox genetic network in the mouse brainstem. PLoS Genet 9:e1003249
17. Di Meglio T, Kratochwil CF, Vilain N et al (2013) Ezh2 orchestrates topographic migration

and connectivity of mouse precerebellar neurons. Science 339:204–207
18. Antonchuk J, Sauvageau G, Humphries RK (2002) HOXB4-induced expansion of adult hematopoietic stem cells ex vivo. Cell 109:39–45
19. Argiropoulos B, Humphries RK (2007) Hox genes in hematopoiesis and leukemogenesis. Oncogene 26:6766–6776
20. McGonigle GJ, Lappin TR, Thompson A (2008) Grappling with the HOX network in hematopoiesis and leukemia. Front Biosci 13:4297–4308
21. Cillo C (2007) Deregulation of the Hox Gene Network and Cancer. In: Papageorgiou S (ed) Hox gene expression. Springer, New York, pp 121–133
22. Paul D, Bridoux L, Rezsohazy R, Donnay I (2011) HOX genes are expressed in bovine and mouse oocytes and early embryos. Mol Reprod Dev 78:436–449
23. Graba Y, Aragnol D, Pradel J (1997) Drosophila Hox complex downstream targets and the function of homeotic genes. Bioessays 19:379–388
24. Akin ZN, Nazarali AJ (2005) Hox genes and their candidate downstream targets in the developing central nervous system. Cell Mol Neurobiol 25:697–741
25. Chen H, Rubin E, Zhang H et al (2005) Identification of transcriptional targets of HOXA5. J Biol Chem 280:19373–19380
26. Lei H, Wang H, Juan AH, Ruddle FH (2005) The identification of Hoxc8 target genes. Proc Natl Acad Sci U S A 102:2420–2424
27. Svingen T, Tonissen KF (2006) Hox transcription factors and their elusive mammalian gene targets. Heredity 97:88–96
28. Makki N, Capecchi MR (2011) Identification of novel Hoxa1 downstream targets regulating hindbrain, neural crest and inner ear development. Dev Biol 357:295–304
29. Donaldson IJ, Amin S, Hensman JJ, Kutejova E, Rattray M, Lawrence N, Hayes A, Ward CM, Bobola N (2012) Genome-wide occupancy links Hoxa2 to Wnt-beta-catenin signaling in mouse embryonic development. Nucleic Acids Res 40:3990–4001
30. Merabet S, Pradel J, Graba Y (2005) Getting a molecular grasp on Hox contextual activity. Trends Genet 21:477–480
31. Merabet S, Sambrani N, Pradel J, Graba Y (2010) Regulation of Hox activity: insights from protein motifs. Adv Exp Med Biol 689:3–16
32. Rezsohazy R (2014) Non-transcriptional interactions of Hox proteins: inventory, facts, and future directions. Dev Dyn 243:117–131
33. Vidalain P, Boxem M, Ge H et al (2004) Increasing specificity in high-throughput yeast two-hybrid experiments. Methods 32:363–370
34. Dreze M, Monachello D, Lurin C et al (2010) High-quality binary interactome mapping. Methods Enzymol 470:281–315
35. Lemmens I, Lievens S, Tavernier J (2010) Strategies towards high-quality binary protein interactome maps. J Proteomics 73: 1415–1420
36. Eyckerman S, Verhee A, der Heyden JV et al (2001) Design and application of a cytokine-receptor-based interaction trap. Nat Cell Biol 3:1114–1119
37. Kodama Y, Hu CD (2012) Bimolecular fluorescence complementation (BiFC): a 5-year update and future perspectives. Biotechniques 53:285–298
38. Hudry B, Viala S, Graba Y, Merabet S (2011) Visualization of protein interactions in living Drosophila embryos by the bimolecular fluorescence complementation assay. BMC Biol 9:5
39. Koh GC, Porras P, Aranda B et al (2012) Analyzing protein-protein interaction networks. J Proteome Res 11:2014–2031
40. Cusick ME, Klitgord N, Vidal M, Hill DE (2005) Interactome: gateway into systems biology. Hum Mol Genet 14(suppl_2):R171–181
41. Lambert B, Vandeputte J, Remacle S et al (2012) Protein interactions of the transcription factor Hoxa1. BMC Dev Biol 12:29
42. Bergiers I, Bridoux L, Nguyen N et al (2013) The Homeodomain Transcription Factor Hoxa2 Interacts with and Promotes the Proteasomal Degradation of the E3 Ubiquitin Protein Ligase RCHY1. PLoS One 8:e80387
43. Vidal M, Braun P, Chen E et al (1996) Genetic characterization of a mammalian protein-protein interaction domain by using a yeast reverse two-hybrid system. Proc Natl Acad Sci U S A 93:10321–10326
44. Vidal M, Brachmann RK, Fattaey A et al (1996) Reverse two-hybrid and one-hybrid systems to detect dissociation of protein-protein and DNA-protein interactions. Proc Natl Acad Sci U S A 93(19):10315–10320
45. James P, Halladay J, Craig EA (1996) Genomic libraries and a host strain designed for highly efficient two-hybrid selection in yeast. Genetics 144:1425–1436
46. Rual JF, Hirozane-Kishikawa T, Hao T et al (2004) Human ORFeome version 1.1: a platform for reverse proteomics. Genome Res 14:2128–2135
47. Wiemann S, Arlt D, Huber W et al (2004) From ORFeome to biology: a functional genomics pipeline. Genome Res 14:2136–2144

48. Rual J-F, Venkatesan K, Hao T et al (2005) Towards a proteome-scale map of the human protein–protein interaction network. Nature 437:1173–1178

49. Lamesch P, Li N, Milstein S et al (2007) hORFeome v3.1: a resource of human open reading frames representing over 10,000 human genes. Genomics 89:307–315

50. Simonis N, Rual JF, Lemmens I et al (2012) Host-pathogen interactome mapping for HTLV-1 and -2 retroviruses. Retrovirology 9:26

51. Orchard S (2012) Molecular interaction databases. Proteomics 12:1656–1662

52. del-Toro N, Dumousseau M, Orchard S et al (2013) A new reference implementation of the PSICQUIC web service. Nucleic Acids Res 41:W601–606

53. Lambert B, Vandeputte J, Desmet PM et al (2010) Pentapeptide insertion mutagenesis of the Hoxa1 protein: mapping of transcription activation and DNA-binding regulatory domains. J Cell Biochem 110:484–496

54. Van Roey K, Orchard S, Kerrien S et al (2013) Capturing cooperative interactions with the PSI-MI format. Database, bat066

55. Aranda B, Blankenburg H, Kerrien S et al (2011) PSICQUIC and PSISCORE: accessing and scoring molecular interactions. Nat Methods 8:528–529

56. van Iersel MP, Pico AR, Kelder T et al (2010) The BridgeDb framework: standardized access to gene, protein and metabolite identifier mapping services. BMC Bioinformatics 11:5

57. Scardoni G, Petterlini M, Laudanna C (2009) Analyzing biological network parameters with CentiScaPe. Bioinformatics 25:2857–2859

58. Bader GD, Hogue CW (2003) An automated method for finding molecular complexes in large protein interaction networks. BMC Bioinformatics 4:2

Chapter 21

Rational Drug Repurposing Using sscMap Analysis in a HOX-TALE Model of Leukemia

Laura M.J. Kettyle, Fabio G. Liberante, and Alexander Thompson

Abstract

Drug discovery and development are often hampered by lack of target identification and clinical tractability. Repurposing of approved drugs to life-threatening diseases such as leukemia is emerging as a promising alternative approach. Connectivity mapping systems link approved drugs with disease-related gene signatures. Relevant preclinical models provide essential tools for system validation and proof-of-concept studies. Herein we describe procedures aimed at generating disease-based gene signatures and applying them to established cross-referencing databases of potential candidate drugs. As a proof of principle, we present the identification of Entinostat as a candidate drug for the treatment of HOX-TALE-related leukemia.

Key words HOXA9, Meis1, Data mining, Gene expression profiles, Connectivity mapping (sscMap), Leukemia, Entinostat, Preclinical model

1 Introduction

1.1 Hox Genes in Leukemia

The evolutionarily conserved Class I *homeobox* (*HOX*) family of genes encode master regulators of developmental processes, including hematopoiesis [1]. Hematopoiesis is the process by which blood cellular components are derived from rare hematopoietic stem cells (HSCs) that reside in the medulla of the bone marrow [2]. HOX proteins form functional complexes with members of the three-amino-loop-extension motif (TALE) family of proteins, including PBX and MEIS isoforms, which together play a primary role in HSC self-renewal and differentiation [3]. Therefore, dysregulation of *HOX* gene expression can dramatically affect cell differentiation, proliferation, and self-renewal, all features associated with neoplastic transformation [1, 4–7].

There is a large body of evidence supporting a particular role for HOX and TALE proteins in leukemogenesis (reviewed by Argiropoulos and Humphries [8]). Indeed, *HOXA9* has been observed to be one of the most consistent indicators of poor

Yacine Graba and René Rezsohazy (eds.), *Hox Genes: Methods and Protocols*, Methods in Molecular Biology, vol. 1196,
DOI 10.1007/978-1-4939-1242-1_21, © Springer Science+Business Media New York 2014

prognosis in acute myeloid leukemia (AML) and acute lymphoid leukemia (ALL) [9–11]. *HOXA9* and *MEIS1* are frequently co-overexpressed in leukemias harboring MLL rearrangements [12, 13] suggesting that there is a specific interaction between these cofactors which is critical for leukemia maintenance [3, 14, 15]. Several techniques and systems have been employed to investigate the role of HOX-TALE in normal and malignant hematopoiesis. This chapter aims to provide an overview and discuss the application of connectivity mapping to gene signatures derived from a *HOXA9 ires Meis1* model of AML within the context of pre-clinical translational research.

1.2 Mining Open-Access Data

The simultaneous measurement of thousands of mRNA transcripts by probe-based microarrays has been used for many years to identify gene expression signatures associated with distinct clinical subtypes of leukemia. These signatures are of particular interest to clinicians and researchers alike and have been used to better understand the diagnosis, prognosis, and pathobiology of the disease [16, 17]. The global Microarray Innovations in Leukaemia (MILE) study, openly available via the Gene Expression Omnibus (GSE13204), was set up in 2005 to assess the feasibility of replacing multiple routine diagnostic methods with a single classifier for 16 subclasses of leukemia, myelodysplastic syndromes (MDSs), and "normal" bone marrow, as determined by gene expression profiling [18]. This repository, together with its European counterpart ArrayExpress, is a great resource for further research into the molecular biology of leukemias (*see* Subheading 3.1.1).

A vast amount of transcriptomic data related to hematological malignancy has been accumulated over the last number of years. Several open-access bioinformatics tools, including HemaExplorer and Differentiation Map (*see* Subheading 3.1.1), have been generated to consolidate and mine this resource.

High-resolution next-generation sequencing (NGS) of complete genomes has led to the creation of disease-specific databases (silos); however their true value in understanding disease and potential in clinical practice has yet to be fully realized [19]. These novel platforms can also be used in cooperation with other techniques, such as chromatin immunoprecipitation to identify low-level protein-DNA interactions across the genome (ChIP-Seq). This has been useful in mapping the binding sites of many transcription factors, including HOX proteins [20]. RNA-seq is another application of the power of NGS; it allows not only quantification of total RNA transcripts but also useful sequence information that does not require a priori knowledge of splice variants or the sequences of noncoding RNAs.

Epigenetics is the terminology used for nonheritable and reversible alterations that provide cues for the expression or the repression of target genes. DNA methylation and histone acetylation status are two of the major processes currently being evaluated in human disease. Growing evidence supports the theory that a disturbance in the epigenetic programming drives or assists leukemogenesis. ChIP-Seq technology is allowing research groups to probe the relationship between various epigenetic states across the genome.

1.3 In Vitro and In Vivo Models of Leukemia

Proof-of-concept studies in relevant cell lines and animal models are required prior to clinical trials. A catalog of well-characterized human leukemia cell lines is available as an excellent resource for selecting appropriate cells for in vitro studies [21]. However, cell culture alone does not provide the ability to examine malignant cells in their appropriate niche [22] and in particular the vital interactions between the leukemia-initiating cell (LIC) and surrounding stroma [23, 24].

Manipulation of *Hox* and *TALE* genes, establishing gain- or loss-of-function models, has been used extensively to study their role in hematopoiesis [8]. A spectrum of phenotypes from relatively mild myeloproliferative disease (*HOXB3* and *HOXB4* [25, 26]) through myeloproliferative neoplasm (*HOXA10* [27]) to full-blown AML (combined *Hoxa9* and *Meis* [28]) have been observed for ectopic overexpression studies. Cytogenetically abnormal AML, primarily resulting from translocations, are associated with altered HOX and TALE expression. However a significant number of AML patients (~50 %) with no underlying chromosomal abnormality, termed cytogenetically normal (CN-AML), are also associated with co-overexpression of *HOXA9* and *MEIS1* [29].

To examine the feasibility of identifying candidate drugs from connectivity mapping in a clinically relevant system, we developed a clonal, conditional (Cre-lox) mouse model of CN-AML by limiting the co-overexpression of *HOXA9* and *Meis1* to levels observed in patient samples, herein called A9M-L2 [30]. The phenotype of the mouse was similar to a previously reported non-conditional counterpart [31].

1.4 Connectivity Mapping and sscMap

Traditionally, rationale based drug design has required a specific and tractable molecular target. While in some cases this has proved very beneficial, as is the case for the archetypal tyrosine-kinase inhibitors being used in Bcr-Abl+ CML [32, 33], success rates are low. The main difficulties are target identification and clinical tractability. Connectivity mapping, a method that links gene signatures to FDA-approved small molecules, has emerged as a way to more rapidly identify candidate drugs for potential clinical repurposing.

The concept behind connectivity mapping is that a particular molecule or "perturbagen" will elicit a similar specific phenotypic or genotypic change to a range of target cells. In practice, differential gene expression profiles generated in experimental cells can be compared to a database of gene signatures generated from an array of perturbagens on a large selection of related cell lines. Ultimately, a connection score is generated, which represents the degree to which a perturbagen led to a similar or an opposite change in gene expression compared to the experimental cells. Members of the Broad Institute were the first to propose and implement this method. They created a public database of perturbed gene expression profiles for over 1,000 different compounds across three different cell lines. In addition, they created an online tool that leverages this database to generate a connection score for a given gene signature. An advanced adaptation of this method, which incorporates a robust measure of statistical significance, has been designed, known as sscMap, or statistically significant connections' map [34, 35]. This Java application calculates a p-value for each connection compared to randomly generated signatures of the same length. By default the application uses the Broad Institute Connectivity Map 02 dataset. In addition, this application allows users to produce custom-built reference profiles, which can be added to the database for future mapping.

1.5 Validation of sscMap

Identification of candidate inhibitors by sscMapping requires subsequent validation in the relevant disease model in order to establish their effects on the disease state. Confirming that the genes present in the signature are actually altered by the compound is a useful first step in validating the process. This can be achieved by treating target cells in vitro with the compound and determining the gene expression of all or a subset of the genes using real-time quantitative PCR (RQ-PCR). Ideally, if the genes originally submitted were downregulated in the model cells compared to control, an increase in expression should be observed and vice versa. In this way, the predicted connection (in silico) can be tested in relevant cells (in vitro) and if confirmed may indicate potential usefulness of the candidate drug in whole-animal studies (in vivo) or clinically.

2 Materials

2.1 General Materials

1. Suspension medium: Dulbecco's Modified Eagle Medium (DMEM), with 10 % fetal calf serum (FCS), 100 IU/ml penicillin, and 100 μg/ml streptomycin (all Invitrogen, Paisley, UK). Store at 4 °C.
2. R10 medium: Roswell Park Memorial Institute (RPMI) 1640 with 10 % FCS, 100 IU/ml penicillin, and 100 μg/ml streptomycin (all Invitrogen). Store at 4 °C.
3. Expansion medium: RPMI 1640 with 10 % FCS (both Invitrogen), 10 ng/ml IL-3, 10 ng/ml IL-6, 10 ng/ml

GM-CSF, and 100 ng/ml SCF (all PeproTech, Rocky Hill, NJ). Store at 4 °C.

4. DMEM with 2 % FCS (Invitrogen). Store at 4 °C.
5. Phosphate Buffered Saline (PBS) (Sigma, St. Louis, MO).
6. Plasticware: 1.5 ml Eppendorf tubes, 15 ml conical base test tube (both Sarstedt, Nümbrecht, Germany) and 6 well plates (Thermo Scientific, Waltham, MA).
7. P200 and P1000 pipettes (Sarstedt).
8. Tissue culture incubator (37 °C and 5 % CO_2).
9. Biological safety cabinet.
10. Hemocytometer and trypan blue solution (Invitrogen).
11. Centrifuges.
12. Vortex.
13. Water bath at 37 °C.
14. A9M-L2 cells were seeded at 1×10^5 cells/ml, maintained in expansion media and passaged every 2/3 days.

2.2 RNA Extraction

1. TRIzol® reagent (Invitrogen). Store at 4 °C.
2. Chloroform (Sigma). Store at 4 °C.
3. Isopropanol (Sigma). Store at 4 °C.
4. Diethylpyrocarbonate (DEPC, Sigma)-treated double-distilled, deionized water (ddH_2O) (0.01 %), autoclaved.
5. Ethanol (Sigma) (75 % v/v, 0.01 % DEPC-treated ddH_2O).
6. A spectrophotometer such as a NanoDrop 1000 spectrophotometer (NanoDrop, DE, USA).

2.3 Complementary DNA (cDNA) Generation

1. Thin walled 0.2 ml PCR tubes (Ambion, Austin, TX).
2. Ampligrade DNase I (Invitrogen), 10× reaction buffer, and 25 mM EDTA. Store all at −20 °C.
3. DEPC ddH_2O (Sigma).
4. Murine Moloney leukemia virus native reverse transcriptase (MMLV-RT), 0.1 M dithiothreitol (DTT), and 5× First-Strand reaction buffer (Invitrogen). Store all at −20 °C.
5. Deoxynucleoside triphosphates (dNTPs) 100 mM (Invitrogen). Store at −20 °C.
6. RNase OUT™ recombinant ribonuclease inhibitor (Invitrogen). Store at −20 °C.
7. Random hexamer primers (300 ng/μl) (Invitrogen). Store at −20 °C.
8. A thermocycler such as a Veriti® 96-Well Fast Thermal Cycler (Applied Biosystems, Foster City, CA).

2.4 Methocult™ Assays

1. Methocult™ M3434 (Stem Cell Technologies). Store at −20 °C.
2. Plasticware: 16–18G blunt-end needles, 1 and 10 ml syringes (both BD Biosciences, Franklin Lakes, NJ) and 35 mm vented NUNC™ tissue culture dishes (Thermo Scientific).
3. Sterile tissue culture water.
4. Grid counter.
5. Microscope with attachment for a digital camera.
6. INT (p-Iodonitrotetrazolium Violet, Sigma). Dilute to a 1 mg/ml working stock in PBS. Filter through a 0.2 μm filter and store protected from light at 4 °C.
7. Digital camera.

2.5 Ex Vivo/In Vivo Assays

1. A9M-L2 leukemias were generated in C57Bl/6 ($CD45.2^+$) mice and transplanted into C57Bl/6 J ($CD45.1^+$) mice (both Charles River Laboratories, Kent, UK).
2. Anesthetic: 100 mg/kg Ketamine (Vétoquinol, Magny-Vernois, France) and 20 mg/kg Xylazine (Bayer HealthCare, Leverkusen, Germany).
3. 300 nM, 1 μM, and 30 mg/kg Entinostat (Selleck, Houston, TX), 7 nM and 25 nM Panobinostat (Selleck, Houston, TX), and DMSO controls.
4. Plasticware: 29G needles and 0.5 ml syringes (both BD Biosciences).
5. Warming device.
6. Paper towel/gauze sponges.
7. Dissection tools: Scissors, blunt-end forceps, scalpels (Harvard Apparatus Kent, UK).
8. Wright Giemsa for cell staining (Sigma Aldrich, St Louis, MO).
9. Nylon Mesh 70 μm (Sefar Nitex Fabric, Lancashire, UK).
10. Red cell lysis buffer (155 mM NH_4Cl, 10 mM $NaHCO_3$, 0.1 mM EDTA).

3 Methods

3.1 Identifying Drugs

3.1.1 Identifying a Gene Signature

In order to use sscMap to identify candidate compounds a gene signature is required. This gene signature is simply a list of differentially expressed genes (DEG) and whether they are up- or down-regulated, regardless of magnitude. A full discussion regarding identification of DEG is beyond the scope of this chapter [36, 37]. However, below are listed some useful resources for identifying genes that represent disease or disease progression related to hematopoiesis and leukemia. Many tools are available for import and

DEG analysis of microarray data, both commercial, such as Partek Genomics Suite http://www.partek.com and GeneSpring GX http://www.agilent.com, and open source, such as dChip https://sites.google.com/site/dchipsoft/ or a multitude of R packages from the Bioconductor stable http://bioconductor.org/.

1. The Gene Expression Omnibus (GEO) provided by the National Center for Biotechnology Information (NCBI) is a public data repository for array- and sequence-based data. It is well curated and very broad in its scope, http://www.ncbi.nlm.nih.gov/geo/. Many of the data sets include both normalized and raw data files. The GEO also provides a number of tools to allow interrogation of the data and identification of DEG without having to download or analyze the raw data. Another online repository is ArrayExpress, which is provided by the European Bioinformatics Institute (EMBL-EBI). This repository, in addition to its own large database of experiments, includes all of the GEO data over 7 days old, http://www.ebi.ac.uk/arrayexpress/. ArrayExpress also includes a suite of tools, perhaps more comprehensive and user friendly, to analyze the data sets directly and identify DEG.
2. Differentiation Map Portal (DMAP) is an online tool based on a data set created by the Broad Institute [38]. It consists of gene expression profiles from cells isolated from each of the canonical stages of hematopoiesis on the basis of cell surface marker expression, www.broadinstitute.org/dmap/. While the raw data is available, the available tools are fairly comprehensive. They include tools to identify DEG, gene correlation, and gene set enrichment analysis (GSEA), comparing each compartment of hematopoiesis.
3. HemaExplorer is similar to DMAP in providing access to data regarding gene expression in the hematopoietic system [39]. However, HemaExplorer is more oriented toward plotting and investigating a specific gene in the cumulative database of collated expression profiles, http://servers.binf.ku.dk/hemaexplorer/. In addition to typical human hematopoiesis it also provides expression information regarding human AML subtypes and the murine hematopoietic system. Direct links to all the data sources it employs are made available and can be downloaded and analyzed to identify DEG.

3.1.2 sscMap

The sscMap tool itself together with the reference database is available for download from the FTP server of the original author. The following address forwards to the current host: http://purl.oclc.org/NET/sscMap.

The tool itself has a relatively simple user-friendly interface and the package comes with a comprehensive user manual. However, below are outlined the basic steps required to generate a list of connection scores based on a gene signature.

1. Create a signature file that lists the probe sets identified as differentially expressed in the relevant analysis (*see* **Note 1**). It should contain only probe sets available in the queried perturbagen (compound)-gene expression reference database. For example, in the case of the database provided with the package, only HG-U133A microarray probe set IDs are allowed. The second column should contain a positive number for upregulated probe sets and a negative number for downregulated probe sets.
2. The software package does not require installation; once downloaded and extracted the sscmap-gui.bat (or sscmap-gui for Unix-based systems) can be executed to launch the user interface.
3. A practically unlimited number of gene signatures can be imported into the software for analysis.
4. Once imported a "query" can be run whereby each signature is analyzed in turn and a corresponding results file is generated. Run times vary based on the size and number of signatures and available computing power. In general, a few hours are typical.
5. The results are presented in the form of a graph plotting connection score against estimated p-value. The threshold for significance is also plotted as a green horizontal line (Fig. 1).

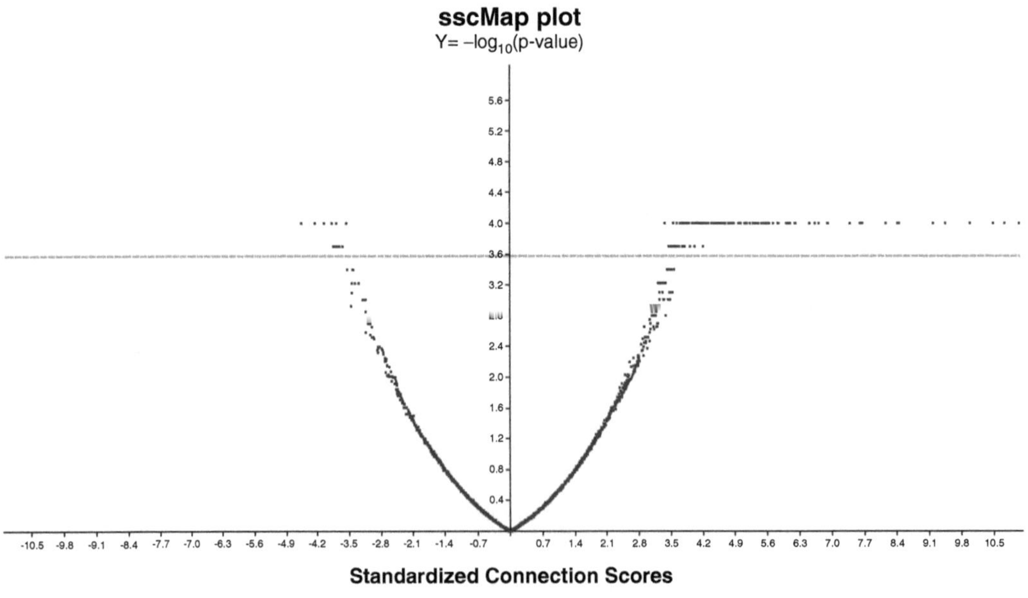

Fig. 1 Representative sscMap graph plotting connection score for each perturbagen/compound against estimated *p*-value for the connection with the gene signature. The threshold for significance is marked as a *horizontal line*. The software default is to position positive connections to the *right* of the origin and negative connections to the *left*

Table 1
An sscMap output demonstrating the analyzed condition (refsetname), the standardized connection score (setscore/stdev), the relative significance of the particular connection "setpvalue," and a score of significance (significance_mark)

Refsetname	setpvalue	setscore/ stdev	significance_ mark
meclozine_8.6muM_HL60	0.0001	−3.54	1
(+)-chelidonine_10.8muM_HL60	0.0001	−3.72	1
(+)-chelidonine_10.8muM_MCF7	0.0002	−3.47	1
(+)-chelidonine_10.8muM_PC3	0.0004	−3.70	0
(+)-isoprenaline_11muM_HL60	0.0016	3.13	0
(+)-isoprenaline_11muM_MCF7	0.8216	0.23	0
(+)-isoprenaline_11muM_PC3	0.2302	−1.19	0
(±)-catechin_13.8muM_HL60	0.2692	−1.10	0
(±)-catechin_13.8muM_MCF7	0.3238	0.99	0
(±)-catechin_13.8muM_PC3	0.2296	−1.19	0
(−)-atenolol_15muM_HL60	0.8588	−0.18	0
metformin_1mM_MCF7	0.6108	0.51	0
metformin_24.2muM_HL60	0.7570	−0.32	0
metformin_24.2muM_MCF7	0.8616	−0.17	0
metformin_24.2muM_PC3	0.6966	−0.38	0
methacholine chloride_20.4muM_MCF7	0.4108	−0.85	0
methacholine chloride_20.4muM_PC3	0.1690	−1.36	0
methanthelinium bromide_9.6muM_HL60	0.4604	−0.76	0
rolipram_14.6muM_HL60	0.0518	1.96	0

A significance mark of 1 indicates that the setpvalue is below the threshold set

6. The results can be exported as a ".tab" file that can be opened in most spreadsheet software for review.
7. The results file contains many columns of information. The first column (refsetname) contains the reference instance that was analyzed. In the default reference set, each instance refers to a particular compound, dose, and cell line. The other columns of interest are the standardized connection score (setscore/stdev), which if negative suggests a reversal of the DEG pattern included in the signature and if positive evokes a similar DEG pattern. The "setpvalue" allows significant connections to be identified on the basis of statistical probability. The significance mark column includes a 1 when the setpvalue is below the threshold set. Table 1 includes an example of the pertinent results columns.

3.2 Validation of sscMap by RQ-PCR

Once a candidate compound has been identified from the sscMap analysis it is important to validate its effect on your target cells of choice with respect to altering the DEG compared to appropriate vehicle controls. In this example, primary A9M-L2 leukemia cells were treated ex vivo with Entinostat or vehicle (0.01 % DMSO) for 24 h (*see* Subheading 3.3) and the expression of a subset of genes ($n = 10$) from the sscMap submitted DEG were quantified by RQ-PCR.

3.2.1 RNA Extraction

1. Collect 5×10^6 cells by centrifugation ($100 \times g$) for 5 min at room temperature (RT).
2. Add 1 ml of TRIzol® or a similar reagent mix by vigorous pipetting (P1000) until sample is homogenized.
3. Incubate for 5 min at RT.
4. Add 200 μl of chloroform.
5. Shake tube vigorously by hand for 15 s and then incubate for 2–3 min at RT.
6. Centrifuge the sample at $12{,}000 \times g$ for 15 min at 4 °C.
7. Gently aspirate the upper (aqueous) phase using a P200 pipette and transfer to a fresh 1.5 ml Eppendorf tube.
8. Add 500 μl of 100 % isopropanol and incubate at RT for 10 min.
9. Centrifuge the sample at $12{,}000 \times g$ for 10 min at 4 °C or store at −20 °C.
10. Remove the supernatant by aspiration.
11. Wash the RNA pellet with 1 ml of 75 % ethanol:25 % DEPC·ddH_2O, mix briefly by inversion, and then centrifuge at $12{,}000 \times g$ for 10 min at 4 °C.
12. Discard the wash and allow the RNA pellet to air-dry for 5–10 min.
13. Resuspend the RNA pellet in 50 μl of DEPC·ddH_2O, incubate for 5 min at 55 °C, and mix by vortex.
14. Spectrophotometrically measure the RNA concentration (absorbance at 260 nm:280 nm for quantification, 260 nm:230 nm for purity).

3.2.2 Complementary DNA (cDNA) Generation

1. Add 1 μl of 10× deoxyribonuclease I (DNase I) reaction buffer and 1 μl DNase I enzyme (100 U) to 2 μg of RNA (plus DEPC·ddH_2O up to 9 μl) in a thin-walled 0.2 ml PCR tube to degrade potentially contaminating genomic DNA.
2. Incubate sample for 10–15 min at RT.

3. Inactivate the DNase I by adding 1 μl of 25 mM EDTA solution (pH 8.0) and heating sample at 65 °C for 10 min.
4. Place sample on ice to prevent formation of secondary structures or store for a long term at −80 °C.
5. For a single reaction to the diluted RNA sample (10 μl) add 4 μl 5× first-strand buffer, 1 μl random primers (300 ng/μl), 1 μl of dNTP (deoxynucleotide triphosphate) mix (10 mM each dATP, dCTP, dGTP, dTTP), 2 μl dithiothreitol (DTT 0.1 M), 1 μl RNaseOUT™(recombinant ribonuclease inhibitor), and 1 μl Moloney Murine Leukemia Virus reverse transcriptase (M-MLV-RT; 200 U/μl, *see* **Note 2**).
6. Incubate samples for 2 h at 37 °C followed by heat inactivation of the enzyme at 70 °C for 10 min (*see* **Note 3**).
7. Use immediately or store cDNA at −20 °C until required for RQ-PCR.

3.2.3 RQ-PCR

For detailed methodology of TaqMan RQ-PCR see Dickson et al. [40]. For this particular analysis SYBR-based RQ-PCR was performed following the manufacturer's (Roche Diagnostics, Basel, Switzerland) procedures briefly:

1. Prepare lyophilized oligonucleotides at 100 μM stock and combine forward and reverse primers (now 50 μM each) into a single Eppendorf tube.
2. For each cDNA sample, make a master mix consisting of 7.5 μl of FastStart SYBR Green master mix, 5.5 μl of DEPC·ddH_2O, and 0.5 μl of cDNA sample.
3. Add 13.5 μl of this master mix in triplicate to a 96-well optical reaction plate on ice.
4. Add 1.5 μl of combined forward and reverse primers to each well (final concentration 5 μM each).
5. Perform RQ-PCR following the standard SYBR cycling and temperatures: 50 °C for 2 min, 95 °C for 15 min, and 40 cycles of 94 °C for 15 s, 55 °C for 1 min, and 72 °C for 30 s.
6. Analyze the data collection step for single melt curve as a measure of specificity; if a single melt curve is not attained redesign primers (Fig. 2).
7. Obtain cycle threshold (C_t) values for each analyte (Fig. 3) and correct for endogenous gene or gene panel (for example *18S*, *Hprt*, *Abl*, *Gapdh*); in this case *18S* rRNA was used. Generate correct C_t (cC_t) for each gene as follows:

$$cC_t = 8 + \bar{C}_t - \bar{C}_t^{18S}$$

where $\bar{C}_t$ is the mean C_t for the subject gene and $\bar{C}_t^{18S}$ is the *18S* mean C_t and where the constant 8 refers to the calculated

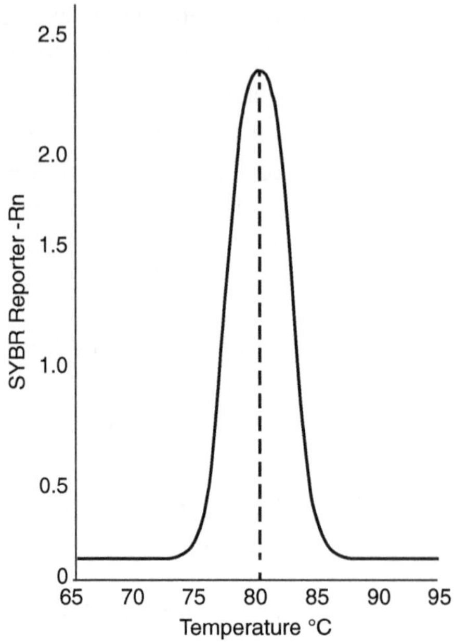

Fig. 2 Representative DNA melt curve for an individual primer set in triplicate, demonstrating specificity of amplification by the presence of a single curve of increased fluorescence at 81 °C

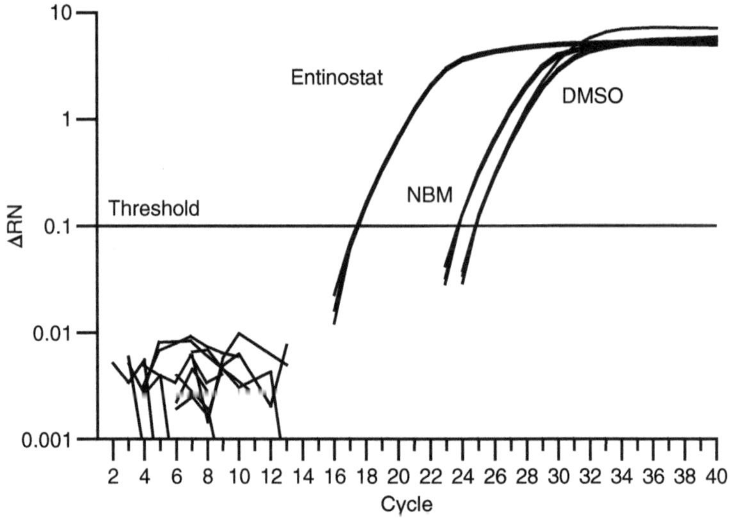

Fig. 3 Representative amplification plots, in triplicate, showing cycle threshold (C_t) values for *Gdf3* in normal bone marrow (NBM) and A9M-L2 bone marrow following treatment with Entinostat or vehicle (DMSO). The lower C_t value for Entinostat [26] compared to DMSO [30] reflects an increase in expression of *Gdf3* following drug treatment. The fold change in expression is calculated as $2^{\Delta\Delta Ct}$ which in this case would be 2^4- or 16-fold if 50 ng equivalents of RNA were used in the initial reaction (as evaluated by an *18SrRNA* C_t value of 8)

C_t value for *18S* representing 50 ng equivalents of RNA in the cDNA sample (as determined by established standard curve analysis).

8. Calculate the change in the expression of a gene by the change in cC_t values (ΔcC_t): A lower C_t value represents higher gene expression:

$$\Delta cC_t = cC_{t,2} - cC_{t,1}$$

In this case, the A9M-L2 v Normal Bone Marrow (NBM) signature that scored a connection with Entinostat was primarily made up of genes with decreased expression. Following Entinostat treatment, we observed an increase in all genes analyzed from the signature. This finding was further supported by the antileukemic effect of Entinostat in the A9M-L2 model.

3.3 Ex Vivo Treatment

Due to the robustness and aggressiveness of the A9M-L2 model, candidate drugs identified from the sscMap can be used to treat primary leukemia cells ex vivo. A dose–response curve should be generated to account for variation between primary cells and the cell lines initially used to generate the connections with specific perturbagens within the cMap database. The ex vivo treatment allows for downstream applications including gene expression profiles, cell viability, western blotting, and colony-forming unit (CFU) assays.

1. Thaw and collect primary A9M-L2 leukemia cells by incubation of the vial in a water bath at 37 °C for 2 min, gentle resuspension of the cells in pre-warmed suspension medium (37 °C), and centrifugation at 100 ×*g* for 5 min at RT.
2. Allow thawed primary leukemia cells to recover overnight in expansion medium 37 °C/5 % CO_2 and then remove dead cells and membranes by collection and centrifugation at 100 ×*g* for 5 min at RT. Resuspend cells and obtain a cell count by hemocytometer/automated counting system.
3. Plate cells (10^5–10^6 per ml) in a 6-well plate with R10 media. Add a range of concentrations of candidate drug over a defined time course, in our case Entinostat (300 nM to 1 μM) and Panobinostat (7–25 nM) along with vehicle controls.
4. Monitor cell viability, apoptosis, and altered cell cycle using multiple applications including Cell Titer Glo™ (*see* **Note 4**), Caspase Glo™ (*see* **Note 5**), flow cytometry (*see* **Note 6**), or equivalent systems. Multiple applications are required to obtain robust data related to cell number, viability, apoptosis, and cell status.
5. Obtain gene expression data for a subset of the gene signature used to generate the connection to the candidate drug in the original sscMap. Ex vivo-treated A9M-L2 can subsequently be used for further analysis including colony formation and transplantation assays.

3.4 Methocult™ Colony-Forming Unit Assay

A principal property of leukemia cells is their ability to form serial colonies in Methocult™ (Methocult™). The semisolid culture system provides niche-like support for the growth of primary hematopoietic cells.

3.4.1 Initial Plating

1. Thaw MethoCult™ at 37 °C in a water bath and mix well by shaking vigorously (or vortexing) for 1–2 min. Allow Methocult™ to stand for 15–20 min or until bubbles have dispersed before aliquoting into labeled Eppendorf tubes using a blunt-end needle (16–18G). Use immediately or store at −20 °C until required.
2. Collect 1×10^4–1×10^5 treated cells by centrifugation ($100 \times g$ for 5 min at RT) and carefully aspirate media using a P200 pipette, leaving ~90 μl for resuspension. Resuspend cells and transfer to Methocult™ (~1 ml, *see* **Note 7**).
3. Mix Methocult™/cell suspension mixture vigorously either by vortex or syringe with blunt-end needle (16–18G). Allow Methocult™ to stand for 15–20 min or until bubbles have dispersed.
4. Aspirate and dispense the Methocult™/cell suspension mixture into a 35 mm vented NUNC tissue culture dish using a syringe with blunt-end needle (16–18G).
5. Incubate Methocult™/cell suspension mixture at 37 °C/5 % CO_2 in a humidified chamber (*see* **Note 8**) for up to 7 days to allow for colony formation. Obtain colony count and subtype microscopically, preferably using a grid counter. Collect colonies at day 5 for replating assays (*see* **Note 9**).

3.4.2 Replating Assays

1. Collect colonies from Methocult™ dishes by addition of 2 ml of pre-warmed suspension medium (37 °C) and resuspension using a P1000 pipette. Transfer the Methocult™/colony suspension mixture to a labeled 15 ml conical test tube and wash dish with an additional 2 ml of suspension medium to collect all cells.
2. Mix the Methocult™/colony suspension vigorously by pipetting up and down with a P1000 and collect cells by centrifugation ($100 \times g$ for 5 min at RT).
3. Repeat **steps 2–5** as described in Subheading 3.4.1.
4. Reduced colony replating ability indicates loss of leukemic repopulating cells.

3.4.3 Staining Methocult™ Colonies

Methocult™ colony images are often difficult to record. P-Iodonitrotetrazolium violet (INT) staining of colonies allows for visualization and digital recording for qualitative and quantitative analyses (Fig. 4, *see* **Note 10**).

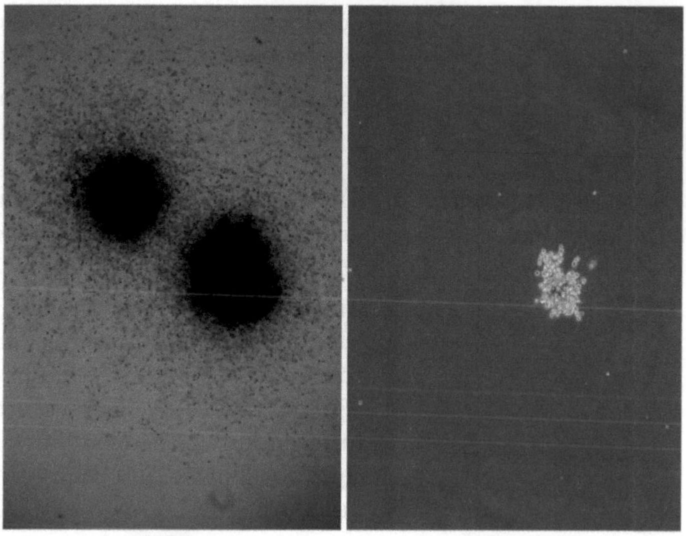

Fig. 4 Representative Methocult™ colony assays from vehicle (DMSO)- or Entinostat-treated leukemia bone marrow cells. Bone marrow-derived hematopoietic colonies grow well in methylcellulose and are used as a surrogate for self-renewal capacity and leukemogenic potential

1. Add 100 μl of INT (final concentration of 100 μg/ml in PBS) directly into each Methocult™ dish containing colonies.
2. Incubate dishes overnight (~16 h) after which time the colonies will appear purple/brown due to the reduction of the tetrazolium dye.
3. Cultures can then be digitally imaged under microscopy to evaluate the effect of the candidate drug(s) on colony-forming ability.

3.5 Transplantation of Ex Vivo-Treated A9M-L2 Cells

1. Collect ex vivo-treated cells by centrifugation ($100 \times g$ for 5 min at RT). Reconstitute cells in DMEM and 2 % FBS at a concentration of $>4 \times 10^6$/ml mix well and adjust volume (<150 μl) to deliver 5×10^5–1×10^6 cells to each recipient mouse using a 29G needle attached to a 0.5 ml syringe (*see* **Note 11**).
2. Anesthetize the mouse using ketamine (100 mg/ml) and xylazine (20 mg/ml) by intraperitoneal injection (100 μl per 20 g body weight). Inject the cells slowly into the retrobulbar sinus (*see* **Note 12**). Remove the needle gently to prevent injury to the eye and apply gentle pressure to the injection site to limit bleeding.
3. Return the mouse to its cage with a protected warming device and monitor its progress through the recovery process (*see* **Note 13**).

4. Mice must be monitored daily for signs of illness and disease development and humanely killed and necropsied and hematopoietic tissue examined at the first signs of ill health.
5. Increased survival or time to morbidity indicates reduction or lack of retention of leukemia-repopulating cells.

3.6 In Vivo Treatment of Leukemic Mice

1. Transplant recipient mice with 1×10^6 A9M-L2 cells as described in Subheading 3.5 and allow leukemia to develop for 14 days.
2. Treat A9M-L2 leukemia mice with candidate drug using the least invasive mode of delivery at levels on or below the maximal tolerated dose [41, 42]. In our case a single dose of Entinostat (30 mg/kg) or vehicle control (DMSO) was administered intravenously (tail vein). Mice should be monitored daily for signs of illness and disease development as described in Subheading 3.5.
3. Increased survival or time to morbidity indicates reduction or lack of retention of leukemia-repopulating cells.

3.7 Disease Monitoring

Mice generally present with groomed fur and a sleek and flexible posture. However mice that are sick will stop grooming, giving matte, ruffled fur and present with a hunched posture. Preleukemic mice can be identified by the presence of blast cells in small peripheral blood samples (~50 μl) obtained from superficial veins. Due to organ infiltration, leukemic mice tend to develop labored breathing and appear lethargic or non-responsive due to the associated anemia. It is upon these signs that mice are sacrificed and investigated.

3.7.1 Gross Pathology

1. Sacrifice mice by a Schedule 1 procedure (Scientific Procedures Act 1986 in the UK) and dissect to gain access to hematopoietic tissue including peripheral blood (cardiac puncture) spleen, thymus, lymph nodes, and long bones (femur, tibia, and pelvis).
2. Necropsied leukemic mice generally have an enlarged spleen (Fig. 5) and pale bone marrow. Leukemic infiltration to other organs such as the liver and lung are also commonly observed.
3. Analyze mice as soon as possible to avoid autolysis.
4. Remove the spleen, liver, kidney, and lung (and lymph nodes if present) under sterile conditions in a biological safety cabinet using a pair of scissors and forceps.
5. Weigh the spleen to assess for splenomegaly (>0.1 g is indicative of leukemia).

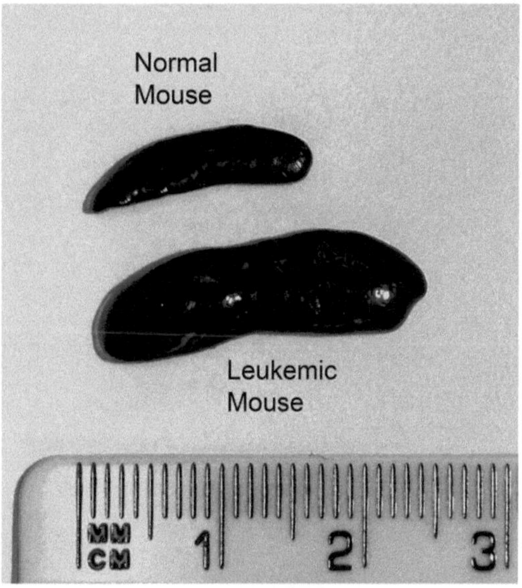

Fig. 5 Representative image showing the size of a leukemic mouse spleen compared to a normal spleen. Human leukemia is associated with an enlarged spleen (splenomegaly) which is reflected within the mouse model. Normal spleen weights are ~0.1 g whereas leukemic spleens for A9M-L2 were generally >3 times the normal size

6. Prepare peripheral blood smears (Fig. 6) and "touch preps" of experimental tissue on glass slides. Touch preps are prepared by touching the dissected side of the organ against a microscope slide to remove a single layer of cells. Allow slides to air-dry prior to staining of leucocytes (modified Wright-Giemsa).

3.7.2 Collection of Leukemic/Normal Hematopoietic Cells

In a Biological Safety Cabinet (Level II):

1. Place the spleen between two pieces of sterile 70 μm nylon mesh in a 50 mm petri dish. Add 3 ml of suspension medium and then mechanically disrupt the spleen; using the back of a pair of blunt-ended forceps transfer cells into a 15 ml conical tube.
2. Remove connective tissue from the outside of the bone(s); using blunt dissection (back of scissors) open both ends of the bone, under sterile conditions, to expose the marrow and flush using 3 ml of suspension medium in a syringe attached to a 25G needle.
3. Collect bone marrow or spleen cells by centrifugation (100 × *g* for 5 min at RT), remove supernatant, resuspend cells in 1 ml of red cell lysis buffer, and incubate on ice for 2 min.
4. After lysis, add 10 ml of suspension medium slowly to the cells, mix gently, and collect by centrifugation (100 × *g* for 5 min at RT) for further studies including cell viability, morphology,

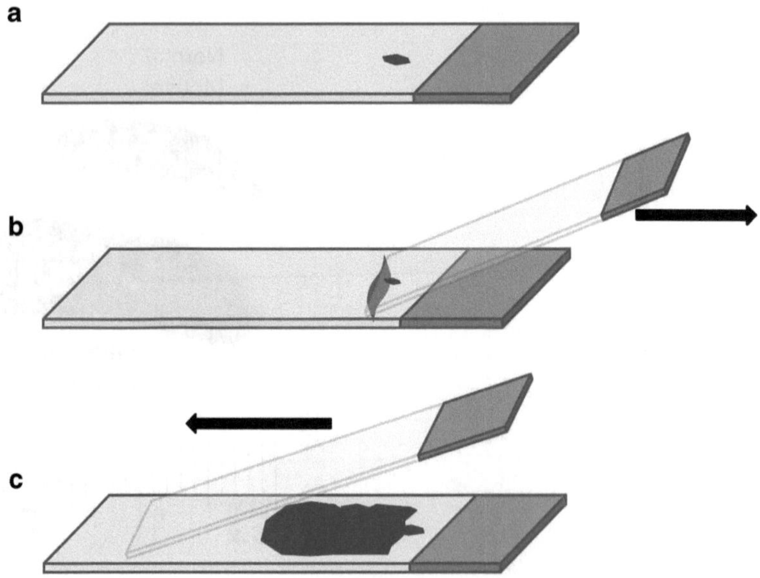

Fig. 6 Diagram showing how to perform blood smears. Blood smears are an important diagnostic test for blood diseases. (**a**) A single drop of patient blood is dropped gently onto a clean microscope slide (slide 1). (**b**) By positioning another microscope slide (slide 2) in contact with the blood drop at <45° angle, the drop will spread over the width of the microscope slide. (**c**) Gentle movement of slide 2 over slide 1 results in distribution of peripheral blood cells along the length of slide 1 and single cells (e.g., leukemic blasts, erythrocytes, and lymphocytes) can be identified at the leading edge of the blood film

flow cytometry, Methocult™ assays, and DNA/RNA/protein analysis. Remaining cells can be frozen down in 90 % FBS/10 % DMSO or re-transplanted into recipient mice.

4 Notes

1. The optimum length of signature is typically between 10 and 100 genes, but is best decided empirically. A signature file is simply a tab-delimited text file labeled with a <.sig> extension. Most spreadsheet software is able to export in the tab-delimited format and the extension can be manually altered afterward. The first column should list the probe set ID (such as 209905_at, present on the HG-U133A chip). The second column should provide the related fold change of that probe set ID, with positive values representing increased expression relative to the control cohort and negative values representing decreased expression. The magnitude of the fold change or the ordering of probe sets is not considered, only the positive/negative dichotomy.

2. For multiple samples it is advised to generate a 2× reverse transcriptase master mix (minus RNA) to decrease sample-to-sample variability.
3. A thermocycler may be programmed to complete both incubations sequentially.
4. The Cell Titer-Glo® Luminescent Cell Viability Assay from Promega determines the number of viable cells in culture by quantifying the amount of ATP present in the sample, indicating the presence of metabolically active cells, and is applicable to high-throughput screening. It is widely used as a surrogate marker of cell number and when appropriate standard curves are generated and can substitute for traditional cell counting assays such as trypan blue exclusion.
5. Caspase Glo® Luminescent Assays from Promega measure the activity of various caspases in the cell culture which are known to have an essential role in apoptosis along with other roles in necrosis and inflammation. Caspase activity is determined by measuring the luminescent signal emitted when the caspase substrate is cleaved. It is widely used for high-throughput screening and as a surrogate for more traditional methods of detecting apoptosis such as Annexin V or TdT staining.
6. Flow cytometry can be used as a means of quantifying the proportion of cells in different phases of the cell cycle by measuring the intensity of fluorescence produced when propidium iodide (PI) is bound to DNA in the cells. The intensity correlates with the amount of DNA that the cells contain.
7. Keep the volume of 1×10^4–1×10^5 cells smaller than 90 μl. Addition of more than 10 % medium to 1 ml Methocult™ may dilute the Methocult™ such that cells are not stationed at a fixed position and do not form nice colonies. If the volume for 1×10^4–1×10^5 cells is larger than 90 μl, spin down the cells ($100 \times g$ for 5 min at RT) and aspirate the supernatant. Resuspend the cells in 90 μl of suspension medium for plating.
8. Filling a dish with water reduces the evaporation of liquid from dishes containing MethoCult™, protecting the samples from drying out.
9. If the pH of the Methocult™ dramatically decreases because of high cell density (color turns from light red to light yellow), replate the cells immediately to avoid starvation and cell death.
10. INT staining is an end-point assay that will affect normal colony formation.
11. A9M-L2 cells were derived by serial transplantation and as a clonal leukemia are highly aggressive (time to leukemia <28 days) and readily transplant to recipient mice without the

need for myeloablation. Tail vein or retro-orbital injection are approved standard transplantation procedures.

12. A mouse weighing 30 g has a blood volume of approximately 2.2 ml; therefore, injecting any more than 200 μl may lead to temporary vascular overload.
13. Protect the mouse from the warming device using paper towel or gauze to prevent thermal injury. The mouse can safely be placed on top of this device until it recovers from the anesthetic.

Acknowledgements

The authors would like to thank the staff within the Biological Resource Unit and Bioinformatics Cores, Queen's University Belfast. Alex Thompson is a recipient of the American Cancer Society for Beginning Investigator Fellowship from the UICC. This work was supported by grants from Leukemia Lymphoma Research, UK (07016 and 09035), the Leukaemia Lymphoma Northern Ireland (LLNI), and Biotechnology and Biological Science Research Council (BBSRC).

References

1. Dickson GJ, Kwasniewska A, Mills KI et al (2009) Hoxa6 potentiates short-term hemopoietic cell proliferation and extended self-renewal. Exp Hematol 37:322–333 e3. doi:10.1016/j.exphem.2008.10.015, S0301-472X(08)00508-0 [pii]
2. Bryder D, Rossi DJ, Weissman IL (2006) Hematopoietic stem cells: the paradigmatic tissue-specific stem cell. Am J Pathol 169:338–346. doi:10.2353/ajpath.2006.060312, S0002-9440(10)62717-4 [pii]
3. Wong P, Iwasaki M, Somervaille TC et al (2007) Meis1 is an essential and rate-limiting regulator of MLL leukemia stem cell potential. Genes Dev 21:2762–2774. doi:10.1101/gad.1602107, gad.1602107 [pii]
4. Lawrence HJ, Sauvageau G, Humphries RK, Largman C (1996) The role of HOX homeobox genes in normal and leukemic hematopoiesis. Stem Cells 14:281–291. doi:10.1002/stem.140281
5. Calvo KR, Sykes DB, Pasillas M, Kamps MP (2000) Hoxa9 immortalizes a granulocyte-macrophage colony-stimulating factor-dependent promyelocyte capable of biphenotypic differentiation to neutrophils or macrophages, independent of enforced meis expression. Mol Cell Biol 20:3274–85
6. Owens BM, Hawley RG (2002) HOX and non-HOX homeobox genes in leukemic hematopoiesis. Stem Cells 20:364–379. doi:10.1634/stemcells.20-5-364
7. Khan SN, Jankowska AM, Mahfouz R et al (2013) Multiple mechanisms deregulate EZH2 and histone H3 lysine 27 epigenetic changes in myeloid malignancies. Leukemia 27:1301–9. doi:10.1038/leu.2013.80
8. Argiropoulos B, Humphries RK (2007) Hox genes in hematopoiesis and leukemogenesis. Oncogene 26:6766–6776, doi:10.1038/sj.onc.1210760, 1210760 [pii]
9. Golub TR, Slonim DK, Tamayo P et al (1999) Molecular classification of cancer: class discovery and class prediction by gene expression monitoring. Science 286:531–537, 7911 [pii]
10. Lawrence HJ, Rozenfeld S, Cruz C et al (1999) Frequent co-expression of the HOXA9 and MEIS1 homeobox genes in human myeloid leukemias. Leukemia 13:1993–1999
11. Soulier J, Clappier E, Cayuela JM et al (2005) HOXA genes are included in genetic and biologic networks defining human acute T-cell leukemia (T-ALL). Blood 106:274–286. doi:10.1182/blood-2004-10-3900, 2004-10-3900 [pii]

12. Armstrong SA, Staunton JE, Silverman LB et al (2002) MLL translocations specify a distinct gene expression profile that distinguishes a unique leukemia. Nat Genet 30:41–47. doi:10.1038/ng765 ng765 [pii]
13. Ferrando AA, Armstrong SA, Neuberg DS et al (2003) Gene expression signatures in MLL-rearranged T-lineage and B-precursor acute leukemias: dominance of HOX dysregulation. Blood 102:262–268. doi: 10.1182/blood-2002-10-3221 2002-10-3221 [pii]
14. So CW, Karsunky H, Wong P et al (2004) Leukemic transformation of hematopoietic progenitors by MLL-GAS7 in the absence of Hoxa7 or Hoxa9. Blood 103:3192–3199. doi:10.1182/blood-2003-10-3722 2003-10-3722 [pii]
15. Somervaille TC, Matheny CJ, Spencer GJ et al (2009) Hierarchical maintenance of MLL myeloid leukemia stem cells employs a transcriptional program shared with embryonic rather than adult stem cells. Cell Stem Cell 4:129–140. doi:10.1016/j.stem.2008.11.015, S1934-5909(08)00617-6 [pii]
16. Haferlach T, Bacher U, Kohlmann A, Haferlach C (2009) Discussion of the applicability of microarrays: profiling of leukemias. Methods Mol Biol 509:15–33. doi:10.1007/978-1-59745-372-1_2
17. Wouters BJ, Lowenberg B, Delwel R (2009) A decade of genome-wide gene expression profiling in acute myeloid leukemia: flashback and prospects. Blood 113:291–298. doi:10.1182/blood-2008-04-153239, blood-2008-04-153239 [pii]
18. Kohlmann A, Kipps TJ, Rassenti LZ et al (2008) An international standardization programme towards the application of gene expression profiling in routine leukaemia diagnostics: the Microarray Innovations in LEukemia study prephase. Br J Haematol 142:802–807. doi:10.1111/j.1365-2141.2008.07261.x, BJH7261 [pii]
19. Kohlmann A, Grossmann V, Haferlach T (2012) Integration of next-generation sequencing into clinical practice: are we there yet? Semin Oncol 39:26–36. doi:10.1053/j.seminoncol.2011.11.008, S0093-7754(11)00297-1 [pii]
20. Huang Y, Sitwala K, Bronstein J et al (2012) Identification and characterization of Hoxa9 binding sites in hematopoietic cells. Blood 119:388–98. doi:10.1182/blood-2011-03-341081
21. Drexler HG (2010) The guide to Leukemia-Lymphoma cell lines, 2nd edn. DSMZ Braunschweig, Germany
22. Shiozawa Y, Havens AM, Pienta KJ, Taichman RS (2008) The bone marrow niche: habitat to hematopoietic and mesenchymal stem cells, and unwitting host to molecular parasites. Leukemia 22:941–50. doi:10.1038/leu.2008.48
23. McMillin DW, Negri JM, Mitsiades CS (2013) The role of tumour-stromal interactions in modifying drug response: challenges and opportunities. Nat Rev Drug Discov 12:217–28. doi:10.1038/nrd3870
24. Sison EAR, Rau RE, McIntyre E et al (2013) MLL-rearranged acute lymphoblastic leukaemia stem cell interactions with bone marrow stroma promote survival and therapeutic resistance that can be overcome with CXCR4 antagonism. Br J Haematol 160:785–97. doi:10.1111/bjh.12205
25. Sauvageau G, Thorsteinsdottir U, Hough MR et al (1997) Overexpression of HOXB3 in hematopoietic cells causes defective lymphoid development and progressive myeloproliferation. Immunity 6:13–22
26. Schmittwolf C, Porsch M, Greiner A et al (2005) HOXB4 confers a constant rate of in vitro proliferation to transduced bone marrow cells. Oncogene 24:561–72. doi:10.1038/sj.onc.1208202
27. Thorsteinsdottir U, Sauvageau G, Hough MR et al (1997) Overexpression of HOXA10 in murine hematopoietic cells perturbs both myeloid and lymphoid differentiation and leads to acute myeloid leukemia. Mol Cell Biol 17:495–505
28. Kroon E, Krosl J, Thorsteinsdottir U et al (1998) Hoxa9 transforms primary bone marrow cells through specific collaboration with Meis1a but not Pbx1b. EMBO J 17:3714–25. doi:10.1093/emboj/17.13.3714
29. Haferlach C, Mecucci C, Schnittger S et al (2009) AML with mutated NPM1 carrying a normal or aberrant karyotype show overlapping biologic, pathologic, immunophenotypic, and prognostic features. Blood 114:3024–3032. doi:10.1182/blood-2009-01-197871, blood-2009-01-197871 [pii]
30. Ramsey JM, Kettyle LMJ, Sharpe DJ et al (2013) Entinostat prevents leukemia maintenance in a collaborating oncogene-dependent model of cytogenetically normal acute myeloid leukemia. Stem Cells 31:1434–45. doi:10.1002/stem.1398
31. Wilhelm BT, Briau M, Austin P et al (2011) RNA-seq analysis of 2 closely related leukemia clones that differ in their self-renewal capacity. Blood 117:e27–e38. doi:10.1182/blood-2010-07-293332, blood-2010-07-293332 [pii]

32. Milojkovic D, Apperley J (2008) State-of-the-art in the treatment of chronic myeloid leukaemia. Curr Opin Oncol 20:112–21. doi:10.1097/CCO.0b013e3282f1fe8a
33. Sullivan C, Peng C, Chen Y et al (2010) Targeted therapy of chronic myeloid leukemia. Biochem Pharmacol 80:584–91. doi:10.1016/j.bcp.2010.05.001
34. Zhang SD, Gant TW (2009) sscMap: an extensible Java application for connecting small-molecule drugs using gene-expression signatures. BMC Bioinformatics 10:236. doi:10.1186/1471-2105-10-236, 1471-2105-10-236 [pii]
35. McArt DG, Zhang S-D (2011) Identification of candidate small-molecule therapeutics to cancer by gene-signature perturbation in connectivity mapping. PLoS One 6:e16382. doi:10.1371/journal.pone.0016382
36. Parmigiani G, Garrett ES, Irizarry RA, Zeger SL (2003) An Overview of Methods and Software. In: The Analysis of Gene Expression Data Statistics for Biology and Health, Springer, New York, pp 1–45. doi:10.1007/0-387-21679-0_1
37. Zhang A (2006) Advanced analysis of gene expression microarray data. World Scientific, Singapore
38. Novershtern N, Subramanian A, Lawton LN et al (2011) Densely interconnected transcriptional circuits control cell states in human hematopoiesis. Cell 144:296–309. doi:10.1016/j.cell.2011.01.004
39. Bagger FO, Rapin N, Theilgaard-Mönch K et al (2013) HemaExplorer: a database of mRNA expression profiles in normal and malignant haematopoiesis. Nucleic Acids Res 41:D1034–9. doi:10.1093/nar/gks1021
40. Dickson GJ, Lappin TR, Thompson A (2009) Complete array of HOX gene expression by RQ-PCR. Methods Mol Biol 538:369–93. doi:10.1007/978-1-59745-418-6_19
41. Srivastava RK, Kurzrock R, Shankar S (2010) MS-275 sensitizes TRAIL-resistant breast cancer cells, inhibits angiogenesis and metastasis, and reverses epithelial-mesenchymal transition in vivo. Mol Cancer Ther 9:3254–66. doi:10.1158/1535-7163.MCT-10-0582
42. Hooker JM, Kim SW, Alexoff D et al (2010) Histone deacetylase inhibitor, MS-275, exhibits poor brain penetration: PK studies of [C] MS-275 using Positron Emission Tomography. ACS Chem Neurosci 1:65–73. doi:10.1021/cn9000268

Index

Yacine Graba and René Rezsohazy (eds.), *Hox Genes: Methods and Protocols*, Methods in Molecular Biology, vol. 1196, DOI 10.1007/978-1-4939-1242-1,

T

U

W

Y

Z

MIX
Papier aus verantwortungsvollen Quellen
Paper from responsible sources
FSC® C105338

If you have any concerns about our products,
you can contact us on
ProductSafety@springernature.com

In case Publisher is established outside the EU,
the EU authorized representative is:
Springer Nature Customer Service Center GmbH
Europaplatz 3, 69115 Heidelberg, Germany

Printed by Libri Plureos GmbH
in Hamburg, Germany